Methods in Molecular Biology

For further volumes:
http://www.springer.com/series/7651

For over 35 years, biological scientists have come to rely on the research protocols and methodologies in the critically acclaimed *Methods in Molecular Biology* series. The series was the first to introduce the step-by-step protocols approach that has become the standard in all biomedical protocol publishing. Each protocol is provided in readily-reproducible step-by-step fashion, opening with an introductory overview, a list of the materials and reagents needed to complete the experiment, and followed by a detailed procedure that is supported with a helpful notes section offering tips and tricks of the trade as well as troubleshooting advice. These hallmark features were introduced by series editor Dr. John Walker and constitute the key ingredient in each and every volume of the *Methods in Molecular Biology* series. Tested and trusted, comprehensive and reliable, all protocols from the series are indexed in PubMed.

Variant Calling

Methods and Protocols

Edited by

Charlotte K. Y. Ng

Department for BioMedical Research, University of Bern, Bern, Switzerland

Salvatore Piscuoglio

Department of Biomedicine, University of Basel, Basel, Switzerland

Humana Press

Editors
Charlotte K. Y. Ng
Department for BioMedical Research
University of Bern
Bern, Switzerland

Salvatore Piscuoglio
Department of Biomedicine
University of Basel
Basel, Switzerland

ISSN 1064-3745 ISSN 1940-6029 (electronic)
Methods in Molecular Biology
ISBN 978-1-0716-2292-6 ISBN 978-1-0716-2293-3 (eBook)
https://doi.org/10.1007/978-1-0716-2293-3

This Humana imprint is published by the registered company Springer Science+Business Media, LLC, part of Springer Nature.
The registered company address is: 1 New York Plaza, New York, NY 10004, U.S.A.

Preface

Genetic diversity is fundamental to species adaptability and survival. The identification of sequence variations provides insights not only into genetic diversity but also to the genetic component of diseases and phenotypes. The advances of next-generation sequencing over the past two decades have offered the opportunity to comprehensively screen and detect various types of variants in large populations. While the Human Genome Project took 13 years to complete, now entire genomes can be screened for genetic variations at single base-pair resolution in a matter of days. The identification of variants in the nearly half-million sequenced species has enabled significant advances in many facets, from medicine to agriculture to industrial processes. The number of known variants will only continue to increase exponentially with the Earth BioGenome Project aiming to sequence and annotate ~1.5 million known eukaryotic species by 2028.

This book was compiled in the midst of one of the great challenges in modern medicine—the SARS-CoV-2 pandemic. Being able to rapidly sequence the pathogen and trace its variants as they spread across the world has been critical to pandemic response planning aiming to contain viral spread. While the initial vaccines developed at breakneck speed have proven effective, the identification of novel variants that could compromise their potency is critical to the further development of vaccines. Next-generation sequencing and variant detection, together with other pandemic response measures, have undoubtedly saved many, many lives.

The evolution of next-generation sequencing has necessitated the continual development of algorithms to address specific questions and to overcome computational challenges. The many types of sequencing protocols and technologies enable the discovery of many different types of variants, each requiring specialist algorithms to accurately model the underlying statistical properties. In this volume, we cover a diverse range of methods relevant to variant calling. In the first part, Chapters 1–11 cover general methods for variant calling, with Chapters 1–5 covering approaches to calling "classical" variants—single nucleotide substitutions, insertions, and deletions—and Chapters 6–11 describing methods for the identification of other, but no less biologically relevant, variant types—copy number variants, structural variants, somatic mitochondrial variants, and splice variants. The second part covers variant calling in specialized data types and use cases. Chapters 12–14 describe three approaches to variant calling in specialized data types—Ion Torrent sequencing data, RNA-seq data, and UMI-tagged paired-end sequencing data. Chapters 15 and 16 describe alignment-free genotyping and SNP calling. Chapters 17 and 18 describe single nucleotide and copy number variant detection in single-cell DNA sequencing data. The third part covers topics relevant and important to variant calling, with Chapters 19 and 20 dedicated to variant annotation—critical to biological interpretation—and Chapter 21 devoted to pre-analytical quality control to ensure successful variant calling. The chapters should be of interest to bioinformatics students and researchers alike looking to broaden and deepen their knowledge on variant calling.

We would like to thank the authors for sharing their methods. We thank Springer Nature and Series Editor Prof. John M Walker for giving us the opportunity to compile this collection and for their support throughout the process.

Bern, Switzerland *Charlotte K. Y. Ng*
Basel, Switzerland *Salvatore Piscuoglio*

Contents

Contributors

ILARIA ALBORELLI • *Institute of Medical Genetics and Pathology, University Hospital Basel, Basel, Switzerland*

RAFAEL ALDANA • *Sentieon® Inc., San Jose, CA, USA*

JISONG AN • *Graduate School of Medical Science and Engineering (GSMSE), Korea Advanced Institute of Science and Technology (KAIST), Daejeon, Republic of Korea*

ARSHI ARORA • *Department of Epidemiology & Biostatistics, Memorial Sloan Kettering Cancer Center, New York, NY, USA*

SAJEDEH BAHONAR • *Department of Bioinformatics, Institute of Biochemistry and Biophysics (IBB), University of Tehran, Tehran, Iran*

GIULIA BERNARDINI • *Centrum Wiskunde & Informatica, Amsterdam, The Netherlands*

NATHALIE BISSONNETTE • *Agriculture and Agri-Food Canada, Sherbrooke, QC, Canada*

ÉLODIE BOHERS • *Department of Pathology, Centre Henri Becquerel, Rouen, France; INSERM U1245, University of Normandie UNIROUEN, Rouen, France*

JEAN-SIMON BROUARD • *Agriculture and Agri-Food Canada, Sherbrooke, QC, Canada*

SHUQI CAO • *Harbin Institute of Technology, Harbin, Heilongjiang, China*

PABLO CINGOLANI • *AstraZeneca, Oncology R&D, Arlington, MA, USA*

DANIEL P. COOKE • *MRC Weatherall Institute of Molecular Medicine, University of Oxford, Oxford, UK*

HÉLÈNE DAUCHEL • *Normandie Univ, UNIROUEN, LITIS EA 4108, Rouen, France*

LUCA DENTI • *Department of Computational Biology, C3BI USR 3756 CNRS, Institut Pasteur, Paris, France*

ANNA FOWLER • *Department of Health Data Science, Institute of Population Health, University of Liverpool, Liverpool, UK*

DONALD FREED • *Sentieon® Inc., San Jose, CA, USA*

ANDREA GAROFOLI • *Institute of Medical Genetics and Pathology, University Hospital Basel, Basel, Switzerland*

WEITAI HUANG • *Laboratory of Computational Cancer Genomics, Genome Institute of Singapore, A*STAR (Agency for Science, Technology and Research), Singapore, Singapore*

FABRICE JARDIN • *Department of Pathology, Centre Henri Becquerel, Rouen, France; INSERM U1245, University of Normandie UNIROUEN, Rouen, France*

PHILIP M. JERMANN • *Institute of Medical Genetics and Pathology, University Hospital Basel, Basel, Switzerland*

SHUANGXI JI • *Department of Bioinformatics and Computational Biology, The University of Texas MD Anderson Cancer Center, Houston, TX, USA*

TAO JIANG • *Harbin Institute of Technology, Harbin, Heilongjiang, China*

YUCHAO JIANG • *Department of Biostatistics, Gillings School of Global Public Health, University of North Carolina, Chapel Hill, NC, USA; Department of Genetics, School of Medicine, University of North Carolina, Chapel Hill, NC, USA; Lineberger Comprehensive Cancer Center, University of North Carolina, Chapel Hill, NC, USA*

YOUNG SEOK JU • *Graduate School of Medical Science and Engineering (GSMSE), Korea Advanced Institute of Science and Technology (KAIST), Daejeon, Republic of Korea*

André Kahles • *Biomedical Informatics Group, Department of Computer Science, ETH Zurich, Zurich, Switzerland; Biomedical Informatics Research, University Hospital Zurich, Zurich, Switzerland; Swiss Institute of Bioinformatics, Lausanne, Switzerland*

Thierry Lecroq • *Normandie Univ, UNIROUEN, LITIS EA 4108, Rouen, France*

Yanbo Li • *School of Computing, Australian National University, Canberra, ACT, Australia*

Yu Lin • *School of Computing, Australian National University, Canberra, ACT, Australia*

Shiqi Liu • *Harbin Institute of Technology, Harbin, Heilongjiang, China*

Philipp Markolin • *Biomedical Informatics Group, Department of Computer Science, ETH Zurich, Zurich, Switzerland; Biomedical Informatics Research, University Hospital Zurich, Zurich, Switzerland; Swiss Institute of Bioinformatics, Lausanne, Switzerland*

Kyoung Il Min • *Graduate School of Medical Science and Engineering (GSMSE), Korea Advanced Institute of Science and Technology (KAIST), Daejeon, Republic of Korea*

Hesam Montazeri • *Department of Bioinformatics, Institute of Biochemistry and Biophysics (IBB), University of Tehran, Tehran, Iran*

Matthew D. Montierth • *Department of Bioinformatics and Computational Biology, The University of Texas MD Anderson Cancer Center, Houston, TX, USA; Quantitative Computational Biology, Baylor College of Medicine, Houston, TX, USA*

Charlotte K. Y. Ng • *Department for BioMedical Research, University of Bern, Bern, Switzerland; SIB Swiss Institute of Bioinformatics, Lausanne, Switzerland*

Hardip Patel • *The John Curtin School of Medical Research, Australian National University, Canberra, ACT, Australia*

Tirso Pons • *Department of Immunology and Oncology, National Center for Biotechnology, Spanish National Research Council (CNB-CSIC), Madrid, Spain*

Marco Previtali • *Department of Informatics, Systems and Communication, University of Milano-Bicocca, Milan, Italy*

Élise Prieur-Gaston • *Normandie Univ, UNIROUEN, LITIS EA 4108, Rouen, France*

Gunnar Rätsch • *Biomedical Informatics Group, Department of Computer Science, ETH Zurich, Zurich, Switzerland; Biomedical Informatics Research, University Hospital Zurich, Zurich, Switzerland; Swiss Institute of Bioinformatics, Lausanne, Switzerland*

Philippe Ruminy • *Department of Pathology, Centre Henri Becquerel, Rouen, France; INSERM U1245, University of Normandie UNIROUEN, Rouen, France*

Vincent Sater • *Normandie Univ, UNIROUEN, LITIS EA 4108, Rouen, France*

Désirée Schnidrig • *Department for BioMedical Research, University of Bern, Bern, Switzerland; SIB Swiss Institute of Bioinformatics, Lausanne, Switzerland*

Venkatraman E. Seshan • *Department of Epidemiology & Biostatistics, Memorial Sloan Kettering Cancer Center, New York, NY, USA*

Ronglai Shen • *Department of Epidemiology & Biostatistics, Memorial Sloan Kettering Cancer Center, New York, NY, USA*

Ngak Leng Sim • *Laboratory of Computational Cancer Genomics, Genome Institute of Singapore, A*STAR (Agency for Science, Technology and Research), Singapore, Singapore*

Anders J. Skanderup • *Laboratory of Computational Cancer Genomics, Genome Institute of Singapore, A*STAR (Agency for Science, Technology and Research), Singapore, Singapore*

Miguel Vazquez • *Genome Informatics Unit, Barcelona Supercomputing Center (BSC-CNS), Barcelona, Spain*

Pierre Vera • *Normandie Univ, UNIROUEN, LITIS EA 4108, Rouen, France; Department of Pathology, Centre Henri Becquerel, Rouen, France*

PIERRE-JULIEN VIAILLY • *Department of Pathology, Centre Henri Becquerel, Rouen, France; INSERM U1245, University of Normandie UNIROUEN, Rouen, France*
MATHIEU VIENNOT • *Department of Pathology, Centre Henri Becquerel, Rouen, France; INSERM U1245, University of Normandie UNIROUEN, Rouen, France*
RUJIN WANG • *Department of Biostatistics, Gillings School of Global Public Health, University of North Carolina, Chapel Hill, NC, USA*
TING-YOU WANG • *The Hormel Institute, University of Minnesota, Austin, MN, USA*
WENYI WANG • *Department of Bioinformatics and Computational Biology, The University of Texas MD Anderson Cancer Center, Houston, TX, USA*
YADONG WANG • *Harbin Institute of Technology, Harbin, Heilongjiang, China*
LIXING YANG • *Ben May Department for Cancer Research, Department of Human Genetics, Comprehensive Cancer Center, University of Chicago, Chicago, IL, USA*
RENDONG YANG • *The Hormel Institute, University of Minnesota, Austin, MN, USA*

Chapter 1

Data Processing and Germline Variant Calling with the Sentieon Pipeline

Rafael Aldana and Donald Freed

Abstract

Public and private genomic sequencing initiatives generate ever-increasing amounts of genomic data creating a need for improved solutions for genomics data processing (Stephens et al.PLoS Biol 13: e1002195, 2015). The Sentieon® Genomics software enables rapid and accurate analysis of next-generation sequence data. In this work, we present a typical use of the Sentieon Genomics software for germline variant calling. The Sentieon germline variant calling pipeline produces more accurate results than other tools on third-party benchmarks (Katherine et al. Front Genet 10:736, 2019; Shen et al. bioRxiv, 885517, 2019) in one tenth the time of comparable pipelines. Parts of this guide come from the official Sentieon Genomics software manual in https://support.sentieon.com/manual (Sentieon. Sentieon Genomics software manual, n.d.) and from the official Sentieon Genomics software application notes in https://support.sentieon.com/appnotes (Sentieon. Sentieon Genomics software application notes, n.d.) and are republished with permission. For additional details and advanced usage instructions of the Sentieon tools, refer to the software manual.

Key words Bioinformatics, DNA sequencing, Genomics, Variant calling, Machine learning

1 Introduction

Next-generation sequencing produces short-read sequences that are derived from a sample of interest. Secondary analysis of next-generation sequence data takes these short sequences and performs alignment to a reference genome, pre-processing through various stages depending on the data type and objective, and variant calling to identify differences between the sample and a reference genome. Downstream, the unique signature of reference matches and differences can be used for the assessment of polygenic risk, identification of known disease-causing variation, inference of ancestry, informing clinical management, or identification of novel associations between specific genetic variants and disease status or other phenotypes.

Charlotte K. Y. Ng and Salvatore Piscuoglio (eds.), *Variant Calling: Methods and Protocols*, Methods in Molecular Biology, vol. 2493, https://doi.org/10.1007/978-1-0716-2293-3_1,

The Sentieon Genomics software is a set of software tools that perform secondary analysis of genomic data obtained from DNA and RNA sequencing. The Sentieon DNAseq®/DNAscope pipelines enable germline variant calling on hundreds of thousands of samples simultaneously. The Sentieon TNseq®/TNscope® pipelines enable accurate calling of somatic variants in paired tumor-normal samples or in an unpaired tumor sample. The Sentieon Genomics software also implements more efficient versions of the most popular sequence alignment tools, BWA for DNAseq data and STAR for RNAseq data. The Sentieon aligners produce the same results as the original versions with a 2×–3× speed improvement. The Sentieon Genomics software implements machine learning-based filtering models to improve variant calling accuracy. Generic models are available which have been calibrated using the GiaB datasets and truth-sets, and the Sentieon Genomics software allows the calibration of user-specific models to achieve better performance.

2 Materials

The Sentieon Genomics software is a license controlled commercial software designed to run on Linux and other POSIX-compatible platforms.

2.1 Platform Requirements

For Linux® systems, it is recommended to use the following Linux distributions or higher: Red Hat®/CentOS 6.5, Debian 7.7, OpenSUSE-13.2, or Ubuntu-14.04. The Sentieon Genomics software has not been tested on older releases and may not work on older distributions.

The Sentieon Genomics software processes file data at speeds around 20–30 MB/s per core using a state-of-the-art server. In order to take advantage of the maximum speed that the software can provide, it is recommended to use high-speed SSD hard drives in the computing system, preferably two identical drives in a high-speed RAID 0 striped volume configuration. If using a RAID 0 configuration, it is advised not to keep any important information on the RAID 0 hard drives, so that the server would have additional storage space for intermediate results.

The Sentieon Genomics software requires that the computing system have either python2.6.x, python2.7.x, or python3.x. It is possible to check the version of the default python in the computing system by running:

```
python --version
```

If neither python2.6.x, python2.7.x, nor python3.x are the default python version in the computing system, it will be required to set an environmental variable to tell the software where an appropriate python binary is located:

```
export SENTIEON_PYTHON=Python_path
```

2.2 Installation Procedure for a Linux-Based System

Install the Sentieon Genomics software in a Linux-based system by following these installation instructions:

1. Obtain the software release package. Contact support@sentieon to obtain the latest version of the software package.
2. Obtain the license file. The license file has extension .lic. Contact support@sentieon to obtain a license file.
3. Upload the software release package and license file to the computing server.
4. Decompress the release package with the command below. This will create a folder named sentieon-genomics-202010.02.

```
tar xvzf sentieon-genomics-202010.02.tar.gz
```

5. Copy the license file to the directory that stores the license files (LICENSE_DIR).
6. Set an environment variable in the computing system to indicate where the license is located (either LICSRVR_HOST: LICSRVR_PORT when using a license server or LICENSE_-DIR/LICENSE_FILE.lic when using a license file).

   It is recommended to add the following line to the shell scripts used to run the Sentieon software:

```
#if using a license server
export SENTIEON_LICENSE=LICSRVR_HOST:LICSRVR_PORT
#if using a license file
export SENTIEON_LICENSE=LICENSE_DIR/LICENSE_FILE.lic
```

3 Methods

3.1 Typical Usage for DNAseq

One of the typical uses of the Sentieon Genomics software is to perform the bioinformatics pipeline for DNA germline analysis using the recommended in the Broad institute best practices described in https://www.broadinstitute.org/gatk/guide/best-practices, using the BWA Mem [1], Picard [2], and GATK [3] mathematics. Figure 1 illustrates such a typical bioinformatics pipeline.

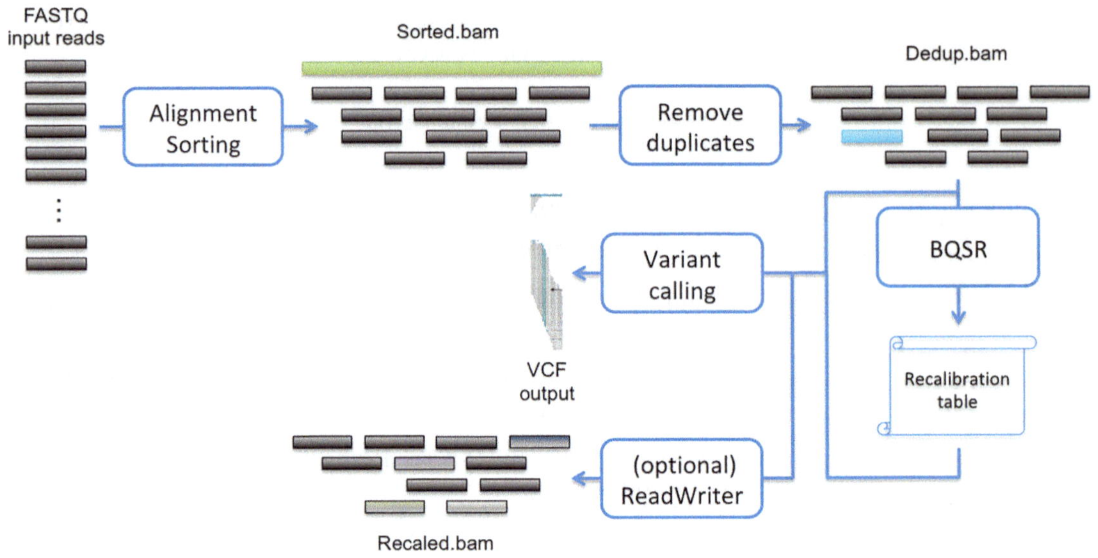

Fig. 1 Recommended bioinformatics pipeline for DNA variant calling analysis

3.1.1 General

This bioinformatics pipeline requires the following inputs:

1. The FASTA file containing the nucleotide sequence of the reference genome for the species corresponding to the sample to be analyzed (*see* **Note 1** for recommended reference files for human samples). The reference data needs to be pre-processed such that the data specified in Table 1 is available to the software.
2. One or multiple FASTQ files containing the nucleotide sequence of the sample to be analyzed. These files contain the raw reads from the DNA sequencing. The software supports inputting FASTQ files compressed using GZIP. The software only supports files containing quality scores in Sanger format (Phred+33).
3. (Optional) The Single-Nucleotide Polymorphism database (dbSNP) data that needs to be included in the pipeline. The data is used in the form of a VCF file; it is possible to use a VCF file compressed with bgzip and tabix indexed.
4. (Optional) Multiple collections of known sites that need to be included in the base quality score recalibration. The data is used in the form of a VCF file; it is possible to use a VCF file compressed with bgzip and tabix indexed.

The following steps compose the typical bioinformatics pipeline:

1. Map reads to the reference: this step aligns the reads contained in the FASTQ files to a reference genome contained in the

Table 1
Data requirements for the reference nucleotide sequence

Data	Description
.fasta	Reference sequence file
.fasta.amb	BWA index file
.fasta.ann	BWA index file
.fasta.bwt	BWA index file
.fasta.pac	BWA index file
.fasta.sa	BWA index file
.fasta.fai	SAMtools fasta index file

FASTA file. This step ensures that the data can be placed in context.

2. Calculate quality metrics: this step produces a statistical summary of the data quality and the pipeline data analysis quality.
3. Remove duplicates: this step detects reads originating from the same DNA molecule. These duplicate reads are not informative and should not be counted as additional evidence.
4. Base quality score recalibration (BQSR): this step modifies the quality scores assigned to individual nucleotide bases of the sequence read data. This modification removes experimental biases caused by variability in the sequencing run.
5. Variant calling: this step identifies the sites where the sample contains variation relative to the reference genome, and calculates sample genotypes at each variant site.

3.1.2 Map Reads to the Reference

A single command is run to efficiently perform the alignment using BWA, and the creation of the BAM file and the sorting using the Sentieon software:

```
(sentieon bwa mem -M -R '@RG\tID:GROUP_NAME\tSM:SAMPLE_NAME
\tPL:PLATFORM' -k 10000000 -t NUMBER_THREADS REFERENCE SAMPLE
[SAMPLE2] || echo -n 'error' )| sentieon util sort -o SORTED_-
BAM -t NUMBER_THREADS --sam2bam -i -
```

Inputs and options for BWA are described in its manual [4].

Alternatively, it is possible to use other aligners that produce a file following the SAM format in stdout and replace the BWA portion of the command. *See* **Note 2** for more information on how the read group information in this command is used by the software during downstream processing.

The following inputs are required for the command:

- **GROUP_NAME**: Readgroup identifier that will be added to the readgroup header line. The RG:ID needs to be unique among all the datasets that will be used, which is important when working with multiple input files, or when performing a Tumor-Normal analysis.
- **SAMPLE_NAME**: name of the sample that will be added to the readgroup header line.
- **PLATFORM**: name of the sequencing platform used to sequence the DNA, which will be used to determine the likely biases in the base quality score recalibration. Possible options are ILLUMINA when the fastq files have been produced in an Illumina™ machine; IONTORRENT when the fastq files have been produced in a Life Technologies™ Ion-Torrent™ machine.
- **NUMBER_THREADS**: the number of computer threads that will be used in the calculation. It is recommended that the number does not exceed the number of computing cores available in the computing system. It is recommended to use the same number of threads for both BWA and for the util binary.
- **REFERENCE**: the location of the reference FASTA file. You should make sure that all additional reference data specified in Table 1 is available in the same location with consistent naming.
- **SAMPLE**: the location of the sample FASTQ file. If the data comes from paired-end sequencing, the second SAMPLE2 corresponding paired FASTQ file will be required.
- **SORTED_BAM**: the location and filename of the sorted mapped BAM output file. A corresponding index file will be created. The Sentieon tools natively and automatically support using smaller sized CRAM files by using a .cram extension (*see* **Note 3**).

BWA will produce slightly different results depending on the number of threads used in the command. This is due to the fact that BWA computes the insert size distribution on a chunk, whose size is dependent on the number of threads. To guarantee that the results are consistent and independent of the number of threads used, it is recommended to fix the chunk size in bases by using the option "-K 10000000."

3.1.3 Calculate Quality Metrics

A single command is run to generate five statistical summaries of the data quality and the pipeline data analysis quality results:

```
sentieon driver -t NUMBER_THREADS -r REFERENCE -i SORTED_BAM
--algo GCBias --summary GC_SUMMARY_TXT GC_METRIC_TXT --algo
MeanQualityByCycle MQ_METRIC_TXT --algo QualDistribution
```

```
QD_METRIC_TXT --algo InsertSizeMetricAlgo IS_METRIC_TXT --algo
AlignmentStat ALN_METRIC_TXT
```

Four commands are run to generate the plots from the statistical summaries:

```
sentieon plot GCBias-o GC_METRIC_PDF GC_METRIC_TXT
sentieon plot MeanQualityByCycle -o MQ_METRIC_PDF MQ_ME-
TRIC_TXT
sentieon plot QualDistribution -o QD_METRIC_PDF QD_METRIC_TXT
sentieon plot InsertSizeMetricAlgo -o IS_METRIC_PDF IS_ME-
TRIC_TXT
```

The following inputs are required for the commands:

- **NUMBER_THREADS**: the number of computer threads that will be used in the calculation. It is recommended that the number does not exceed the number of computing cores available in the computing system.
- **REFERENCE**: the location of the reference FASTA file. You should make sure that the reference is the same as the one used in the mapping stage.
- **SORTED_BAM**: the location where the previous mapping stage stored the result.
- **GC_SUMMARY_TXT**: the location and filename of the GC bias metrics summary results output file.
- **GC_METRIC_TXT**: the location and filename of the GC bias metrics results output file.
- **MQ_METRIC_TXT**: the location and filename of the mapping quality metrics results output file.
- **QD_METRIC_TXT**: the location and filename of the quality/depth metrics results output file.
- **IS_METRIC_TXT**: the location and filename of the insertion size metrics results output file.
- **ALN_METRIC_TXT**: the location and filename of the alignment metrics results output file.
- **GC_METRIC_PDF**: the location and filename of the GC bias metrics report output file.
- **MQ_METRIC_PDF**: the location and filename of the mapping quality metrics report output file.
- **QD_METRIC_PDF**: the location and filename of the quality/depth metrics report output file.
- **IS_METRIC_PDF**: the location and filename of the insertion size metrics report output file.

3.1.4 Remove Duplicates

Two individual commands are run to remove duplicates on the BAM file after alignment and sorting. The first command collects read information, and the second command performs the deduplication.

```
sentieon driver -t NUMBER_THREADS -i SORTED_BAM --algo Locus-
Collector --fun score_info SCORE_TXT
sentieon driver -t NUMBER_THREADS -i SORTED_BAM --algo Dedup
--rmdup --score_info SCORE_TXT --metrics DEDUP_METRIC_TXT DE-
DUP_BAM
```

The following inputs are required for the commands:

- **NUMBER_THREADS**: the number of computer threads that will be used in the calculation. It is recommended that the number does not exceed the number of computing cores available in the computing system.
- **SORTED_BAM**: the location where the previous mapping stage stored the result.
- **SCORE_TXT**: the location and filename of the temporary score output file. It is important that the same file is used for both commands, as the file contains information from the first command on which reads should be marked as duplicate.
- **DEDUP_METRICS_TXT**: the location and filename of the deduplication metrics results output file.
- **DEDUP_BAM**: the location and filename of the deduped BAM output file. A corresponding index file will be created. The Sentieon tools natively and automatically support using smaller sized CRAM files by using a .cram extension (*see* **Note 3**).

3.1.5 Base Quality Score Recalibration (BQSR)

A single command is run to calculate the required modification of the quality scores assigned to individual nucleotide bases of the sequence read data; the actual recalibration is applied during the variant calling stage.

```
sentieon driver -t NUMBER_THREADS -r REFERENCE -i DEDUP_BAM --
algo QualCal [-k KNOWN_SITES] RECAL_DATA.TABLE
```

Three commands are run to apply the recalibration and create a report on the base quality sore recalibration. The first command applies the recalibration to calculate the post calibration data table and optionally applies the recalibration on the BAM file, the second creates the data for plotting; the third command plots the calibration data tables, both pre and post, into graphs in a pdf.

```
sentieon driver -t NUMBER_THREADS -r REFERENCE -i DEDUP_BAM -q
```

```
RECAL_DATA.TABLE --algo QualCal [-k KNOWN_SITES] RECAL_DATA.
TABLE.POST [--algo ReadWriter RECALIBRATED_BAM]
sentieon driver -t NUMBER_THREADS --algo QualCal --plot --
before RECAL_DATA.TABLE --after RECAL_DATA.TABLE.POST RECAL_-
RESULT.CSV
sentieon plot QualCal -o BQSR_PDF RECAL_RESULT.CSV
```

The following inputs are required for the command:

- **NUMBER_THREADS**: the number of computer threads that will be used in the calculation. It is recommended that the number does not exceed the number of computing cores available in the computing system.
- **REFERENCE**: the location of the reference FASTA file. You should make sure that the reference is the same as the one used in the mapping stage.
- **DEDUP_BAM**: the location where the previous deduplication stage stored the result.
- **RECAL_DATA.TABLE**: the location and filename of the recalibration table.
- **RECAL_DATA.TABLE.POST**: the location and filename of the temporary post recalibration table.
- **RECAL_RESULT.CSV**: the location and filename of the temporary recalibration results output file used for plotting.
- **BQSR_PDF**: the location and filename of the BSQR results output file.

The following inputs are optional for the command:

- **KNOWN_SITES**: the location of the VCF file used as a set of known sites. You can include multiple collections of known sites by repeating the -k KNOWN_SITES option.
- **RECALIBRATED_BAM**: the location and filename of the recalibrated BAM output file. A corresponding index file will be created. This output is optional as Sentieon variant callers can perform the recalibration on the fly using the before recalibration bam plus the recalibration table. The Sentieon tools natively and automatically support using smaller sized CRAM files by using a .cram extension.

3.1.6 *Variant Calling*

A single command is run to call variants and additionally apply the BQSR calculated before.

```
sentieon driver -t NUMBER_THREADS -r REFERENCE -i DEDUP_BAM -q
RECAL_DATA.TABLE --algo Haplotyper [-d dbSNP] VARIANT_VCF
```

Alternatively, it is possible to use a previously recalibrated BAM file; in this case, it is not necessary to re-apply the BQSR, so the recalibration table option should not be used when using the previously recalibrated BAM:

```
sentieon driver -t NUMBER_THREADS -r REFERENCE -i RECALIBRA-
TED_BAM --algo Haplotyper [-d dbSNP] VARIANT_VCF
```

Using the recalibrated BAM or the non-recalibrated BAM plus the recalibration data table will give identical results. It is very important to be careful not to use the recalibration data table together with the already recalibrated BAM, as that would apply the recalibration twice leading to incorrect results.

The following inputs are required for the command:

- **NUMBER_THREADS**: the number of computer threads that will be used in the calculation. It is recommended that the number does not exceed the number of computing cores available in the computing system.
- **REFERENCE**: the location of the reference FASTA file. You should make sure that the reference is the same as the one used in the mapping stage.
- **DEDUP_BAM**: the location where the previous deduplication stage stored the result.
- **RECAL_DATA.TABLE**: the location where the previous BQSR stage stored the result.
- **RECALIBRATED_BAM**: the location of the recalibrated BAM file.
- **VARIANT_VCF**: the location and filename of the variant calling output file. A corresponding index file (.idx) will be created. The tool will output a compressed file by using .gz extension.

The following inputs are optional for the command:

- **dbSNP**: the location of the Single-Nucleotide Polymorphism database (dbSNP) that will be used to label known variants. You can only use one dbSNP file.

3.2 Typical Usage for DNAscope

The Sentieon Genomics software includes DNAscope, an improved algorithm to perform the variant calling step of germline DNA analysis. DNAscope utilizes a machine learning model to perform variant calling with higher accuracy and improves both candidate variant detection and filtering (*see* **Note 4**). The pre-processing pipeline used for DNAscope is similar to the DNAseq one described in Subheading 3.1, but some steps are not executed.

3.2.1 General

This bioinformatics pipeline requires the following inputs:

1. The FASTA file containing the nucleotide sequence of the reference genome corresponding to the sample to be analyzed (*see* **Note 1** for recommended reference files for human samples).
2. One or multiple FASTQ files containing the nucleotide sequence of the sample to be analyzed. These files contain the raw reads from the DNA sequencing. The software supports inputting FASTQ files compressed using GZIP. The software only supports files containing quality scores in Sanger format (Phred+33).
3. (Optional) The Single-Nucleotide Polymorphism database (dbSNP) data that can be included in the pipeline. The data is used in the form of a VCF file; it is possible to use a VCF file compressed with bgzip and tabix indexed.
4. A machine learning model file. Sentieon can provide a model for diploid samples [13] that has been trained using a subset of the data from the GiaB truth-set [6]. The model was created by processing samples HG001 and HG005 through a pipeline consisting of Sentieon BWA-mem alignment and Sentieon deduplication, and using the variant calling results to calibrate a model to fit the GiaB truth-set for these samples. In addition, Sentieon can assist in the creation of models using user data, which will calibrate the specifics of the sequencing and bio-informatics processing.

The following steps compose the typical bioinformatics pipeline for improved accuracy germline variant calling with Sentieon DNAscope:

1. Pre-process the sample using a DNA seq pipeline like the one introduced in Subheading 3.1.2, with the following stages:
   (a) Map reads to the reference.
   (b) Calculate quality metrics.
   (c) Remove duplicates.
2. Variant calling using DNAscope: This step identifies the sites where the sample contains variation relative to the reference genome and calculates sample genotypes at each variant site.

3.2.2 Map Reads to the Reference, Calculate Quality Metrics, and Remove Duplicates

For the mapping, calculation of quality metrics, and duplicate removal stages, refer to Subheading 3.1 for detailed usage instructions.

3.2.3 Germline Variant Calling Using a Machine Learning Model

Two individual commands are run to call variants and to apply the machine learning model. The input BAM file should come from a pipeline where only alignment and deduplication have been performed, to match the model creation methodology. In addition, it is important to add option --pcr_indel_model NONE when running DNAscope if the data being used is PCR-Free. Depending on whether PCR is involved, DNAscope will use different priors for finding significant INDEL variants and the default --pcr_indel_model setting is appropriate for PCR samples.

```
PCRFREE=true #PCRFREE=true means the sample is PCRFree, change
it to false for PCR samples.
if [ "$PCRFREE" = true ] ; then
 pcr_option="--pcr_indel_model none"
else
 pcr_option=""
fi
sentieon driver -t NUMBER_THREADS -r REFERENCE -i DEDUPED_BAM
--algo DNAscope --model MODEL_FILE [ -d dbSNP ] $pcr_option
TMP_VARIANT_VCF
sentieon driver -t NUMBER_THREADS -r REFERENCE --algo DNAMo-
delApply -v TMP_VARIANT_VCF --model MODEL_FILE VARIANT_VCF
```

The following inputs are required for the command:

- **NUMBER_THREADS**: the number of computer threads that will be used in the calculation. It is recommended that the number does not exceed the number of computing cores available in the computing system.
- **REFERENCE**: the location of the reference FASTA file. You should make sure that the reference is the same as the one used in the mapping stage.
- **DEDUP_BAM**: the location where the previous deduplication stage stored the result.
- **MODEL_FILE:** the location of the machine learning model file; the model will also determine the correct settings to be used.
- **TMP_VARIANT_VCF**: the location and filename of the variant calling output file from DNAscope. This is a temporary file.
- **VARIANT_VCF**: the location and filename of the variant calling output file. A corresponding index file (.idx) will be created. The tool will output a compressed file by using .gz extension.

The following inputs are optional for the command:

- **dbSNP**: the location of the Single-Nucleotide Polymorphism database (dbSNP) that will be used to label known variants. You can only use one dbSNP file.

3.3 Limitations of the Machine Learning Model

When using DNAscope with a machine learning model, most of the options for DNAscope should not be used, even though the software will not give an error if the options are present. In particular, the following options should be avoided:

- --emit_mode GVCF: The use of DNAscope with a machine learning model is incompatible with the generation of GVCF outputs.
- --var_type BND: The use of DNAscope with a machine learning model is incompatible with structural variant calling.

In addition, using an input BAM file created using a pipeline with additional stages such as INDEL realignment or BQSR is not recommended and will likely degrade the performance as the impact of those stages was not fit during model training.

3.4 Working with Multiple Input Files from the Same Sample

When there are more than one set of input fastq files for the sample, it is recommended to perform an individual alignment for each set of input fastq files, and then use the multiple sorted BAM files as input to the next stage (Fig. 2). The next stage will take care of merging all the BAM files into one. It is important to make sure that each set of input fastq files are aligned using the same SM sample name, but a different RG readgroup name, so that the rest of the stages properly handle the readgroups.

```
#Run alignment for 1st input file set
(sentieon bwa mem -M -R '@RG\tID:GROUP_NAME_1 \tSM:
SAMPLE_NAME \tPL:PLATFORM' -t NUMBER_THREADS REFERENCE SAM-
PLE_1 [SAMPLE_1_2] || echo -n 'error' )| sentieon util sort -o
SORTED_BAM_1 -t NUMBER_THREADS --sam2bam -i -
#Run alignment for 2nd input file set
(sentieon bwa mem -M -R '@RG\tID:GROUP_NAME_2 \tSM:
SAMPLE_NAME \tPL:PLATFORM' -t NUMBER_THREADS REFERENCE SAM-
PLE_2 [SAMPLE_2_2] || echo -n 'error' )| sentieon util sort -o
SORTED_BAM_2 -t NUMBER_THREADS --sam2bam -i -
#Run dedup on both BAM files
sentieon driver -t NUMBER_THREADS -i SORTED_BAM_1 -i SORTED_-
BAM_2 --algo LocusCollector --fun score_info SCORE_TXT
sentieon driver -t NUMBER_THREADS -i SORTED_BAM_1 -i SORTED_-
BAM_2 --algo Dedup --rmdup --score_info SCORE_TXT --metrics
DEDUP_METRIC_TXT DEDUP_BAM
```

Alternatively, it is possible to input the results of multiple BWA alignments onto a single util sort command to merge the results onto a single sorted bam files as soon as possible. It is still important to make sure that each set of input fastq files are aligned using the

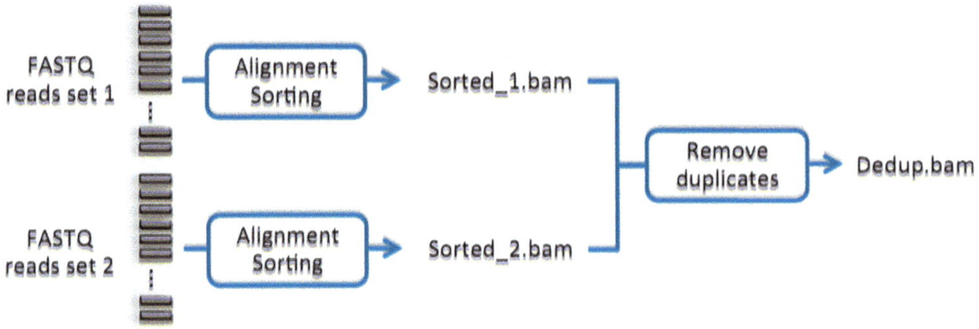

Fig. 2 Bioinformatics pipeline using multiple input files from the same sample

same SM sample name, but a different RG readgroup name, so that the rest of the stages properly handle the readgroups.

```
#Run alignment for both input file sets sequentially
(sentieon bwa mem -M -R '@RG\tID:GROUP_NAME_1 \tSM:
SAMPLE_NAME \tPL:PLATFORM' -t NUMBER_THREADS REFERENCE SAM-
PLE_1 [SAMPLE_1_2] && sentieon bwa mem -M -R '@RG\tID:
GROUP_NAME_2 \tSM:SAMPLE_NAME \tPL:PLATFORM' -t NUMBER_THREADS
REFERENCE SAMPLE_2 [SAMPLE_2_2] || echo -n 'error' )| sentieon
util sort -o SORTED_COMBINED_BAM -t NUMBER_THREADS --sam2bam
-i -
#Run dedup on the combined BAM file
sentieon driver -t NUMBER_THREADS -i SORTED_COMBINED_BAM_1 --
algo LocusCollector --fun score_info SCORE_TXT
sentieon driver -t NUMBER_THREADS -i SORTED_COMBINED_BAM_1 --
algo Dedup --rmdup --score_info SCORE_TXT --metrics DEDUP_ME-
TRIC_TXT DEDUP_BAM
```

3.5 Complete Pipeline for Processing Whole-Genome Sequencing Data

The complete pipeline recommended to process whole-genome samples can be found in https://github.com/Sentieon/sentieon-scripts/tree/master/example_pipelines [5].

3.6 Processing Data Coming from Exome Sequencing

When processing Whole-Exome Sequencing (WES) data, it is important to restrict certain steps of the pipeline to only the exonic regions of the genome, so that off-target reads are excluded from the analysis. Off-target reads are byproducts of the WES capture process and might not accurately represent variation at the off-target loci. On the other hand, it is also important not to use an interval file in any BAM/CRAM generating step, to prevent the loss of reads from the outputs.

Thus, when processing WES data, the exome capture BED file should be used via the –interval option in the non-BAM generating

QualCal step of the pipeline and the variant calling, i.e., Haplotyper or DNAscope, step of the pipeline, but it is important to make sure the –interval option is not used in other steps such as Dedup or ReadWriter. This may require separating BAM generating command from non-BAM generating commands so that they are not run in the same command line.

The complete pipeline recommended to process whole-exome samples can be found in https://github.com/Sentieon/sentieon-scripts/blob/master/example_pipelines/germline/wes-interval.sh [5].

4 Benchmark Results

In this section, we present the accuracy performance of both DNAseq and DNAscope when processing some samples for which there exists a NIST GiaB truth-set [6]; parts of these results come from the official Sentieon GitHub page [7, 8] and are republished with permission.

4.1 Performance for DNAseq

4.1.1 Input Data Files

For this evaluation, we used both HG001 and HG002 with depth of about 50x from the PrecisionFDA truth challenge [9]. We used reference b37 (hs37d5.fa) for this benchmark with the standard DNAseq processing pipeline outlined in Subheading 3.1 to process the following input files:

- HG001 (50x): HG001-NA12878-50x_1.fastq.gz and HG001-NA12878-50x_2.fastq.gz.
- HG002 (50x): HG002-NA24385-50x_1.fastq.gz and HG002-NA24385-50x_2.fastq.gz.

The truthset of HG001 and HG002 can be found at the GiaB latest release page [10]. To obtain the performance numbers below, the following truth-sets were used:

- HG001 VCF: HG001_GRCh37_GIAB_highconf_CG-IllFB-IllGATKHC-Ion-10X-SOLID_CHROM1-X_v.3.3.2_highconf_PGandRTGphasetransfer.vcf.gz.
- HG001 BED: HG001_GRCh37_GIAB_highconf_CG-IllFB-IllGATKHC-Ion-10X-SOLID_CHROM1-X_v.3.3.2_highconf_nosomaticdel.bed.
- HG002 VCF: HG002_GRCh37_GIAB_highconf_CG-IllFB-IllGATKHC-Ion-10X-SOLID_CHROM1-22_v.3.3.2_highconf_triophased.vcf.gz.
- HG002 BED: HG002_GRCh37_GIAB_highconf_CG-IllFB-IllGATKHC-Ion-10X-SOLID_CHROM1-22_v.3.3.2_highconf_noinconsistent.bed.

4.1.2 Accuracy Benchmarking Results

We use RTG's vcfeval [11] and hap.py [12] for variant evaluation, with the same comparison methodology as used in precisionFDA Truth Challenge. The following commands were run to obtain the performance results:.

```
HAPPY="/opt/hap.py/bin/hap.py"
OUTPREFIX="happy_eval"
$HAPPY truth.vcf query.vcf -f truth.bed -o $OUTPREFIX -r
hs37d5.fa --engine=vcfeval --engine-vcfeval-template hs37d5.
sdf
```

The results of the benchmark are shown in Table 2.

4.2 Performance for DNAscope

4.2.1 Input Data Files

For this evaluation, we used only HG002 with depth of about 50x from the PrecisionFDA truth challenge [9], this is because the machine learning model used in the processing used some of HG001 as training data which may bias the evaluation.

We used reference b37 (hs37d5.fa) for this benchmark with the standard DNAscope processing pipeline outlined in Subheading 3.2 and using Sentieon DNAscope Machine Learning Model version 0.5 [13] to process the following input files:

- HG002 (50x): HG002-NA24385-50x_1.fastq.gz and HG002-NA24385-50x_2.fastq.gz.

The truthset of HG001 and HG002 can be found at the GiaB latest release page [10]. To obtain the performance numbers below, the following truth-sets were used:

- HG002 VCF: HG002_GRCh37_GIAB_highconf_CG-IllFB-IllGATKHC-Ion-10X-SOLID_CHROM1-22_v.3.3.2_highconf_triophased.vcf.gz.
- HG002 BED: HG002_GRCh37_GIAB_highconf_CG-IllFB-IllGATKHC-Ion-10X-SOLID_CHROM1-22_v.3.3.2_highconf_noinconsistent.bed.

4.2.2 Accuracy Benchmarking Results

We use RTG's vcfeval [11] and hap.py [12] for variant evaluation, with the same comparison methodology as used in precisionFDA Truth Challenge. The following commands were run to obtain the performance results:.

```
HAPPY="/opt/hap.py/bin/hap.py"
OUTPREFIX="happy_eval"
$HAPPY truth.vcf query.vcf -f truth.bed -o $OUTPREFIX -r
hs37d5.fa --engine=vcfeval --engine-vcfeval-template hs37d5.
sdf
```

The results of the benchmark are shown in Table 3.

Table 2
Performance of DNAseq

Sample	Type	TP	FN	FP	Recall	Precision	F1_Score
HG001	INDEL	359926	3112	10,133	0.9914	0.9726	0.9819
	SNP	2785549	1741	7236	0.9994	0.9974	0.9984
HG002	INDEL	462614	806	1085	0.9983	0.9977	0.9980
	SNP	3046197	1640	5339	0.9995	0.9983	0.9989

Table 3
Performance of DNAscope. *PrecisionFDA Truth Challenge Result taken from* https:/precision.fda.gov/challenges/truth/results

Type	TP	FN	FP	Recall	Precision	F1_score	PrecisionFDA truth challenge winning Fscore*
SNP	3046378	**1459**	**700**	0.999521	0.99977	**99.9646%**	99.9587%
INDEL	463754	**1010**	**685**	0.997827	0.998585	**99.8206%**	99.4009%

5 Notes

1. Human next-generation sequencing samples are most commonly processed using reference genome build GRCh37 (hg19), first released in 2009, or GRCh38 (hg38), first released in 2013. Many groups have published derivate versions of the human reference genome that are optimized for specific applications. For the Sentieon mapping and variant calling pipelines, we recommend using hs37d5 when aligning to build GRCh37 (available from ftp://ftp-trace.ncbi.nih.gov/1000genomes/ftp/technical/reference/phase2_reference_assembly_sequence/hs37d5.fa.gz) and GRCh38 without alternate contigs when aligning to build GRCh38 (available from ftp://ftp.ncbi.nlm.nih.gov/genomes/all/GCA/000/001/405/GCA_000001405.15_GRCh38/seqs_for_alignment_pipelines.ucsc_ids/GCA_000001405.15_GRCh38_no_alt_analysis_set.fna.gz). Our recommendations follow the advice of BWA author Heng Li [14].
2. The readgroup information that is added to the file header during alignment can be used to control how multiple sets of reads are grouped during different stages of the pipeline. Importantly, the software requires that reads with the same ID tag have consistent readgroup attributes across files. The LB tag is used to identify reads from the same sequencing

library during the duplicate removal stage. The PU tag is used to identify reads from the same sequencer unit during BQSR. The SM tag is used to identify reads from the same sample during variant calling. More details on the Sentieon software's handling of readgroup attributes can be found in Sentieon's "Recommendations on Read Groups" application note [15].

3. The Sentieon software has built-in support for reading and writing aligned sequence reads in either BAM or CRAM format. Uncompressed BAM format is recommended for temporary or intermediate files due to its faster reading and writing speed while CRAM is recommended for long-term storage of aligned reads due to its smaller file size. The option "--bam_-compression 0" can be added to stages that output temporary or intermediate BAM files for even faster writing while "--cram_write_options version=3.0, compressor=rans" can be added to stages that output CRAM files for long-term storage for improved file compression.
4. Sentieon DNAscope uses a similar architecture to Sentieon Haplotyper (and the GATK's HaplotypeCaller) with separate steps for active region detection, local assembly, and read-likelihood calculation (Pair-HMM) for the production of variant candidates. DNAscope functions similarly to the Sentieon Haplotyper during these stages, but DNAscope implements additional enhancements such as an improved local assembler and will output additional variant annotations. In the typical usage for DNAscope, variant candidates are then genotyped using a machine learned model, rather than a Bayesian statistical model. All of these changes improve variant calling accuracy; the improved local assembler and annotations output more complete information on variant candidates while the machine learning model improves variant filtering and genotyping.

References

1. Li H (2013) Aligning sequence reads, clone sequences and assembly contigs with BWA-MEM. arXiv:1303.3997v2. https://arxiv.org/abs/1303.3997
2. "Picard Toolkit." 2019. Broad institute, GitHub repository. http://broadinstitute.github.io/picard/, Broad Institute
3. DePristo M, Banks E, Poplin R, Garimella K, Maguire J, Hartl C, Philippakis A, del Angel G, Rivas MA, Hanna M, McKenna A, Fennell T, Kernytsky A, Sivachenko A, Cibulskis K, Gabriel S, Altshuler D, Daly M (2011) A framework for variation discovery and genotyping using next-generation DNA sequencing data. Nat Genet 43:491–498. https://doi.org/10.1038/ng.806
4. BWA manual. http://bio-bwa.sourceforge.net/bwa.shtml
5. Sentieon GitHub example scripts repository. https://github.com/Sentieon/sentieon-scripts/tree/master/example_pipelines
6. NIST Genome in A Bottle. https://github.com/genome-in-a-bottle
7. Sentieon GitHub page for DNAseq benchmark. https://github.com/Sentieon/sentieon-matching-performance/tree/master/DNAseq

8. Sentieon GitHub page for DNAscope benchmark. https://github.com/Sentieon/sentieon-dnascope-ml
9. PrecisionFDA truth challenge. https://precision.fda.gov/challenges/truth
10. NIST Genome in A Bottle latest benchmark release. https://github.com/genome-in-a-bottle/giab_latest_release
11. vcfeval. https://github.com/RealTimeGenomics/rtg-tools
12. hap.py. https://github.com/Illumina/hap.py
13. DNAscope model version 0.5. https://console.cloud.google.com/storage/browser/_details/sentieon-dnascope-model/models/SentieonModelBeta0.5.model
14. Which human reference genome to use? https://lh3.github.io/2017/11/13/which-human-reference-genome-to-use
15. Recommendations on Read Groups. https://support.sentieon.com/appnotes/read_groups/

Chapter 2

MuSE: A Novel Approach to Mutation Calling with Sample-Specific Error Modeling

Shuangxi Ji, Matthew D. Montierth, and Wenyi Wang

Abstract

Accurate detection of somatic mutations in genetically heterogeneous tumor cell populations using next-generation sequencing remains challenging. We have developed MuSE, *Mutation calling using a Markov Substitution model for Evolution*, a novel approach for modeling the evolution of the allelic composition of tumor and normal tissue at each reference base. It adopts a sample-specific error model to depict inter-tumor heterogeneity, which greatly improves the overall accuracy. Here, we describe the method and provide a tutorial on the installation and application of MuSE.

Key words Next-generation DNA sequencing, Cancer evolution, Tumor heterogeneity, Single-nucleotide variants, Bayesian model

1 Introduction

Cancer is primarily a genetic disease caused by accumulation of DNA mutations. Next-generation sequencing (NGS) technology is commonly used to describe genetic alterations in cancer, thanks to the massive amount of reads it can generate. Methods to detect somatic point mutations, *i.e.*, single-nucleotide variants (SNVs), have rapidly evolved to address the analytical challenges in recent years [1–6]. The decreasing cost of NGS and the increasing interest to thoroughly interrogate the cancer genome further motivated an expansion from using whole-exome sequencing (WES) to using whole-genome sequencing (WGS) technology. The WGS technology introduces more challenges to somatic mutation calling as compared to WES. WGS mutation calling has to meet with a higher standard for specificity. A false positive rate (FPR) of just 1 per megabase pair (Mbp) results in a high number of 3000 false calls for WGS. Furthermore, the sequencing depth decreases from 100–200× for WES to 30–60× for WGS. This results in a lower signal-to-noise ratio, making it more difficult for accurate

Charlotte K. Y. Ng and Salvatore Piscuoglio (eds.), *Variant Calling: Methods and Protocols*,
Methods in Molecular Biology, vol. 2493, https://doi.org/10.1007/978-1-0716-2293-3_2,

mutation calling, especially for subclonal mutations that are presented at a low variant allele frequency (VAF).

Another key challenge in somatic mutation calling lies in the need to account for inter- and intra-tumor heterogeneity (ITH). This effect varies across samples that contain differing amounts of normal cells and tumor subclones [7, 8]. It is therefore important to report tier-based SNV call sets and dynamically identify sample-specific cutoffs for all samples. In contrast, applying one single cutoff across samples is commonly done among existing methods. Furthermore, within one sample, a tiered ranking system for variant calls provides options for users to reach their desired balance between sensitivity and specificity, especially for variants with a low VAF. Such tiered system was not fully utilized in most mutation calling methods. Our method, *Mu*tation calling using a Markov *S*ubstitution model for *E*volution, or MuSE, addresses these key issues by first modeling the transition from reference allele to the variant allele at each locus, and then identifying sample-specific cutoffs on the model output summary metric which reflect variations in tumor heterogeneity across samples [9]. We benchmarked MuSE using both synthetic and real data, and demonstrated its reliable performance with high sensitivity and specificity [9–11].

2 Software

MuSE is written in C++ computer language and the source code is freely available at https://github.com/wwylab/MuSE and https://bioinformatics.mdanderson.org/public-software/muse/ under a GPL-2.0 open source license.

2.1 Pre-Processing

Before running MuSE, the raw WES/WGS data need to be pre-processed with the following list of software tools, as further outlined in Fig. 1.

1. Biobambam (https://github.com/gt1/biobambam) to convert to FASTQ, if the input is a binary sequence alignment map (BAM) file.
2. cutadapt (https://cutadapt.readthedocs.io/en/stable/) for adapter and low-quality bases trimming.
3. bwa (http://bio-bwa.sourceforge.net/) for alignment against the reference genome.
4. samtools (http://www.htslib.org/) to sort and index BAM files.
5. picard (https://broadinstitute.github.io/picard/) to mark duplicates.

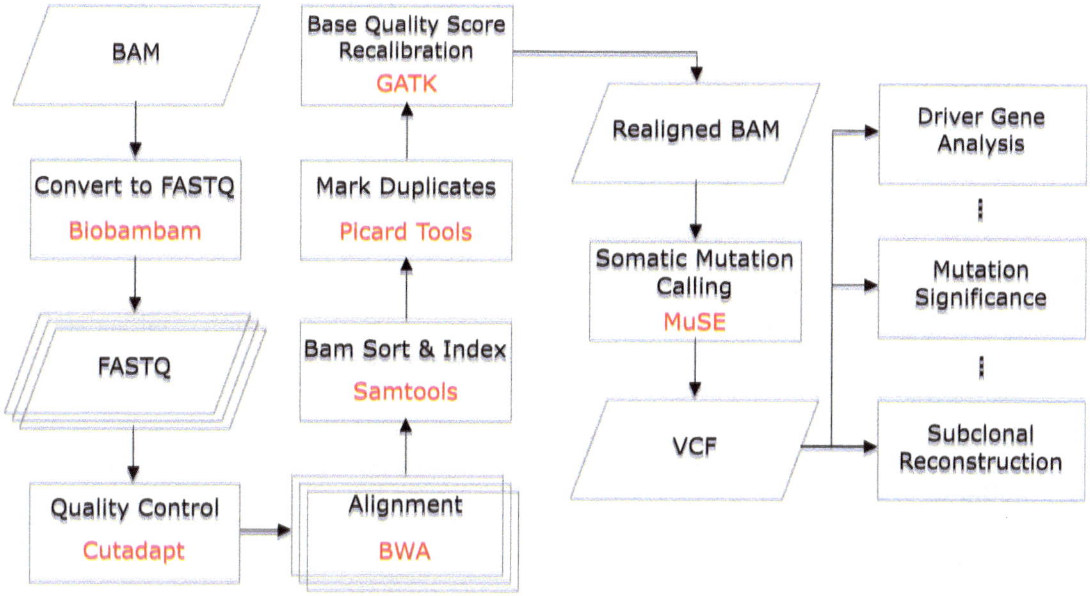

Fig. 1 Somatic mutation calling workflow for MuSE

6. GATK 4.2 (https://github.com/broadgsa/gatk) for base quality score recalibration.

2.2 Download and Install MuSE

The latest source code of MuSE can be cloned with the following `git` command:

```
git clone https://github.com/wwylab/MuSE
cd MuSE
git checkout tags/v1.0-rc
```

The compiling process is done with `make` command:

```
make
```

You can find the executable file in the current directory: `MuSE`. You may also add this directory to your `$PATH` to avoid using the full path.

For Windows, please install Cygwin (https://www.cygwin.com/) first, which provides functionality similar to a Linux distribution on Windows.

2.3 Input Requirements

MuSE requires the following files as input:

1. Indexed reference genome FASTA file.
2. BAM formatted sequence data from the pair of tumor and normal DNA samples that have gone through the pre-processing step.

3. dbSNP variant call format (VCF) file that should be bgzip compressed, tabix indexed, and based on the same reference genome as File (1).

Files (1) and (3) can be obtained from publicly available sources. For example: The human GRCh38 FASTA reference file can be downloaded at https://console.cloud.google.com/storage/browser/_details/genomics-public-data/resources/broad/hg38/v0/Homo_sapiens_assembly38.fasta, The corresponding dbSNP reference VCF can be found at https://console.cloud.google.com/storage/browser/_details/genomics-publicdata/resources/broad/hg38/v0/Homo_sapiens_assembly38.dbsnp138.vcf.

3 Methods

3.1 Running MuSE

MuSE runs in two steps (*see* **Note 1**). The first step, "`MuSE call`," takes as input files (1) and (2). This step carries out pre-filtering and calculating position-specific summary statistics using the Markov substitution model (*see* **Note 2**). The full argument list of "`MuSE call`" can be obtained by running in the terminal:

```
./MuSE call
```

They include four optional parameters and the full paths for `tumor.bam` and `matched_normal.bam`.

`-f`, path of the indexed reference genome FASTA file.

`-r`, a region where somatic mutations are called in format chromosome:start position-end position (`chr:pos-pos`).

`-l`, path of a BED format (https://m.ensembl.org/info/website/upload/bed.html) file storing the list of regions where somatic mutations are called, with one region (`chr:pos-pos`) per line.

`-O`, name of the output file. The suffix ".MuSE.txt" is automatically added.

Example:

```
./MuSE call -O Output.Prefix -f Reference.Genome Tumor.bam
Matched.Normal.bam
```

The second step, "`MuSE sump`," takes as input the output file from "`MuSE call`" and File (3). This step computes tier-based cutoffs from a sample-specific error model. The full argument list of "`MuSE sump`" can be obtained by running in the terminal:

```
./MuSE sump
 -E, error model for WES data.
 -G, error model for WGS data.
 -I, the single file generated by 'MuSE call'.
 -D, the path of the dbSNP vcf file.
 -O, the name of the output file for mutation calling.
```

Example:

```
./MuSE sump -I Output.Prefix.MuSE.txt -G -O Output.Prefix.
vcf -D dbsnp.vcf.gz
```

3.2 Output

The final output of MuSE is a VCF file (v4.1) that lists the identified somatic variants along with tiered rankings (in the FILTER field): "PASS" is the highest confidence category, followed by Tiers 1–5, with Tier 5 being the tier with the lowest confidence (*see* **Notes 3** and **4**). The INFO field of the VCF file is always "SOMATIC." For each variant, the FORMAT field specifies the genotype, read depth, number of reads supporting the reference and alternate alleles, average base quality for these reads, and the variant status of tumor and normal samples. The variant status can be 0 (wild-type), 1 (germline), 2 (somatic), 3 (loss of heterozygosity), 4 (post-transcriptional modification), or 5 (unknown), which is shown in the tumor section of the FORMAT field. The default status is 2. Other options are for use when further annotation is performed. For the normal section, the variant status is always denoted by a period symbol (".").

4 Notes

1. Typically, running MuSE on WES data from a pair of tumor and matched normal samples takes ~2–3 h, while on WGS data it takes ~20 h. These time estimates are based on a computer cluster equipped with 28 Intel(R) Xeon cores and 256GB RAM.
2. To speed up "MuSE call," we recommend splitting the WGS data into small blocks (<50 Mb) by using the provided option either "-r" or "-l", and concatenating all the output files by the Linux command "cat".
3. MuSE calls have been extensively evaluated during the Pan-Cancer Analysis of Whole Genomics (PCAWG) and The Cancer Genome Atlas (TCGA) studies [10, 11]. We recommend, as default and a starting point, for the post-processing of the variants called: for WES data, filter out mutations in the

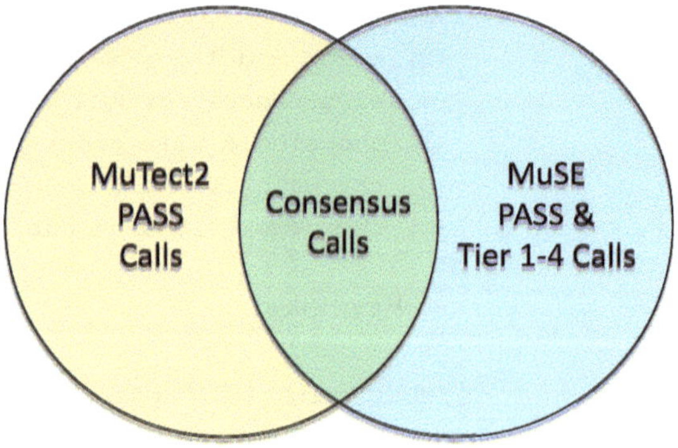

Fig. 2 Consensus calling using an intersection of mutation calls of MuSE and MuTect2

"Tier 5" category; for WGS data, keep all mutations including "Tier 5."

4. We further recommend using MuSE in conjunction with one or more mutation callers and taking a consensus approach for mutation calling, as was previously done in the PCAWG and TCGA studies [10, 11]. For individual laboratories with limited computing resources, we recommend using MuSE and MuTect2 (in GATK v4) and taking the intersect of the mutation call from both callers. Figure 2 illustrates the intersect of high-confidence calls between MuSE and MuTect2 for WES data. Such Venn diagrams may also be used to modify tier selection in the given dataset in order to reach a high percent overlap of the consensus call set with individual callers for a given study.

References

1. Koboldt DC, Zhang Q, Larson DE, Shen D, McLellan MD, Lin L, Miller CA, Mardis ER, Ding L, Wilson RK (2012) VarScan 2: somatic mutation and copy number alteration discovery in cancer by exome sequencing. Genome Res 22(3):568–576. https://doi.org/10.1101/gr.129684.111
2. Reumers J, De Rijk P, Zhao H, Liekens A, Smeets D, Cleary J, Van Loo P, Van Den Bossche M, Catthoor K, Sabbe B, Despierre E, Vergote I, Hilbush B, Lambrechts D, Del-Favero J (2011) Optimized filtering reduces the error rate in detecting genomic variants by short-read sequencing. Nat Biotechnol 30(1):61–68. https://doi.org/10.1038/nbt.2053
3. Roth A, Ding J, Morin R, Crisan A, Ha G, Giuliany R, Bashashati A, Hirst M, Turashvili G, Oloumi A, Marra MA, Aparicio S, Shah SP (2012) JointSNVMix: a probabilistic model for accurate detection of somatic mutations in normal/tumour paired next-generation sequencing data. Bioinformatics 28(7):907–913. https://doi.org/10.1093/bioinformatics/bts053
4. Saunders CT, Wong WS, Swamy S, Becq J, Murray LJ, Cheetham RK (2012) Strelka: accurate somatic small-variant calling from sequenced tumor-normal sample pairs. Bioinformatics 28(14):1811–1817. https://doi.org/10.1093/bioinformatics/bts271
5. Cibulskis K, Lawrence MS, Carter SL, Sivachenko A, Jaffe D, Sougnez C, Gabriel S, Meyerson M, Lander ES, Getz G (2013) Sensitive detection of somatic point mutations in impure and heterogeneous cancer samples. Nat Biotechnol 31(3):213–219. https://doi.org/10.1038/nbt.2514

6. Ewing AD, Houlahan KE, Hu Y, Ellrott K, Caloian C, Yamaguchi TN, Bare JC, P'ng C, Waggott D, Sabelnykova VY, participants I-TDSMCC, Kellen MR, Norman TC, Haussler D, Friend SH, Stolovitzky G, Margolin AA, Stuart JM, Boutros PC (2015) Combining tumor genome simulation with crowdsourcing to benchmark somatic single-nucleotide-variant detection. Nat Methods 12(7):623–630. https://doi.org/10.1038/nmeth.3407
7. Meyerson M, Gabriel S, Getz G (2010) Advances in understanding cancer genomes through second-generation sequencing. Nat Rev Genet 11(10):685–696. https://doi.org/10.1038/nrg2841
8. Gerlinger M, Rowan AJ, Horswell S, Math M, Larkin J, Endesfelder D, Gronroos E, Martinez P, Matthews N, Stewart A, Tarpey P, Varela I, Phillimore B, Begum S, McDonald NQ, Butler A, Jones D, Raine K, Latimer C, Santos CR, Nohadani M, Eklund AC, Spencer-Dene B, Clark G, Pickering L, Stamp G, Gore M, Szallasi Z, Downward J, Futreal PA, Swanton C (2012) Intratumor heterogeneity and branched evolution revealed by multiregion sequencing. N Engl J Med 366(10):883–892. https://doi.org/10.1056/NEJMoa1113205
9. Fan Y, Xi L, Hughes DST, Zhang JJ, Zhang JH, Futreal PA, Wheeler DA, Wang WY (2016) MuSE: accounting for tumor heterogeneity using a sample-specific error model improves sensitivity and specificity in mutation calling from sequencing data. Genome Biol 17:178. https://doi.org/10.1186/s13059-016-1029-6
10. ICGC/TCGA Pan-cancer Analysis of Whole Genomes Consortium (2020) Pan-cancer analysis of whole genomes. Nature 578(7793):82–93. https://doi.org/10.1038/s41586-020-1969-6
11. Ellrott K, Bailey MH, Saksena G, Covington KR, Kandoth C, Stewart C, Hess J, Ma S, Chiotti KE, McLellan M, Sofia HJ, Hutter C, Getz G, Wheeler D, Ding L, Grp MW, Network CGAR (2018) Scalable Open Science approach for mutation calling of tumor exomes using multiple genomic pipelines. Cell Syst 6(3):271–281. https://doi.org/10.1016/j.cels.2018.03.002

Chapter 3

Octopus: Genotyping and Haplotyping in Diverse Experimental Designs

Daniel P. Cooke

Abstract

Haplotype-based variant callers have become the de facto choice for genotyping from next-generation sequencing (NGS) as they are able to resolve read-mapper alignment errors and implicitly phase heterozygous variants. Here, I describe how the haplotype-based variant calling tool Octopus can be used for genotyping and haplotyping in several common experimental designs.

Key words Variant calling, Genotyping, Haplotyping, Bioinformatics, Sequencing

1 Introduction

Genotyping from sequencing now plays a fundamental role in research and healthcare across a number of fields. Accurate software tools that are able to model complex sources of uncertainty from differing experimental designs are crucial for this task, particularly in a clinical setting. Haplotype-based methods, which implicitly phase heterozygous variant sites, are established as the state-of-the-art for bulk diploid genotyping [1–6], but similarly powered methods have been lacking for other designs, such as tumor, microbial, and single-cell sequencing.

Octopus [1] is a haplotype-based variant calling tool that is able to genotype biological samples from sequencing data generated from divergent experimental designs using a unified haplotype-based framework. Octopus accepts sequencing reads mapped to a linear reference genome and genotyping the study samples with respect to the reference. Like most other mapping-based variant callers, Octopus consists of three major steps: (1) candidate discovery, (2) genotyping, (3) variant filtering. However, a distinctive feature of Octopus is that the genotyping step can be optimized for certain experimental designs to improve calling accuracy and provide experiment-specific inferences. In addition, because these

Charlotte K. Y. Ng and Salvatore Piscuoglio (eds.), *Variant Calling: Methods and Protocols*,
Methods in Molecular Biology, vol. 2493, https://doi.org/10.1007/978-1-0716-2293-3_3,

calling models exist within a unified haplotype framework, calls are automatically phased and reported in a consistent format (VCF).

Currently, there are six calling models available:

1. Individual: calls germline variants in a bulk sample with known ploidy (including polyploids [7]).
2. Population: calls germline variants jointly in multiple bulk samples with known ploidy.
3. Trio: calls germline and de novo variants in a parent-offspring trio.
4. Cancer: calls germline and somatic variants in bulk tumor samples from a patient.
5. Polyclone: calls variants in a pooled sample of haploid organisms with unknown composition.
6. Cell: calls germline and somatic variants in a population of single cells.

2 Software

Octopus is freely available under the Massachusetts Institute of Technology (MIT) open-source license.

The source code is located at https://github.com/luntergroup/octopus.

Octopus Docker images can be found at https://hub.docker.com/r/dancooke/octopus.

Octopus Conda packages are available from https://anaconda.org/bioconda/octopus.

2.1 Requirements

In order to build and run Octopus, the following tools and packages must be installed in a Linux or Unix environment:

- A C++14 compiler with SSE2 support and standard library. Both GCC and LLVM have been tested. Generally speaking, LLVM is preferrable on MacOS and GCC on Linux systems. Use the most recent compiler where possible.
- Boost (1.75+) [https://www.boost.org].
- htslib (1.4+) [https://github.com/samtools/htslib].
- GMP (5.1.0+) [https://gmplib.org].
- CMake (3.9+) [https://cmake.org].
- Git (2.5+) [https://github.com/git/git].
- Python3 and the distro package.

All of these packages are available via Homebrew (https://github.com/Homebrew/brew) on Linux and MacOS.

2.2 Installation

The latest release version can be cloned with git:

```
$ git clone -b master https://github.com/luntergroup/octopus.
git
```

Alternatively, release versions can be downloaded directly from https://github.com/luntergroup/octopus/releases.

The simplest way to install Octopus is via the provided Python installer script:

```
$ octopus/scripts/install.py --dependencies --forests
```

This command will automatically download and install all required dependencies and pre-trained random forest models. However, if all dependencies are already installed then the `--dependencies` option may be omitted.

In addition to these two options, there are several others that may be of interest:

- `--clean`: clear all previous build history before building. This can sometimes resolve build problems caused by caching.
- `--prefix`: where to install the binary.
- `--architecture`: sets the CPU architecture to build for. By default, compilation will be optimized for the building machine. However, this can mean that the binary will not work on other systems.
- `--static`: use static linking where possible. The default is to use dynamic linking.
- `--threads`: sets the number of CPU threads to use for building.
- `--sanitize`: add compiler sanitizers to the binary, which can be very useful for debugging.

Successful installation may be quicky tested by running:

```
$ octopus --version
```

Which should print version and build information..

3 Methods

To see command line options available for all calling models, run:

```
$ octopus --help
```

To see command line options available to a specific calling model use:

```
$ octopus --caller MODEL --help
```

where `MODEL` is replaced with the name of the calling model of interest (individual, population, trio, cancer, polyclone, or cell).

3.1 Basic Usage

The basic usage for all Octopus commands is:

```
$ octopus \
--reference genome.fa \
--reads in1.bam [in2.bam [...]] or --reads-file bams.txt \
[options]
```

The mandatory options—the same for all Octopus calling models—are:

- `--reference`: the FASTA format reference genome. There must also be the corresponding fasta index (.fai) file in the same directory.
- `--reads`: the input reads to use for variant calling. Files can be in either BAM or CRAM format and must have associated index files (.bai and .crai, respectively). All reads should be mapped to the reference genome given in `--reference` and must include the @RG tag in the header file defining sample names. Samples split over several input files are automatically merged by Octopus.
- `--reads-file`: a text file containing a list (newline separated) of BAM/CRAM file paths. This option may be used in place of, or in addition to, `--reads`.

3.1.1 General Options

General options available to all calling models include:

- `--output`: the file to output results to. The output file format is automatically deduced from the extension provided. Supported formats are VCF (.vcf), compressed VCF (.vcf.gz), and BCF (.bcf). If the given output filename is not one of these formats, then VCF is assumed. If no output is specified, then calls are output to standard output in VCF. Octopus reports VCF 4.3 including use of the * allele (*see* **Note 1**).
- `--working-directory`: set the working directory. All relative paths will be with respect to this.
- `--samples`: a list of samples to call. The default behavior is to call all samples present in the input read files provided.
- `--samples-file`: a file containing a list (one per line) of samples to call.

- `--regions`: a list of genomic regions to analyze, in the format <chr>[:start[–[end]]] (e.g., chrX, chrX:100, chrX:100-, chrX:100–200). The default behavior is to call all contigs present in the reference genome index, which is usually not recommended (*see* **Note 2**).
- `--regions-file`: a file containing a list of genomic regions to call. The format can be either BED or the same format as `--regions` (one per line).
- `--skip-regions`: a list of genomic regions not to call. In the same format as `--regions`.
- `--skip-regions-file`: the opposite `--regions-file`.
- `--threads`: the number of CPU threads to use for analysis. If the option is specified without an argument (i.e., just `--threads`), then the number of threads is automatically determined based on the number of CPU cores available. Default is 1.
- `--temp-directory-prefix`: the prefix of the temporary directory created by Octopus.
- `--max-open-read-files`: the maximum number of read files that can be kept open at one time. Default is 250.
- `--target-read-buffer-memory`: the maximum amount of memory that should be used for buffering read date. This is a non-binding request. Default is 6 GB.
- `--max-reference-cache-memory`: the maximum memory that can be used for caching the reference genome. Default is 500 MB.
- `--ignore-unmapped-contigs`: do not raise an error if there is a mismatch between contigs in the reference file and input reads file (so long as all active contigs in the reads are in the reference).

3.1.2 Read Pre-processing

Octopus has built-in read pre-processing that either removes or transforms problematic reads before genotyping, which can improve calling accuracy and runtime. The default setting should work well for most users. However, some study designs may call for adjustments. Pre-processing options, available for all calling models, include:

- `--disable-read-preprocessing`: turns off all read pre-processing.
- `--mask-soft-clipped-bases`: set all soft-clipped base qualities to 0.
- `--disable-adapter-masking`: do not mask of putative adapter sequencing, determined by examining read pair overlap.

- `--disable-overlap-masking`: do not mask bases from one strand of overlapping read pairs.
- `--mask-inverted-soft-clipping`: mask soft-clipped sequence that is an inverted copy of a proximate sequence—a common artifact of MDA.
- `--mask-3prime-shifted-soft-clipped-heads`: mask soft-clipped read head sequence that is a copy of a proximate 3′ sequence—a common artifact of MDA.
- `--consider-unmapped-reads`: do not filter reads with the SAM unmapped flag set.
- `--min-mapping-quality`: minimum read mapping quality. Default is 5.
- `--allow-qc-fails`: do not filter reads with the SAM QC flag set.
- `--allow-marked-duplicates`: use reads with the SAM duplicate flag set.
- `--allow-octopus-duplicates`: do not apply Octopus' own read deduplication algorithm.
- `--allow-secondary-alignments`: use reads with the SAM secondary alignment flag set.
- `--allow-supplementary-alignments`: use reads with the SAM `supplementary` alignment flag set.
- `--allow-secondary-alignments`: use reads with the SAM secondary alignment flag set.
- `--no-reads-with-unmapped-segments`: filter reads with the SAM 0x2 or 0x8 flag bits set.
- `--no-reads-with-distant-segments`: filter reads with pairs on different contigs.
- `--no-adapter-contaminated-reads`: filter reads with putative adapter contamination.
- `--max-decoy-supplementary-alignment-mapping-quality`: filter reads with supplementary alignments to decoy contigs (those with suffix "_decoy") with mapping quality above this. Default is 5.
- `--max-unplaced-supplementary-alignment-mapping-quality`: filter reads with supplementary alignments to `unplaced` contigs (those with prefix "chrU_") with mapping quality above this. Default is 5.
- `--max-unlocalized-supplementary-alignment-mapping-quality`: filter reads with supplementary alignments to `unlocalized` contigs (those with suffix "_random") with mapping quality above this. Default is 5.
- `--no-adapter-contaminated-reads`: filter reads.

- `--min-read-length`: the minimum allowed read length. Default is 0.
- `--max-read-length`: the maximum allowed read length. Default is 10,000.
- `--split-long-reads`: split reads longer than `--max-read-length` into linked fragments.
- `--disable-downsampling`: turn off read downsampling.
- `--downsample-above`: downsample reads in regions where the depth is above this. Default is 1000.
- `--downsample-target`: when downsampling is applied, try to get the average depth in the region to this amount.
- `--use-same-read-profile-for-all-samples`: rather than generating per-sample read statistics for downstream use, just use one sample for all samples.

3.1.3 Variant Discovery Options

Variant discovery options, available to all calling models, include:

- `--disable-denovo-variant-discovery`: turn off all read-based candidate variant discovery generators.
- `--disable-pileup-candidate-generator`: do not discover candidate variants from raw input alignments.
- `--disable-assembly-candidate-generator`: do not discover candidate variants from local reassembly.
- `--disable-repeat-candidate-generator`: do not discover variants from realigned input alignments.
- `--source-candidates`: a VCF file containing variants to include as candidates. This does *not* mean that all of these candidates will be regenotyped (*see* **Note 3**).
- `--source-candidates-file`: a file containing a list (one per line) of VCF file paths to use as candidates.
- `--use-filtered-source-candidates`: include filtered variants as candidates. Default is only to use PASS variants.
- `--disable-denovo-variant-discovery`: turn off all read-based variant discovery; only use variants provided by the user with `--source-candidates` or `--source-candidates-file`.
- `--force-pileup-candidates`: use candidates discovered by the pileup generator, even if the reads appear to be misaligned.
- `--max-variant-size`: the maximum variant region size (with respect to the reference) that can be discovered.
- `--kmer-sizes`: a list of k-mer sizes to use for local reassembly. Default is 10 15 20.

- `--num-fallback-kmers`: the maximum number of additional k-mer sizes to try if all of the k-mers given in `--kmer-sizes` fail to produce a valid assembly graph. Default is 10.
- `--fallback-kmer-gap`: interval between fallback k-mers. Default is 10.
- `--max-region-to-assemble`: the maximum reference region length that can be locally assembled. Default is 600.
- `--max-assemble-region-overlap`: the maximum overlap between independent assembly regions. Default is 200.
- `--assemble-all`: force all input regions to be assembled, rather than just regions deemed likely to contain variation.
- `--min-bubble-score`: the minimum score of bubbles found in the assembly graph to include as candidates. Default is 2.

3.1.4 Haplotype Generating Options

Octopus uses an iterative haplotype proposal mechanism where haplotypes previously evaluated on incomplete read data are selectively extended in order to avoid windowing artefacts. Options that control haplotype proposal include:

- `--max-haplotypes`: the maximum number of haplotypes that can be considered by the genotype model. Default is 200.
- `--haplotype-holdout-threshold`: the maximum number of haplotypes that can be generated before some variants are held out. Default is 2500.
- `--haplotype-overflow`: regions with more haplotypes than this will be skipped. Default is 200,000.
- `--max-holdout-depth`: the maximum number of variants that can held out at one time. If more than this are required, then the region will be skipped. Default is 20.
- `--min-protected-haplotype-posterior`: haplotypes with posterior probability less than this may be pruned on each iteration. Default is 1e-10.
- `--extension-level`: aggressiveness of haplotype extension on each iteration [MINIMAL, CONSERVATIVE, MODERATE, AGGRESSIVE, UNLIMITED]. Default is MODERATE.
- `--lagging-level`: aggressiveness of haplotype lagging on each iteration [NONE, CONSERVATIVE, MODERATE, OPTIMISTIC, AGGRESSIVE]. Default is MODERATE.
- `--backtrack-level`: aggressiveness of haplotype backtracking [NONE, MODERATE, AGGRESSIVE]. Default is NONE.
- `--dont-protect-reference-haplotype`: allow the reference haplotype to be pruned.

3.1.5 General Variant Calling Options

Options that are applicable to more than one—but not necessarily all—calling models include:

- `--organism-ploidy`: the default ploidy for all contigs in the input samples.
- `--contig-ploidies`: the ploidy for individual contigs or sample contigs. The argument format is `[sample:]contig=-ploidy`. For example, `--contig-ploidies X=1` will set all the ploidy of the X contig of all samples to 1 while `NA24385:X=1` will just set the ploidy for `NA24385`. The default value for this option is `Y=1 chrY=1 MT=1 chrM=1`. All contigs not included in this option are assumed to have ploidy `--organism-ploidy` (default 2).
- `--min-variant-posterior`: the minimum variant posterior (in Phreds) to call. Default is 0.1.
- `--refcall`: call homozygous reference genotypes. Also accept an optional argument value [POSITIONAL, BLOCKED].
- `--refcall-block-merge-quality`: merge adjacent homozygous reference calls with probability difference less than this.
- `--min-refcall-posterior`: the minimum reference call probability (in Phreds) to report. Default is 0.
- `--max-refcall-posterior`: cap the reference call probability (in Phreds).
- `--snp-heterozygosity`: the SNP heterozygosity parameter to use for modeling germline variants.
- `--indel-heterozygosity`: the baseline indel heterozygosity parameter to use for modeling germline variants.
- `--max-uniform-genotype-priors`: use a uniform prior for genotypes; posteriors will be genotype likelihoods.
- `--max-genotypes`: the maximum number of genotypes that may be evaluated by the model. Default depends on calling model.
- `--max-genotype-combinations`: the maximum number of genotype combinations that may be considered by the model for joint calling. Default is 100,000.
- `--use-uniform-genotype-priors`: use a uniform prior for genotypes; posteriors will be genotype likelihoods.
- `--use-independent-genotype-priors`: assume sample genotype is independent for joint calling.
- `--disable-inactive-flank-scoring`: do not adjust haplotype likelihoods for potential windowing artefacts.
- `--dont-model-mapping-quality`: do not consider mapping quality for haplotype likelihoods.

- `--max-indel-errors`: the maximum number of sequencing indel errors to model. Default is 16.
- `--use-wide-hmm-scores`: use 32-bit rather than 16-bit scores for pair HMM probabilities.
- `--read-linkage`: molecular linkage between the input reads [UNLINKED, PAIRED, LINKED]. Default is PAIRED.
- `--sequence-error-model`: the sequence error model to use for calling. The argument is either the name of a built-in model or a file path to a custom model (*see* **Note 4**). The name format of built-in models is `[library-preparation][:platform]` where `library-preparation` is one of `PCR`, `PCRF`, or `10X`, and `platform` is one of `HISEQ-2000`, `HISEQ-2500`, `HISEQ-4000`, `X10`, `NOVASEQ`, `BGISEQ-5000`.
- `--max-vb-seeds`: the maximum number of seeds to use for the variational Bayes models. Default is 12.
- `--min-phase-score`: the minimum phase score (Phred scaled) allowed for a phase set; phase sets with scores less than this will be split. Default is 5.
- `--phasing-policy`: aggressiveness of applying phasing algorithm [AUTO, CONSERVATIVE, AGGRESSIVE]. Default is AUTO.
- `--bad-region-tolerance`: tolerance for skipping regions that are considered unlikely to be callable [LOW, NORMAL, HIGH, UNLIMITED]. Default is NORMAL.

3.1.6 Variant Filtering Options

Octopus optionally applies filtering to called variants to improve accuracy. Currently, variants can be filtered either using hand-tune threshold filters or a random forest classifier. Both the methods use a common set of variant annotations, which can be listed using:

```
$ octopus --help --annotations
```

Filter expressions are constructed using hard thresholds—inequalities between annotations and numerical constants. Multiple thresholds can be specified using the OR logical operator ("|"). For example, "AD < 1 | SB > 0.99" specifies a filter that fails a sample genotype if either any called allele depth is zero *or* there is substantial strand bias. Notably, annotations are automatically aggregated across alleles or haplotypes into a single numeric value when used in a filter expression.

Options used for filtering include:

- `--disable-call-filtering`: do not apply filtering to variant calls; report unfiltered calls.

- `--filter-expression`: expression to apply for filtering variant calls.
- `--annotations`: a list of filtering annotations to report in the final VCF output.
- `--forest-model`: the random forest model to use for variant filtering. This will result in the output VCF being annotated with a FORMAT/RFGQ—the per-sample genotype quality score (Phred scaled) assigned by the random forest model, and INFO/RFGQ_ALL—the geometric mean of all RFGQs. If no forest model is provided, then calls will be filtered using filter expressions.
- `--min-forest-quality`: the minimum passing random forest score. RFGQ_ALL is used to filter variant calls: sites with RFGQ_ALL less than `--min-forest-quality` are filtered with the "RF" FILTER; otherwise, sites are PASSed. Individual genotype fields are also independently filtered using each RFGQ using the same threshold. Filtered calls have FOTMAT/FT=RF and passing calls have FORMAT/FT=PASS. Default is 3.

3.2 Calling Germline Variants in Individuals

Calling germline variants in an individual with known ploidy from bulk sequencing data is the simplest study design that Octopus can analyze and is achieved using the **individual** calling model. There is no need to explicitly specify the individual calling model as it is the default model for single sample calling. An example command including recommend options is:

```
$ octopus \
 --reference hs38DH.fa \
 --reads NA24385.bam \
 --contig-ploidies chrX=1 \
 --regions-file hs38DH_chromosomes.bed \
 --sequence-error-model PCR.NovaSeq \
 --forest-model germline.forest \
 --threads 20 \
 --output NA24385.vcf.gz
```

Because Octopus assumes samples are diploid by default, for polyploid calling, it is also necessary to specify the `--organism-ploidy` option. This model can also be used for haploid calling (`--organism-ploidy 1`); however, this should only be used if the sample is clonal; if polyclonal samples are suspected, then the polyclone calling model should be used.

3.3 Calling Variants in Cohorts

It is usually preferable to call multiple samples jointly, as this ensures consistent variant representation across samples and can improve calling accuracy in some cases. If samples are unrelated, the **population** calling model is most appropriate. This is the default model for multi-sample calling so there is no need to specify it explicitly. An example including recommended options is:

```
$ octopus \
 --reference hs38DH.fa \
 --reads NA12878.bam NA24385.bam NA24631.bam \
 --contig-ploidies NA12878:chrY=0 NA24385:chrX=1 NA24631:chrX=1 \
 --regions-file hs38DH_chromosomes.bed \
 --sequence-error-model PCR.NovaSeq \
 --forest-model germline.forest \
 --threads 20 \
 --output cohort.vcf.gz
```

For calling larger cohorts (>10 samples), it may be preferable to first call smaller batches of samples, before joint calling all samples together but restricting candidate variants to those previously called using the `--disable-denovo-variant-discovery` and `--source-candidates` options. However, Octopus has not currently been tested at calling large cohorts.

3.4 Calling Germline and De Novo Mutations in Trios

If the input samples are a parent-offspring trio, then it is better to use the **trio** calling model than the population calling model. To invoke this model, it is just required to provide a pedigree (PED) file to the `--pedigree` option defining the trio. An example command is with recommended options is:

```
$ octopus \
 --reference hs38DH.fa \
 --reads NA24143.bam NA24149.bam NA24385.bam \
 --contig-ploidies NA24143:chrY=0 NA24149:chrX=1 NA24385:chrX=1 \
 --pedigree trio.ped \
 --regions-file hs38DH_chromosomes.bed \
 --sequence-error-model PCR.NovaSeq \
 --forest-model germline.forest \
 --threads 20 \
 --output trio.vcf.gz
```

Variants called as de novo mutations in the offspring will be annotated with the DENOVO INFO field. In addition, Mendelian violations that do not result in a novel allele are marked with the REVERSION INFO tag.

- INFO/PP: the posterior probability (Phred scaled) that the non-reference alleles segregate in the sample **and** are the call classification (germline/de novo) is correct.
- INFO/MP: the posterior probability (Phred scaled) that the model used for calling is correct. Low MP values indicate that the assumptions made by the calling model have been violated (e.g., the sample ploidy) and may indicate false calls.

Additional options that are specific to the trio calling model are:

- `--denovo-snv-prior`: the prior probability of a de novo SNV mutation in the sample.
- `--denovo-indel-prior`: the prior probability of a de novo indel mutation in the sample.
- `--min-denovo-posterior`: the minimum posterior probability (in Phreds) to call a de novo mutation. If a variant is not called de novo, a germline variant call may still be made. Default is 3.

3.5 Calling Germline and Somatic Mutations in Bulk Tumors

Octopus can call germline and somatic variants in bulk tumor data, with or without a normal sample. In both the cases, the **cancer** calling model is used. Several tumor samples may be jointly called, but all samples must be pertaining to the same individual or patient, such as with multi-region tumor sampling. Command line options relevant to this model:

- `--normal-samples`: normal sample names for cancer calling.
- `--somatic-snv-prior/--somatic-indel-prior`: SNV/indel somatic mutation priors (default 1e-4 and 1e-6, respectively).
- `--somatic-forest-model`: similar to `--forest-model` but used for SOMATIC mutations (*see* **Note 5**).
- `--max-somatic-haplotypes`: the maximum number of somatic haplotypes that the model may consider. Default is 2.
- `--min-credible-somatic-frequency`: only somatic haplotypes that have credible frequency above than this will be called. The credible region mass is specified by `--somatic-credible-mass`. Also specifies the minimum VAF for variants discovered by the pileup candidate generator. Default is 0.005.
- `--min-expected-somatic-frequency`: the minimum somatic haplotype frequency that is expected in the sample. Used to calculate somatic segregation probabilities. Default is 0.01.
- `--min-somatic-posterior`: the minimum posterior probability required to report a somatic mutation. Default is 0.5.

Values lower than this are not recommended due to this risk of misclassifying germline variation.

3.5.1 Paired Tumor–Normal Samples

To invoke the cancer calling model with paired samples, it is simply necessary to specify the name of the normal sample with the `--normal-samples` option or, because no other option has the prefix "normal-sample" (*see* **Note 6**), just `--normal-sample`:

```
$ octopus \
 --reference hs38DH.fa \
 --reads normal.bam tumor.bam [tumor2.bam [...]] or --reads-
file bams.txt \
 --normal-sample NORMAL \
 [options]
```

Note that, for the cancer calling model, only one normal sample is permitted. An example command with recommended options specified is:

```
$ octopus \
 --reference hs38DH.fa \
 --reads normal.bam tumorA.bam tumorB.bam \
 --normal-sample NORMAL \
 --contig-ploidies chrX=1 \
 --regions-file hs38DH_chromosomes.bed \
 --somatic-snv-prior 1e-7 \
 --somatic-indel-prior 1e-9 \
 --max-somatic-haplotypes 3 \
 --sequence-error-model PCR.NovaSeq \
 --forest-model germline.forest \
 --somatic-forest-model somatic.forest \
 --threads 20 \
 --output calls.vcf.gz
```

3.5.2 Reporting Phased Somatic Variants

Both germline and somatic variants are included in the output VCF file—somatic variants are annotated with the SOMATIC INFO tag. Calls are also annotated with INFO/PP.

- INFO/PP: the posterior probability that the alleles are segregating, **and** the classification (germline/somatic) is correct.
- FORMAT/HSS: the somatic status of each haplotype (one for each entry in the GT field): 1 for somatic, 0 for germline.
- FORMAT/HPC: the posterior Dirichlet concentration parameters for each called haplotype.
- FORMAT/MAP_HF: the maximum a posteriori frequency estimate of each called haplotype.

```
##fileformat=VCFv4.3
##INFO=<ID=SOMATIC,Number=0,Type=Flag,Description="Record includes a somatic mutation">
##FORMAT=<ID=GT,Number=1,Type=String,Description="Genotype">
##FORMAT=<ID=PS,Number=1,Type=String,Description="Phase set">
##FORMAT=<ID=HSS,Number=.,Type=Integer,Description="Somatic status for each haplotype">
#CHROM POS ID REF ALT QUAL FILTER INFO FORMAT NORMAL TUMOUR
1 50  . G  A   . . SOMATIC GT:PS:HSS  0|0:50:0,0   1|0|0:50:1,0,0
1 100 . A  C   . . SOMATIC GT:PS:HSS  0|0:100:0,0  0|0|0|1:100:0,0,1,1
1 150 . G  T   . . .       GT:PS:HSS  1|0:100:0,0  1|0|1|0:100:0,0,1,1
1 200 . A  AC  . . SOMATIC GT:PS:HSS  0|0:100:0,0  0|0|1|0:100:0,0,1,1
1 250 . TA T   . . .       GT:PS:HSS  1|1:100:0,0  1|1|1|1:100:0,0,1,1
2 300 . C  A,T . . SOMATIC GT:PS:HSS  0|2:300:0,0  1|0|2:300:1,0,0
2 350 . G  T   . . .       GT:PS:HSS  0|1:300:0,0  0|0|1:300:1,0,0
2 400 . T  C   . . SOMATIC GT:PS:HSS  0|0:300:0,0  1|0|0:300:1,0,0
3 100 . A  C   . . SOMATIC GT:PS:HSS  0|0:100:0,0  0|0|1|1:100:0,0,1,1
3 150 . C  CC  . . SOMATIC GT:PS:HSS  0|0:100:0,0  0|0|1|0:100:0,0,1,1
3 200 . G  T   . . .       GT:PS:HSS  0|1:100:0,0  0|1|1|1:100:0,0,1,1
4 100 . C  T   . . .       GT:PS:HSS  0|1:100:0,0  0|1|1|1:100:0,1,1,0
4 150 . A  G   . . SOMATIC GT:PS:HSS  0|0:100:0,0  0|0|1|0:100:0,1,1,0
4 200 . A  G   . . SOMATIC GT:PS:HSS  0|0:100:0,0  0|1|0|0:100:0,1,1,0
```

Fig. 1 Example Octopus VCF output, with the records colored by phase set

- FORMAT/HF_CR: frequency credible intervals for each called haplotype, where the credible mass is given by the command line option `--somatic-credible-mass`.

In order to report germline and somatic variants with phasing information, Octopus encodes unique somatic haplotypes in the GT field. In the VCF example shown in Fig. 1, records are colored by phase set. The first phase set 1:50 includes a simple somatic mutation; it is not phased with any other variants. Downstream of this is another phase block starting at 1:100 that includes two germline and two somatic variants. The first somatic mutation in this phase set at 1:100 is phased onto the germline haplotype including the reference allele at the germline variant at 1:150. The second somatic mutation in this phase set is phased with the alternate allele of this germline variant. In the third phase set stating at 2:300, there is a somatic mutation that segregates with a germline variant at the same position, and another somatic mutation which is phased onto the same germline haplotype—the reference. The somatic allele in the multiallelic record is determined by looking at the HSS flags (1 indicates somatic), so "C>A" is the somatic mutation in this case. The third phase set starting at 3:100 includes two somatic mutations phased onto the same germline haplotype, but the first somatic mutation was inferred to segregate with both the germline allele *and* somatic allele at 3:150, which suggests linear progression of the C>CC somatic mutation. The last phase set starting at 4:100 includes two somatic mutations phased onto the same germline haplotype, but not with each other.

3.5.3 Tumor-Only Samples

When a normal sample is not available, Octopus is still able to call germline and somatic variants in the tumor, with reduced classification accuracy (*see* **Note 7**). Because a normal sample is not present, the `--caller` option must be used to specify the cancer calling model:

```
$ octopus \
 --caller cancer \
 --reference hs38DH.fa \
 --reads normal.bam tumor.bam [tumor2.bam [...]] or --reads-
file bams.txt \
 [options]
```

If the cancer calling model was not specified, then the population calling model would be used. An example command with recommended options specified is:

```
$ octopus \
--caller cancer \
--reference hs38DH.fa \
--reads tumorA.bam tumorB.bam \
--contig-ploidies chrX=1 \
--regions-file hs38DH_chromosomes.bed \
--sequence-error-model PCR.NovaSeq \
--forest-model germline.forest \
--somatic-forest-moodel somatic.forest \
--threads 20 \
--output calls.vcf.gz
```

The same format of VCF output described above for paired mode is used for tumor-only calls.

3.6 Calling Variants in Haploid Polyclonal Samples

It is common for microbial or viral samples to contain several distinct strains of a species, either due to evolution or mixed infection. Most often, the number of strains is unknown a priori and must be inferred. Octopus can achieve this with the **polyclone** calling model. An example command with recommended options specified is:

```
$ octopus \
 --caller polyclone \
 --reference H37Rv.fa \
 --reads ecoli.bam \
 --max-clones 5 \
 --sequence-error-model PCR.NovaSeq \
 --forest-model polyploid.forest \
 --threads 20 \
 --output calls.vcf.gz
```

The genotype (GT) ploidies in the output VCF file indicates the locally inferred number of strains, in a similar way to how somatic haplotypes are reported in the cancer calling model. For final inference regarding the number of strains in the sample, it is important to consider whether distinct haplotypes may have resulted from variant calling errors that have been filtered (*see* **Note 8**).

3.7 Calling Germline and Somatic Variants in Single Cells

Octopus supports joint genotyping of single cells that have undergone whole-genome amplification, such as with multiple displacement amplification (MDA), with the **cell** calling model. The model also supports the inclusion of bulk samples and normal samples (i.e., samples that are ancestors to all other samples). Options specific to the cell calling model include:

- `--max-copy-gain`: the maximum number of copy chromosome gains to model.
- `--max-copy-loss`: the maximum number of copy chromosome losses to model.
- `--somatic-cnv-prior`: the prior probability of a somatic CNV event.
- `--dropout-concentration`: the default Dirichlet concentration parameter for all samples.
- `--sample-dropout-concentrations`: Dirichlet concentration prior on haplotype mixture frequencies for specified samples. Higher values make the likelihood more binomial and can therefore be used to encode bulk samples that are less impacted by amplification stochasticity (*see* **Note 9**).
- `--phylogeny-concentration`: the Dirichlet prior concentration parameter on phylogeny clone assignments.

An example command with recommended options specified for MDA single-cell datasets is:

```
$ octopus \
 --caller cell \
 --reference hs38DH.fa \
 --reads-file cell_bams.txt \
 --normal-samples CONTROL \
 --sample-dropout CONTROL=100 \
 --somatic-snv-prior 1e-7 \
 --somatic-indel-prior 1e-9 \
 --max-clones 4 \
 --max-vb-seeds 50 \
 --contig-ploidies chrX=1 \
 --regions-file hs38DH_chromosomes.bed \
 --sequence-error-model PCR.NovaSeq \
 --mask-inverted-soft-clipping \
 --mask-3prime-shifted-soft-clipped-heads \
 --forest-model germline.forest \
 --threads 20 \
 --output cells.vcf.gz
```

In addition to SOMATIC flags in, there are several VCF fields reported in the output unique to the cell calling model:

- INFO/PY: The called cell phylogeny.
- INFO/PPP: The posterior probability of the called phylogeny.
- INFO/PSPP: Posterior probabilities that the phylogeny has 0, 1, . . . nodes.
- FORMAT/PNAP: Posterior probabilities that the sample is assigned to node 0, 1, . . . in the called phylogeny.

The phylogeny reported in INFO/PY is given in bracket notation using the sterilization procedure:

```
def serialise(result="", node=ROOT):
 if (node != NULL):
 result += '(' + str(node.id)
 for child in node:
 serialise(result, child)
 result += ')
```

3.8 Using Config Files

As an alternative to specifying options explicitly on the command line, Octopus can accept configuration files that provide all, or a subset of option values. The format of configuration files is one option value per line separated with the=sign (i.e., OPTION=VALUE). Options that accept multiple values must be separated over multiple lines. For example, the single cell command in Subheading 3.6 could be modified to:

```
$ octopus \
 --config single_cell.config \
 --reads-file cell_bams.txt \
 --normal-samples CONTROL \
 --sample-dropout CONTROL=100 \
 --contig-ploidies chrX=1 \
 --threads 20 \
 --output cells.vcf.gz
```

where the config file is:

```
# single_cell.config
caller=cell
reference=hs38DH.fa
somatic-snv-prior=1e-7
somatic-indel-prio=1e-9
max-clones=4
max-vb-seeds=50
regions-file=hs38DH_chromosomes.bed
sequence-error-model=PCR.NovaSeq
mask-inverted-soft-clipping=yes
mask-3prime-shifted-soft-clipped-heads=yes
forest-model=germline.forest
```

3.9 Making Realigned Evidence BAMs

Octopus can output realigned and haplotype-tagged BAMs that provide visual evidence for calls. In order to use this feature:

1. All input BAM files must be single sample.
2. VCF output (--output) must be a file.

Evidence BAMs are requested with the `--bamout` command line option that accepts a single path argument. If a single sample is being called, the path is the full output BAM path, e.g.:

```
$ octopus \
 --reference hs38DH.fa \
 --reads NA12878.bam \
 --output calls.vcf.gz \
 --bamout NA12878_realigned.bam
```

If multiple samples are being called, then the value is a *directory* to output evidence BAMs for each sample—the name of the output BAM is the same as the input BAM, for each sample, e.g.:

```
$ octopus \
 --reference hs38DH.fa \
 --reads normal.bam tumorA.bam tumorB.bam \
 --normal-sample NORMAL \
 --output calls.vcf.gz \
 --bamout evidence_bams
```

Octopus adds several tags to realigned reads:

- HP: A list (, separated) of haplotype IDs that the read is inferred to originate from. A haplotype ID, which is zero-indexed, corresponds to column in the GT field of the affiliated phased VCF. A haplotype ID indicates that the read was unambiguously assigned to the haplotype, while multiple values indicate that the read could equally well be assigned to any of the listed haplotype.
- MD: Reference free alignments. As defined in the SAM specification.
- md: Like MD but alignment is relative to the inferred haplotype rather than the reference (i.e., mismatches are inferred sequencing errors).
- hc: The CIGAR alignment to the inferred haplotype.

By default, only reads supporting regions containing called variation are realigned. However, Octopus can also copy reads overlapping regions where no variation was called using the --bamout-type FULL command. Only primary reads are used for BAM realignment.

3.10 Calling Long Haplotypes

By default, Octopus does not aim to maximize haplotype length because this can result in long runtimes with limited genotyping accuracy improvement. However, tuning parameters for longer haplotypes—in all calling models—is reasonably straightforward. Relevant options include: `--max-haplotypes`, `--lagging-level`, `--extension-level`, `--backtrack-level`, `--min-protected-haplotype-posterior`, and `--read-linkage`.

The following command may be useful for calling HLA loci with short-read sequencing, for example:

```
$ octopus \
 --reference hs38DH.fa \
 --reads NA24385.bam \
 --regions-file HLA.bed \
 --sequence-error-model PCRF.NovaSeq \
 --forest-model germline.forest \
 --max-haplotypes 500 \
 --backtrack-level AGGRESSIVE \
 --output NA24385_HLA.vcf.gz
```

Another strategy for getting even longer haplotypes is to perform one round of calling with normal haplotype lengths, and then only use filtered calls for a second round, the options set for long haplotypes. For example, taking the output of the example from the individual calling model (Subheading 3.2), a second round may be achieved like:

```
$ octopus \
 --reference hs38DH.fa \
 --reads NA24385.bam \
 --contig-ploidies chrX=1 \
 --regions-file hs38DH_chromosomes.bed \
 --sequence-error-model PCR.NovaSeq \
 --forest-model germline.forest \
 --threads 20 \
 --source-candidates NA24385.vcf.gz \
 --disable-denovo-variant-discovery \
 --max-haplotypes 500 \
 --lagging-level OPTIMISTIC \
 --backtrack-level AGGRESSIVE \
 --min-protected-haplotype-posterior 1e-3 \
 --output NA24385.longhaps.vcf.gz
```

The advantage of this approach is that the candidate variants used should be high-quality so there is less time wasted evaluating long combinations of false variants. The same technique may be applied to calls from any calling model.

4 Notes

1. Octopus reports variants in VCF v4.3 (https://samtools.github.io/hts-specs/VCFv4.3.pdf).
   (a) The * allele is used to indicate sites that overlap with an explicitly called allele in another record. Note that unlike some other callers, Octopus uses * for both upstream and downstream overlapping alleles, and the interacting allele can be of any type (not just deletions). This means that any instance of 0 in the GT field can be treated as an explicit reference call.
   (b) All genotypes (GT) are reported using the *phased* notation (alleles are separated with | rather than /). The extent of phasing is indicated by phase set (PS) field, which gives the position (POS) of the first record in the phase set. Therefore, records where PS=POS should be treated as if they were unphased.

2. Some reference genomes, including human references, include decoy contigs to clean up reads that would otherwise be mis-mapped. These contigs are rarely of interest but can take up substantial genotyping compute time, so are best ignored during calling.

3. Regenotyping is usually taken to mean that genotypes will be reported for every variant in the input set. However, Octopus does not guarantee that all sites will be reported.

4. Custom sequencing error models can be generated using the bundled Python script "profiler.py" in the "scripts" directory.

5. If `--forest-model` is used but not `--somatic-forest-model`, then all calls (including SOMATIC ones) will be assessed by the model specified in `--forest-model`. This may be acceptable if the model given to `--forest-model` has been trained on both germline and somatic variants, but will likely result in poor performance if only germline variants have been used for training. It is recommended to use separate forests trained on germline and somatic calls independently, as is the case with the pre-trained provided forests.

6. The full option name is --normal-samples as other calling models (e.g., cell) may permit more than one normal sample; the cancer calling model (currently) only permits one normal. More generally, any command line option name may be shortened so long as the shortened name is not a prefix of another option. So --normal-samples can be shortened to --normal-sample or just --normal-sample as "normal-sample" is not a prefix to any other option.

7. For tumor-only samples, it is highly recommended to refine calls based on either INFO/PP or the classification confidence (INFO/CC)—the probability that the germline/somatic *classification* is correct, conditional on the variant being true. This annotation is not included by default but can be requested with the --annotations command. In particular, high-frequency (clonal) somatic mutations will often be misclassified as germline variants, so germline variant calls with low PP should be treated with caution.

8. In some cases, a false-positive variant call can result in calling a novel haplotype, and this will be reported for all sites that are phased with the problem variant. In particular, if a variant is filtered, then it is important to assess whether that variant caused the creation of a novel haplotype before making a final determination of the number of strains in the sample. Octopus can do this automatically by recalling the sample but only using previously called (PASS) variants as input, in the same was as was done for calling long haplotypes (Subheading 3.9). This

process could even be iterated until the call set only contains PASS variants.

9. The concentration to use requires some trial-and-error, but as a guiding principle—the more cells in the bulk sample, the higher the concentration can be.

References

1. Cooke D et al (2021) A unified haplotype-based method for accurate and comprehensive variant calling. Nat Biotechnol 39(7):885–892. https://doi.org/10.1038/s41587-021-00861-3
2. Rimmer A et al (2014) Integrating mapping-, assembly- and haplotype-based approaches for calling variants in clinical sequencing applications. Nat Genet 46:912–918
3. Kim S et al (2018) Strelka2: fast and accurate calling of germline and somatic variants. Nat Methods 15:591–594
4. Poplin R et al (2018) A universal SNP and small-indel variant caller using deep neural networks. Nat Biotechnol 36:983–987
5. Garrison, E and Math, G (2012). Haplotype-based variant detection from short-read sequencing. Preprint at. https://arxiv.org/abs/1207.3907
6. Poplin, R. et al (2017). Scaling accurate genetic variant discovery to tens of thousands of samples. Preprint at bioRxiv https://doi.org/10.1101/201178
7. Cooke, D. et al (2021). Benchmarking small-variant genotyping in polyploids. Preprint at bioRxiv https://doi.org/10.1101/2021.03.29.436766

Chapter 4

Accurate Ensemble Prediction of Somatic Mutations with SMuRF2

Weitai Huang, Ngak Leng Sim, and Anders J. Skanderup

Abstract

Accurate identification of somatic mutations is crucial for discovery and identification of driver mutations in cancer tumors. Here, we describe the updated **S**omatic **Mu**tation calling method using a **R**andom **F**orest (SMuRF2), an ensemble method that combines the predictions and auxiliary features from individual mutation callers using supervised machine learning. SMuRF2 provides an efficient workflow to predict both somatic point mutations (SNVs) and small insertions/deletions (indels) in cancer genomes and exomes. We describe the latest method and provide a detailed tutorial for running SMuRF2.

Key words Somatic mutation calling, Next-generation sequencing, Supervised machine-learning, Bioinformatics tools, Cancer genomics

1 Introduction

Somatic mutations are acquired DNA alterations in somatic cells throughout their lifetime. These mutations often go undetected and do not give rise to malignancies. However, somatic mutations accumulate when they evade DNA replication and repair checkpoints. Genetic alterations that occur at critical genes or DNA regulatory elements are capable of providing a selective advantage for the cancer cell to survive and proliferate, leading to tumorigenesis. Somatic mutations may trigger the constitutive activation of signaling pathways and disrupt feedback mechanisms that give rise to the hallmarks of cancer [1]. Identifying these key driver mutations is an important step for cancer research and stratified treatments.

Somatic mutation calling is the process of identifying somatic genetic alterations at the local tumor site of a patient. This is commonly performed through a comparative analysis of the DNA in the tumor tissue and normal blood sample extracted from the same patient. The analysis could be performed over a period of time

Charlotte K. Y. Ng and Salvatore Piscuoglio (eds.), *Variant Calling: Methods and Protocols*,
Methods in Molecular Biology, vol. 2493, https://doi.org/10.1007/978-1-0716-2293-3_4,

to monitor the changes in the patient tumor genome before and after cancer treatment as well as mapping the genomic profiles of metastatic lesions detected at a later stage. Here, we will focus on the identification and analysis of single-nucleotide variants (SNVs) and short insertions and deletions (indels) in matched tumor and normal samples.

We developed Somatic Mutation calling using a Random Forest (SMuRF), an ensemble mutation caller that integrates the predictions from multiple callers using supervised machine learning (Fig. 1 and **Note 1**). We have shown that SMuRF accurately predicts SNVs and indels at low allele frequencies and performs well when tested on multiple cancer types [2]. Previously, we incorporated multiple SNV and indel callers including: MuTect2 [3], VarDict [4], VarScan2 [5], and FreeBayes (ArXiv: https://arxiv.org/abs/1207.3907) into the prediction model [6]. In our latest version, SMuRF version 2.0 (SMuRF2), we also included the Strelka2 caller [7] to provide higher sensitivity and overall accuracy to the somatic mutation calls. SMuRF2 also leverages on efficient data processing by utilizing the newest GATK4 (https://github.com/broadinstitute/gatk) in the bcbio-nextgen 1.1.5+ pipeline. New sequence and variant features from Strelka2 are incorporated into the SMuRF2 prediction models, improving its sensitivity. Overall, SMuRF2 improves performance over existing individual callers and outperforms the previous version of SMuRF (Table 1 and Fig. 2). Benchmarks using independent data from the DREAM Somatic Mutation Challenge [8] also showed that SMuRF2 performed better than the previous version (Fig. 3).

In this chapter, we describe the steps required to set up SMuRF2 for somatic mutation calling, understanding the required input and output file formats and how to customize SMuRF2 parameters for your analysis.

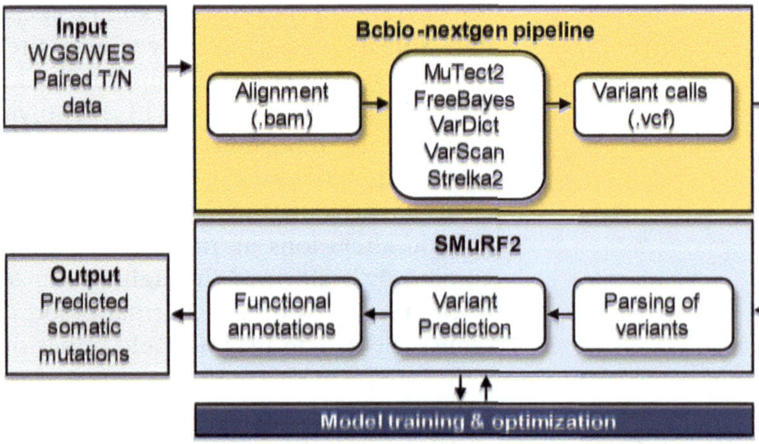

Fig. 1 Workflow of SMuRF2

Table 1
Performance of SMuRF and callers on chronic lymphocytic leukemia (CLL) and medulloblastoma (MBL) whole genomes

	Caller	Precision	Recall	F1
CLL	SMuRF2	0.965	0.926	0.945
	SMuRFv1.6	0.928	0.884	0.906
	Strelka2	0.551	0.917	0.688
	MuTect2	0.435	0.902	0.587
	FreeBayes	0.193	0.854	0.316
	VarDict	0.214	0.873	0.344
	VarScan	0.030	0.889	0.059
MBL	SMuRF2	0.983	0.854	0.914
	SMuRFv1.6	0.896	0.822	0.857
	Strelka2	0.380	0.854	0.526
	MuTect2	0.388	0.861	0.535
	FreeBayes	0.098	0.733	0.174
	VarDict	0.063	0.749	0.116
	VarScan	0.014	0.762	0.028

2 Materials

2.1 Environment

SMuRF2 is implemented as an R package (requires R version ≥3.3.1) (https://cran.rproject.org) and runs on a standard computer (4 CPUs, 16GB RAM). The source code for SMuRF2 is available at Github (https://github.com/skandlab/SMuRF). SMuRF2 processes both whole-genome (~80–100× coverage) and whole-exome (~200–400× coverage) sequencing samples with a typical runtime between 3 and 10 min. It is recommended to run SMuRF2 on systems with higher RAM (minimum 4 GB) for whole-genome sequencing samples.

To get started, install the package from Github with the following commands in R:

```
# From Github directly
install.packages("devtools")
library(devtools)
install_github("skandlab/SMuRF", subdir="smurf")

# Download package from Github and install
install.packages("my/current/directory/smurf",
repos = NULL, type = "source")
```

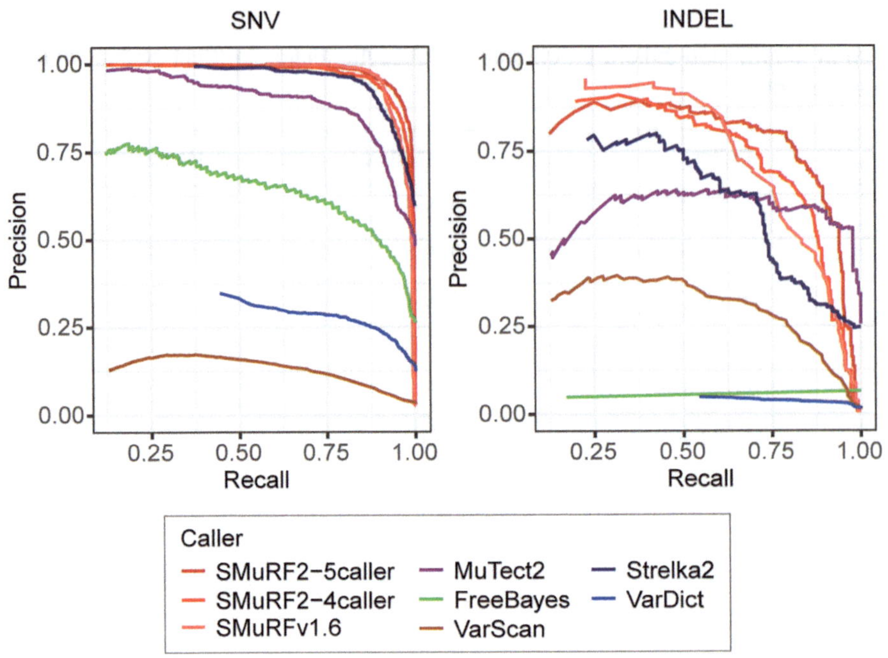

Fig. 2 Performance of SMuRF2. Precision-recall profiles for individual somatic mutation callers and SMuRF are evaluated on SNV and indels using 20% withheld test data. SMuRF models were trained on 80% matched tumor–normal WGS data from a chronic lymphocytic leukemia patient and a medulloblastoma patient. To generate variations in sequencing coverage and tumor purity, noise was injected into the training and testing data; (1) normal samples down-sampled to 30× coverage, (2 and 3): lowered purity tumor samples spiked in with either 20× or 40× normal reads, respectively (70% and 55% purity)

SMuRF2 requires the following dependencies: bioconductor:: *VariantAnnotation* (R-3.3 & R-3.4) or BiocManager:: *VariantAnnotation* (≥R-3.5), *data.table* and *h2o* (v3.26.0.7 and above). These dependencies will be installed automatically the first time SMuRF2 is installed.

2.2 Input Requirements

SMuRF2 performs its prediction based on a set of input somatic mutations from the following five mutation callers: MuTect2, VarDict, VarScan, FreeBayes, and Strelka2 (*see* **Note 2**). The required Variant Call Format (VCF) files from these five callers can be generated by the *variant2* bcbio-nextgen somatic mutation calling workflow (https://bcbio-nextgen.readthedocs.io). To ensure full compatibility with SMuRF2, make sure you run the workflow with the following bcbio-nextgen YAML configuration parameters (Table 2).

The variant calls can alternatively be generated by independently running each of the five callers, but we recommend using *bcbio-nextgen* for ease of use and full compatibility with SMuRF2 and its input data requirements.

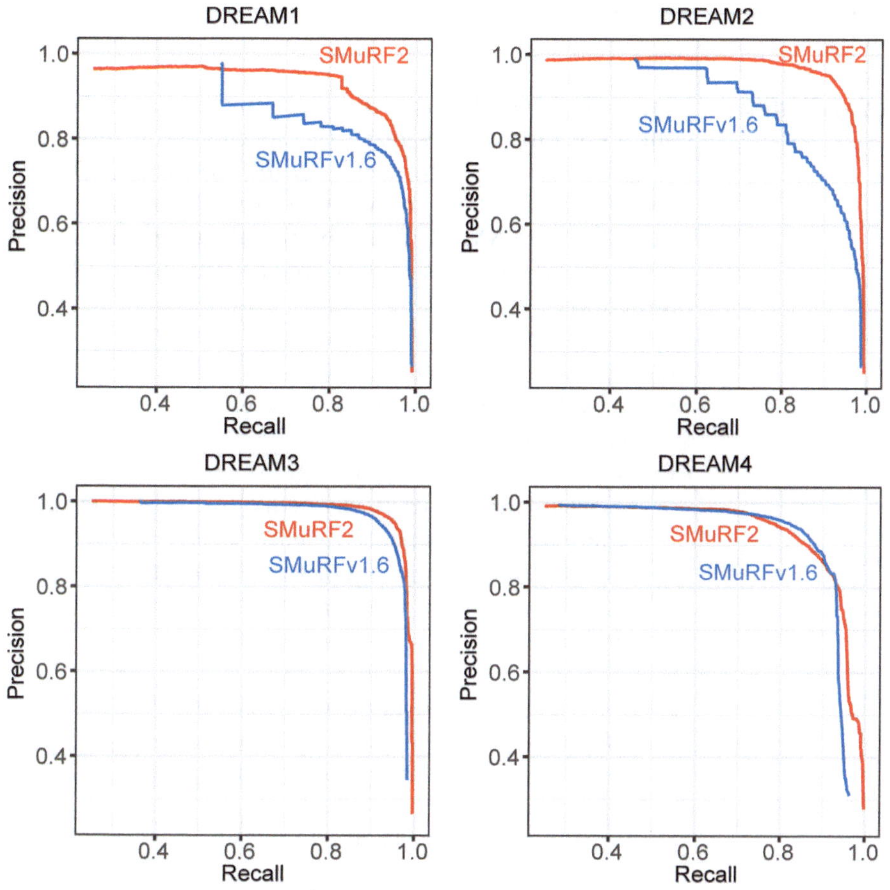

Fig. 3 Benchmark of SMuRF2 using independent data from the DREAM Somatic Mutation Challenge [8]

Table 2
Bcbio-nextgen parameters required for SMuRF2

bcbio-nextgen parameter	Value
mark_duplicates	True
Recalibrate	False
Realign (GATK4)	False
Variantcaller	[mutect2,freebayes,vardict,varscan,strelka2]

2.3 Test Dataset

SMuRF2 is pre-loaded with a test dataset that comprises input VCF files generated from a subset of a chronic lymphocytic leukemia (CLL)-matched tumor and normal sample (European Genome-phenome Archive (EGA) accession number EGAS00001001539). Below, we will demonstrate the use of SMuRF2 based on the analysis of this dataset.

3 Methods

3.1 Running SMuRF2

Minimally, SMuRF2 requires the following parameters: (1) path to your VCF inputs (***directory***), (2) specified run mode (***mode***), and (3) genomic assembly build (***build***) to begin. Other SMuRF2 running parameters allow you to customize your settings (Table 3). Run SMuRF2 on the test data provided in the package using the commands below to obtain both SNV and indel predictions:

```
library("smurf") #load package
smurf() #check version and parameters
# "SMuRF2.0.9 (16th Apr 2021)"
smurf(directory=NULL, mode=NULL, nthreads = -1,
     annotation=F, output.dir=NULL, parse.dir=NULL,
     snv.cutoff = 'default', indel.cutoff = 'default',
     build=NULL, change.build=F, find.build=F,
     t.label=NULL, re.tabIndex=F,
     check.packages=T, file.exclude=NULL)

mydir <- paste0(find.package("smurf"), "/data") #test data
root dir
mydir <- "my/path/to/smurf/data" #custom dir to smurf package
data
setwd(mydir) #set target directory as working directory

myresults <- smurf(directory=mydir, mode="combined",
build="hg19") #run SMuRF
```

SMuRF2 requires five caller VCF (vcf.gz) files as input stated under the ***directory*** parameter. Both caller.vcf.gz (compressed) and caller.vcf are accepted. Provide a path to a directory containing all five caller VCF files tagged with the names of each caller (e.g., sample1-mutect2.vcf.gz, sample1-freebayes.vcf.gz, sample1-vardict.vcf.gz, sample1-varscan.vcf.gz, and sample1-strelka2.vcf.gz, *see* **Note 3**). The tab-indexed (.tbi) files for each caller are required for the parsing step. If the required files are missing, specify using *re.tabIndex=T* on SMuRF2 to generate them from the vcf files.

```
# Running on Windows OS (nthreads = 1)
myresults <- smurf(directory = paste0(find.package("smurf"),
"/data"),
     mode ="combined", nthreads = 1, build = 'hg19',
     re.tabIndex = T)  #generate .tbi files in directory
#"Generating .tbi files in directory..."
```

Table 3
SMuRF2 running parameters

Arguments	Description
directory	Choose directory where the variant caller format(VCF) files are located
output.dir	Path to output directory (if saving files as. txt)
parse.dir	Specify if changing SMuRF2 default cutoffs. Path to the location of existing snv-parse.txt and indel-parse.txt files generated by SMuRF2
mode	Choose "snv," "indel," or "combined" (snv+indel). "Combined" provides a separate list of SNVs and indels
annotation	TRUE or FALSE (default). Provide gene annotations for each variant call
nthreads	Number of cores used for RandomForest prediction. Default (−1) for maximum number of cores. *For 32-bit Windows, only 1 core is allowed (nthreads = 1)*
t.label	(Optional) Provide the sample name for your tumor sample to ease the identification of the normal and tumor sample names in your vcf
file.exclude	(Optional) Additional keywords in file directory names to be filtered
build	Specify your human genome build: build="hg19" or build="hg38"
change.build	TRUE or FALSE (default). For conversion of your genomic coordinates
Find.Build	TRUE or FALSE (default). Additional genome build check for the annotation step
snv.cutoff	Default SNV model cutoff, unless a number between 0 and 1 is stated
indel.cutoff	Default indel model cutoff, unless a number between 0 and 1 is stated
re.tabIndex	TRUE or FALSE (default). Set to TRUE to create tab-indexed (.tbi) files for each vcf
check.packages=T	Developer mode

```
# If the vcf files are in different directories, you may
manually specify each directory in this order:
dir.list <- list(mutect='/path1/to/mutect2.vcf.gz',
                 freebayes='/path2/to/freebayes.vcf.gz',
                 vardict='/path3/to/vardict.vcf.gz',
                 varscan='/path4/to/varscan.vcf.gz',
                 strelka='/path5/to/strelka2.vcf.gz')

myresults <- smurf(directory=dir.list,
      mode="combined", build='hg19')
```

3.2 Retrieving Gene Annotation Information

The *cdsannotation* function annotates mutations found in the coding sequence (CDS) region using SnpEff [9] (generated from the bcbio-nextgen pipeline).

The genome build must be defined (either *build='hg19'* or *build='hg38'*) to ensure gene annotations are added correctly.

SMuRF2 checks the defined genome build with the one used in the original input VCF file. If unsure of the genome build used in the analysis, specify *find.build=T*. The genome build will be corrected to the original build. To convert the SMuRF2 results to a different build from the input VCF, specify *change.build=T*.

```
# Detecting genome build
myresults <- smurf(directory = paste0(find.package("smurf"),
"/data"),
     mode ="combined", nthreads = 1, annotation = T,
     build = 'hg38', #wrong build stated
     find.build = T) #if unsure of genome build

# Genome build stated in SMuRF2:"hg38"
# Ref genome used in vcf:"/bcbio/genomes/Hsapiens/GRCh37/seq/
GRCh37.fa"
# "Warning: build provided does not match ref genome used in
vcf. SMuRF2 CDS annotation may not run properly if genome build
is incorrect."
# "Changing build variable provided" (change.build=T)
# "hg38 -> hg19"
# "Final genome build used for analysis: hg19" (follows
original ref)

# Changing genome build
myresults <- smurf(directory = paste0(find.package("smurf"),
"/data"),
     mode ="combined", nthreads = 1, annotation = T,
     build = 'hg19',
     change.build = T) #genome build conversion

# Genome build stated in SMuRF2:"hg19"
# "Ref genome used in vcf:"/bcbio/genomes/Hsapiens/GRCh37/seq/
GRCh37.fa"
# "Final genome build used for analysis: hg19"

# "Compiling annotations"
# "Changing annotations from hg19 to hg38"
```

3.3 Tweaking the Precision and Recall of SMuRF2

SMuRF2 is calibrated for both high accuracy and sensitivity. The SNV and indel cut-offs in our model are defined to achieve 99% sensitivity in our test dataset. The user may want to adjust the SMuRF2 confidence score threshold to balance the precision or the recall (sensitivity) on their specific datasets. Use parameters *snv.cutoff* or *indel.cutoff* to adjust these thresholds (*see* **Note 4**). To re-adjust the cut-off value of an existing SMuRF2 output, simply provide the directory (*parse.dir*) to the snv-parse and indel-parse files for re-processing (*see* **Note 5**).

```
# Start with default cutoffs
myresults <- smurf(directory = paste0(find.package("smurf"),
"/data"),
     mode="combined",
     snv.cutoff='default', indel.cutoff='default',
     output.dir = 'C:/Users/admin/myresults')

# Modify cutoff from existing SMuRF2 parse files
myresults <- smurf(directory = paste0(find.package("smurf"),
"/data"),
     mode="combined",
     snv.cutoff=0.2, indel.cutoff=0.1, #specify new cutoffs
     parse.dir = 'C:/Users/admin/myresults',
     output.dir = 'C:/Users/admin/myresults2') #new output
```

3.4 Interpreting and Saving SMuRF2 Output

SMuRF2 output will be stored as a list object with the following information:

1. Variant statistics (stats)
   General statistics showing the total number of somatic mutations that passed each caller as well as SMuRF2.
2. Predicted mutations (predicted)
   *The predicted file is a data frame containing information on the predicted SNVs/indels, including their SMuRF2 scores (*Table 4*).*
3. Predicted mutations with annotations (annotated)
   *The annotation file consists of 15 columns (*Table 5*). The information in the annotation file is extracted from SnpEff (http://snpeff.sourceforge.net/SnpEff_manual.html#input).*
4. Parsed-raw file (parse)
   The original list of somatic mutations passed by at least 1 individual mutation caller.
5. Time taken (time).
   Information on run-time.

To export the SMuRF2 output in tab-delimited file format, specify the *output.dir* as the targeted directory (*see* **Note 6**).

```
# Saving SMuRF2 output as .txt files

myresults <- smurf(directory = paste0(find.package("smurf"),
```

Table 4
Description for each column in the SMuRF2 output file

Column Name	Description
Chr	Chromosome number
START_POS_REF/ END_POS_REF	Start and End nucleotide position of the somatic mutation
REF/ALT	Consensus Ref and Alt nucleotide changes of the highest likelihood
REF_MFVdVs/ ALT_MFVdVs	Reference and Alternative nucleotide changes from each caller; Mutect2 (M), Freebayes (F), Vardict (Vd), Varscan (Vs), and Strelka2 (not abbreviated to preserve column name)
FILTER	Passed (TRUE) or Reject (FALSE) mutation calls from the individual callers
Sample_Name	Sample name is extracted based on your labeled samples in the vcf files
Alt_Allele_Freq	Mean variant allele frequency calculated from the tumor reads of the callers
Depth ref/alt N/T	Mean read depth from the N/T sample for ref/alt alleles
SMuRF_score	SMuRF2 confidence score of the predicted mutation

Table 5
Description for gene annotation features

Annotation info	Description
Allele	Alternate allele (ALT)
Annotation	Effect based on Sequence Ontology terms
Impact	Estimated level of impact (high, moderate, low, modifier)
Gene name	HGNC gene name
Gene ID	Gene ID
Feature type	Sequence type
Feature ID	Sequence ID
Transcript biotype	Ensembl biotypes (coding, non-coding, etc.)
Rank	Exon or intron rank
HGVS.c	HGVS cDNA notation
HGVS.p	HGVS protein notation
cDNA.pos	Position in cDNA and transcript cDNA length
CDS.pos	Position and number of coding bases
AA.pos	Position and number of amino acids
Distance	Distance to nearest feature

```
"/data"),
    mode="combined",
     output.dir = 'C:/Users/admin/myresults') #path to output
directory
```

The summary statistics for each run can be retrieved using "myresults$smurf_snv$stats_snv" and "myresults$smurf_indel $stats_indel" for SNVs and indels, respectively.

```
myresults$smurf_indel$stats_indel
#              Passed_Calls
# Strelka2              466
# Mutect2               232
# FreeBayes             306
# VarDict               483
# VarScan              1273
# Atleast1             2431
# Atleast2              278
# Atleast3               43
# Atleast4                7
# All5                    1
# SMuRF_INDEL            88

myresults$smurf_snv$stats_snv
#            Passed_Calls
# Strelka2           1362
# Mutect2            1539
# FreeBayes           216
# VarDict             239
# VarScan            1734
# Atleast1           4017
# Atleast2            928
# Atleast3             60
# Atleast4             48
# All5                 37
# SMuRF_SNV          1043
```

We added gene annotations using SnpEff, and SMuRF2 extracts the coding annotations from the canonical transcripts with the highest functional impact (Table 6). The parsed-raw output allows the user to examine all variant caller variables used in the SMuRF2 training model.

```
myresults$smurf_indel$parse_indel
myresults$smurf_snv$parse_snv
myresults$time.taken
# <!-- Time difference of 20.52405 secs -->
```

Table 6
SMuRF2 predicted SNVs with CDS annotations (example)

No.	1	2
Chr	9	3
START_POS_REF	119582946	137807254
END_POS_REF	119582946	137807254
REF	C	C
ALT	T	T
REF_MFVdVs	C/NA/NA/NA/C	C/NA/NA/NA/C
ALT_MFVdVs	T/NA/NA/NA/T	T/NA/NA/NA/T
FILTER_Mutect2	TRUE	TRUE
FILTER_Freebayes	FALSE	FALSE
FILTER_Vardict	FALSE	FALSE
FILTER_Varscan	FALSE	FALSE
FILTER_Strelka2	TRUE	TRUE
Sample_Name	icgc_cll_tumour	icgc_cll_tumour
Alt_Allele_Freq	0.472	0.5
N_refDepth	45	39
N_altDepth	1	1
T_refDepth	56	14
T_altDepth	50	14
Allele	T	T
Annotation	synonymous_variant	synonymous_variant
Impact	LOW	LOW
Gene_name	ASTN2	DZIP1L
Gene_ID	ENSG00000148219	ENSG00000158163
Feature_Type	Transcript	Transcript
Feature_ID	ENST00000313400	ENST00000327532
Transcript_BioType	protein_coding	protein_coding
Rank	12/23	6/16
HGVS.c	c.2157G>A	c.966G>A
HGVS.p	p.Thr719Thr	p.Glu322Glu
cDNA.pos	2258/4747	1329/3493
CDS.pos	2157/4020	966/2304

(continued)

Table 6 (continued)

AA.pos	719/1339	322/767
Distance	.	.
REGION	CDS	CDS
SMuRF_score	0.964683105	0.916847514

4 Notes

1. We have previously trained the SMuRF SNV and indel models on a set of gold standard somatic mutation calls that were manually curated by the International Cancer Genome Consortium (ICGC) [2].
2. Although SMuRF2 is generally robust to variation in sequencing coverage, we recommend ensuring a tumor and normal coverage of at least 60× and 30×, respectively.
3. It is recommended to place the input VCF files in the same directory. VCF file names should contain the unique caller names: "mutect," "freebayes," "vardict," "varscan," and "strelka." Additionally, non-unique VCF file names (e.g., Sample1-mutect.vcf.gz and Sample1-mutect-germline.vcf.gz) can be filtered by stating the keyword using the *file.exclude* parameter (*file.exclude='germline'*).
4. Higher SMuRF2 cut-off scores provide a smaller, accurate set of calls. This may be useful when analyzing whole-genome sequencing data where a large set of non-coding mutations are detected.
5. To further analyze all mutations (passed by at least one caller), look into the parse-raw output under "myresults$smurf_indel $parse_indel" and "myresults$smurf_snv$parse_snv".
6. SMuRF2 output is saved as a tab-delimited .txt file (not VCF). The output consists of columns that differ from the conventional VCF format (*see* Table 4).

References

1. Hanahan D, Weinberg Robert A (2011) Hallmarks of cancer: the next generation. Cell 144(5):646–674. https://doi.org/10.1016/j.cell.2011.02.013
2. Huang W, Guo YA, Muthukumar K, Baruah P, Chang MM, Skanderup AJ (2019) SMuRF: portable and accurate ensemble prediction of somatic mutations. Bioinformatics (Oxford, England) 35(17):3157–3159. https://doi.org/10.1093/bioinformatics/btz018
3. Cibulskis K, Lawrence MS, Carter SL, Sivachenko A, Jaffe D, Sougnez C, Gabriel S,

Meyerson M, Lander ES, Getz G (2013) Sensitive detection of somatic point mutations in impure and heterogeneous cancer samples. Nat Biotechnol 31:213. https://doi.org/10.1038/nbt.2514

4. Lai Z, Markovets A, Ahdesmaki M, Chapman B, Hofmann O, McEwen R, Johnson J, Dougherty B, Barrett JC, Dry JR (2016) VarDict: a novel and versatile variant caller for next-generation sequencing in cancer research. Nucleic Acids Res 44(11):e108. https://doi.org/10.1093/nar/gkw227
5. Koboldt DC, Zhang Q, Larson DE, Shen D, McLellan MD, Lin L, Miller CA, Mardis ER, Ding L, Wilson RK (2012) VarScan 2: somatic mutation and copy number alteration discovery in cancer by exome sequencing. Genome Res 22(3):568–576. https://doi.org/10.1101/gr.129684.111
6. Huang W, Guo YA, Chang MM, Skanderup AJ (2020) Ensemble-based somatic mutation calling in cancer genomes. In: Boegel S (ed) Bioinformatics for cancer immunotherapy: methods and protocols. Springer, US, New York, NY, pp 37–46. https://doi.org/10.1007/978-1-0716-0327-7_3
7. Kim S, Scheffler K, Halpern AL, Bekritsky MA, Noh E, Källberg M, Chen X, Kim Y, Beyter D, Krusche P, Saunders CT (2018) Strelka2: fast and accurate calling of germline and somatic variants. Nat Methods 15(8):591–594. https://doi.org/10.1038/s41592-018-0051-x
8. Ewing AD, Houlahan KE, Hu Y, Ellrott K, Caloian C, Yamaguchi TN, Bare JC, P'ng C, Waggott D, Sabelnykova VY, participants I-TDSMCC, Xi L, Dewal N, Fan Y, Wang W, Wheeler D, Wilm A, Ting GH, Li C, Bertrand D, Nagarajan N, Chen Q-R, Hsu C-H, Hu Y, Yan C, Kibbe W, Meerzaman D, Cibulskis K, Rosenberg M, Bergelson L, Kiezun A, Radenbaugh A, Sertier A-S, Ferrari A, Tonton L, Bhutani K, Hansen NF, Wang D, Song L, Lai Z, Liao Y, Shi W, Carbonell-Caballero J, Dopazo J, CCK L, Guinney J, Kellen MR, Norman TC, Haussler D, Friend SH, Stolovitzky G, Margolin AA, Stuart JM, Boutros PC (2015) Combining tumor genome simulation with crowdsourcing to benchmark somatic single-nucleotide-variant detection. Nat Methods 12:623. https://doi.org/10.1038/nmeth.3407
9. Cingolani P, Platts A, Wang le L, Coon M, Nguyen T, Wang L, Land SJ, Lu X, Ruden DM (2012) A program for annotating and predicting the effects of single nucleotide polymorphisms, SnpEff: SNPs in the genome of Drosophila melanogaster strain w1118; iso-2; iso-3. Fly 6(2): 80–92. https://doi.org/10.4161/fly.19695

Chapter 5

Detecting Medium and Large Insertions and Deletions with transIndel

Ting-You Wang and Rendong Yang

Abstract

Insertions and deletions (indels) are primarily detected from DNA sequencing (DNA-seq) data, but their transcriptional consequences remain unexplored due to challenges in distinguishing medium- and large-sized indels from RNA splicing events in RNA-seq data. We introduce transIndel, a splice-aware algorithm that parses the chimeric alignments predicted by a short read aligner and reconstructs the mid-sized insertions and large deletions based on the linear alignments of split reads from DNA-seq or RNA-seq data. Here, we describe the method and provide a tutorial on the installation and application of transIndel.

Key words Indel detection, DNA-seq, RNA-seq, Split reads, Chimeric alignment

1 Introduction

A key aspect of genomic research is to determine the genetic difference among individuals and phenotypic differences between healthy and disease states, including cancers. The recent development of next-generation sequencing (NGS) including DNA-seq and RNA-seq has enabled the detection of genomic variations at single-nucleotide resolution. DNA-seq data is generally used as the primary source of genetic mutations while RNA-seq data is used for evaluating gene expression changes. However, with the generation of massive multi-omics data, there is a consensus that the integrative analysis of them may provide researchers with a deeper and comprehensive understanding of disease etiology and the functional consequences and interactions between them. For example, integrative analysis of DNA-seq and RNA-seq data could determine whether DNA variants are expressed, identify alterations in genomic DNA when altered DNA fragments escape hybrid capture in whole exome sequencing applications, or identify non-canonical RNA splicing events that are caused by underlying DNA alterations [1].

Charlotte K. Y. Ng and Salvatore Piscuoglio (eds.), *Variant Calling: Methods and Protocols*,
Methods in Molecular Biology, vol. 2493, https://doi.org/10.1007/978-1-0716-2293-3_5,

In order to detect single-nucleotide variants, small insertions and deletions (indels, <10 bp), and structural variations (or gene fusions) from RNA-seq data, computational algorithms have been developed [2–7]; however, the scientific community currently lacks an effective and efficient method for indel identification, especially for mid-sized and large indels, from RNA-seq data. Detecting indels in RNA-seq data is a challenging task for three reasons. First, RNA-seq splice-aware aligners fail to map short reads which contain mid-sized insertions and large deletions, as these will be treated as splicing junctions [8]. Second, most existing indel callers were designed for indel discovery using DNA-seq data, so utilization with RNA-seq data results in an extremely high false-positive rate of medium- and large-sized indel calls due to inability to account for splicing events [8]. Third, it is difficult to distinguish genomic indels from non-canonical splicing events, such as microexons and exitrons [9].

In this chapter we introduce transIndel [10], an algorithm designed flexibly to detect indels from DNA-seq or RNA-seq data. transIndel employs the chimeric alignments from short read aligners such as BWA-MEM [11] and reconstructs the mid-sized insertions and large deletions based on the linear alignments of chimeric reads. In addition, transIndel makes use of several filters to distinguish deletions with RNA splicing events when analyzing RNA-seq data.

In general, transIndel takes two steps to call indels: (1) indel reconstruction, and (2) detect indels using samtools pileup function [12]. The identified indels are reported in standard VCF format.

In indel reconstruction step, an indel is detected based on the genomic location, strand, and type of the two linear alignments reported from a chimeric alignment. Then, the CIGAR string of this chimeric alignment is re-defined as a linear alignment where the inferred indel event is included. The linear alignments are categorized into MS, SM, and O classes, where MS stands for left matched alignment and right soft-clipping (e.g., 40M60S as the CIGAR string); SM stands for left soft-clipping and right matched alignment (e.g., 40S60M as the CIGAR string) and O stands for other alignments that do not belong to MS or SM class. The indels are inferred based on the genomic location and types of the two linear alignments in a chimeric alignment (Fig. 1). Eventually, a new BAM file with a corrected CIGAR string will be generated for the subsequent indel calling step using pileup function.

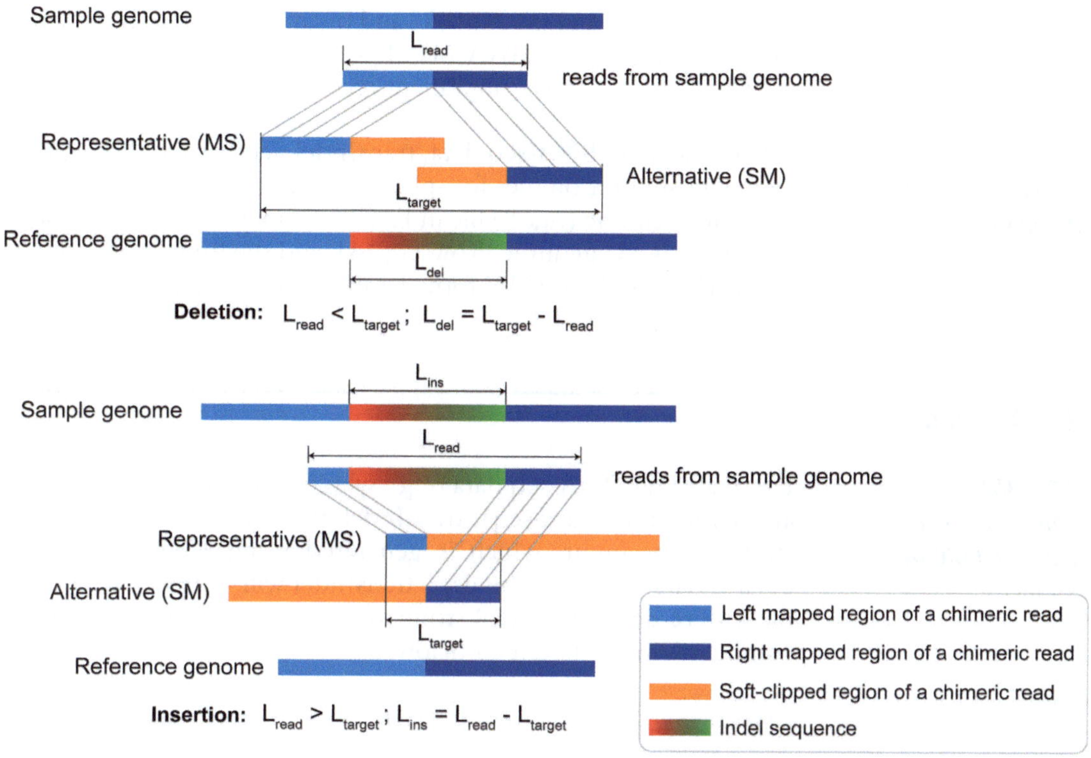

Fig. 1 Conceptual overview of transIndel algorithm. Deletions and insertions are recovered from representative and alternative alignments of chimeric reads. Size of the indels is determined based on the difference of read length and target offset

2 Software

2.1 Environment

transIndel is implemented using Python language (required Python version ≥3.6) (https://www.python.org). Python is a platform-independent language, so transIndel can run on a standalone computer with any OS installed. The source code of transIndel is available at GitHub (https://github.com/cauyrd/transIndel).

To get started, download and install the package from GitHub with the following command:

```
git clone git://github.com/cauyrd/transIndel.git
```

transIndel requires the following dependencies: SamTools (version ≥1.0) [12], pysam (http://pysam.readthedocs.io), and HTSeq [13]. These dependencies need to be installed before executing transIndel.

2.2 Input Data

transIndel detects indels based on DNA-seq or RNA-seq data from different sequencing platforms. It takes aligned short reads data in BAM format as input. Both paired-end reads and single-end reads are acceptable. transIndel requires the input BAM file to be aligned by BWA-MEM [11] (*see* **Note 1**) and the read length to be 75 bp or longer, which allows BWA-MEM to report chimeric alignments with an "SA" tag in the alignment records.

2.3 Reference Genome and Gene Annotation Files

When analyzing RNA-seq data, transIndel needs reference genome and gene annotation files for splice site determination. The human reference genome (hg38) file in FASTA format can be downloaded from the UCSC genome browser [14], and the human gene annotation (hg38) file in GTF format can be downloaded from GENCODE project [15].

3 Methods

3.1 BAM File Refinement with Corrected CIGAR String

For analyzing DNA-seq data (e.g., whole genome-seq, exome-seq, and targeted capture-seq), transIndel expects a coordinate sorted BAM file as input, which is generated using BWA-MEM with default parameters by the user. transIndel will output a new BAM file with re-defined CIGAR string in this step (*see* **Note 2**). Example command line is described below:

```
transIndel_build_DNA.py -i input.bam -o output_build.bam
```

Here we list all parameters used by transIndel_build_DNA.py script (Table 1).

When analyzing RNA-seq data using transIndel, the reference genome file in FASTA format and the gene annotation file in GTF format are needed. RNA-seq raw reads are first aligned with BWA-MEM, and transIndel is used to infer indel events from the aligned BAM. Deletion candidates are subjected to several filters to discriminate bona fide genomic deletion event from RNA splicing events. The filtering criteria include the removal of deletions neighboring canonical splicing motifs or annotated splice sites. These filtered reads meeting these criteria will be labeled as junction reads (*see* **Note 3**). The example command line to process RNA-seq data is showed below:

Table 1
Parameters for indel calling by transIndel_build_DNA.py

Parameter	Description
input/i	Input BAM file
output/o	Output BAM file
length/l	Maximum deletion length to detect (default: 1e6)
mapq/m	Minimum MAPQ of read to support Indel (default: 15)

```
transIndel_build_RNA.py -i input.bam -o output_build.bam -g
annotations.gtf -r reference.fasta
```

Here we list all parameters used by transIndel_build_RNA.py script (Table 2).

3.2 Indel Detection

In this step, transIndel uses the refined BAM file with corrected CIGAR string generated from the previous step as input to report final indel calls. The example command line to execute this step is showed below:

```
transIndel_call.py -i output_build.bam -o prefix_of_vcf
```

Here we list all parameters used by transIndel_call.py script (Table 3).

Table 2
Parameters for indel calling by transIndel_build_RNA.py

Parameter	Description
input/i	Input BAM file
output/o	Output BAM file
ref/r	Reference genome FASTA file
gtf/g	Gene annotation GTF file
splice_bin/s	Splice site half bin size (default: 20)
length/l	Maximum deletion length to detect (default: 1e6)
mapq/m	Minimum MAPQ of read to support Indel (default: 15)

Table 3
Parameters for indel calling by transIndel_call.py

Parameter	Description
input/i	Input BAM file with redefined CIGAR string
output/o	Prefix of output VCF file
ao/c	Minimum observation count for indel (default: 4)
depth/d	Minimum depth to call indel (default: 10)
vaf/f	Minimum variant allele frequency (default: 0.1)
length/l	Minimum Indel length to report (default: 10)
mapq/m	Minimum MAPQ of read to call indel (default: 15)
t	Limit analysis to targets listed in the BED format file or a SamTools region string (chr:start-end) (*see* **Note 4**)

Table 4
Description for INFO fields in the transIndel VCF output file

Name	Value
SVTYPE	The type of event, INS or DEL
SVLEN	Difference in length between REF and ALT alleles
END	End position of the indel
AO	Alternate allele observations
DP	Total read depth at the locus
AB	Estimated allele frequency, representing the ratio of reads showing the alternative allele to all reads

3.3 TransIndel Output Interpretation

transIndel final output is a standard VCF file [16] that reports the information of identified indels. Table 4 illustrates the detail description of each item in the INFO field included in the VCF file (*see* **Note 5**).

4 A Running Example

To demonstrate our tool, we analyze RNA-seq data of LNCaP (ATCC, #CRL-1740) prostate cancer cell line from Cancer Cell Line Encyclopedia (CCLE) [17] to discover indels with transIndel.

4.1 Data Preparation

Illumina RNA-seq data of LNCaP can be downloaded from NCBI SRA (Accession: SRX5414853). The reads raw data in FASTQ format can be retrieved using the SRA toolkit [18]. The RNA-seq reads are aligned to hg38 human genome assembly (https://hgdownload.cse.ucsc.edu/goldenpath/hg38/bigZips/hg38.fa.gz) using BWA-MEM followed by sorting BAM file with SamTools and duplicated reads removal with Picard tools (http://broadinstitute.github.io/picard/).

The main steps of mapping and postprocessing are shown below:

```
fastq-dump --outdir fastq --skip-technical --read-filter pass
--dumpbase --split-3 -clip SRR8615547

bwa mem -M -R "@RG\tID:LNCaP\tLB:lib1\tPL:illumina\tSM:LNCaP
\tPU:unit1" -t 8 hg38.fa fastq/SRR8615547_pass_1.fastq fastq/
SRR8615547_pass_2.fastq |
samtools sort -@ 8 -O BAM -o LNCaP.bwa.bam -

picard MarkDuplicates I=LNCaP.bwa.bam O=LNCaP.bam M=LNCaP.
marked_dup_metrics.txt CREATE_INDEX=true REMOVE_DUPLICATES=-
true
```

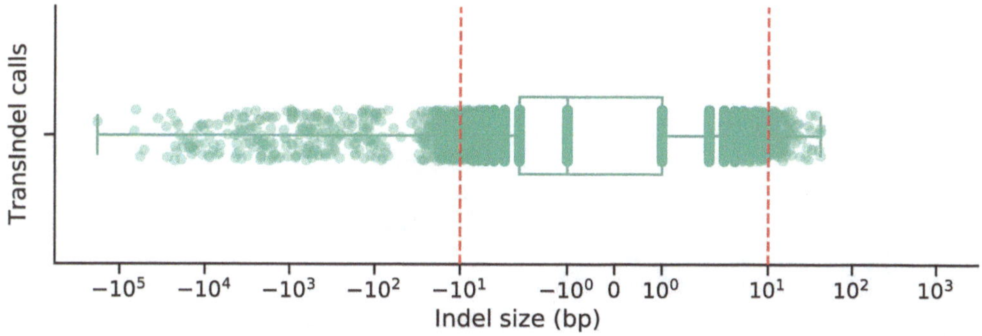

Fig. 2 The size distribution of indels identified in RNA-seq of prostate cancer LNCaP cells with transIndel algorithm

4.2 Running transIndel

transIndel can be used to identify indels using the BAM file generated by BWA-MEM with two steps: (1) build new BAM file with redefined CIGAR string and (2) indel calling.

The two steps of building and calling are executed as following:

```
transIndel_build_RNA.py -i LNCaP.bam -o LNCaP.build.bam -r
hg38.fa -g gencode.v21.annotation.gtf
transIndel_call.py -i LNCaP.build.bam -l 1 -o LNCaP
```

The human reference gene annotation (hg38) can be downloaded from GENCODE project (ftp://ftp.ebi.ac.uk/pub/databases/gencode/Gencode_human/release_21/gencode.v21.annotation.gtf.gz).

After accomplishing the two steps using the transIndel pipeline, a VCF file (LNCaP.indel.vcf) is generated as the output. The output VCF file includes 23,786 indels in total, including 711 medium- and large-sized indels ($\geq$10 bp) (Fig. 2). Two identified indels are shown in Fig. 3.

5 Notes

1. Any aligner that is able to output chimeric alignments (with SA tags) could be used to prepare the input file for transIndel.
2. The original representative alignment will store in the redefined BAM file with an OA tag. The format for OA tag is OA: position, CIGAR string.
3. The splicing junction reads inferred from gene annotation GTF or canonical splice site motifs will store in the redefined BAM file with JM tags.
4. If parameter "`-t`" is not set by users, then transIndel will process all chromosomes in the redefined BAM file one by one.

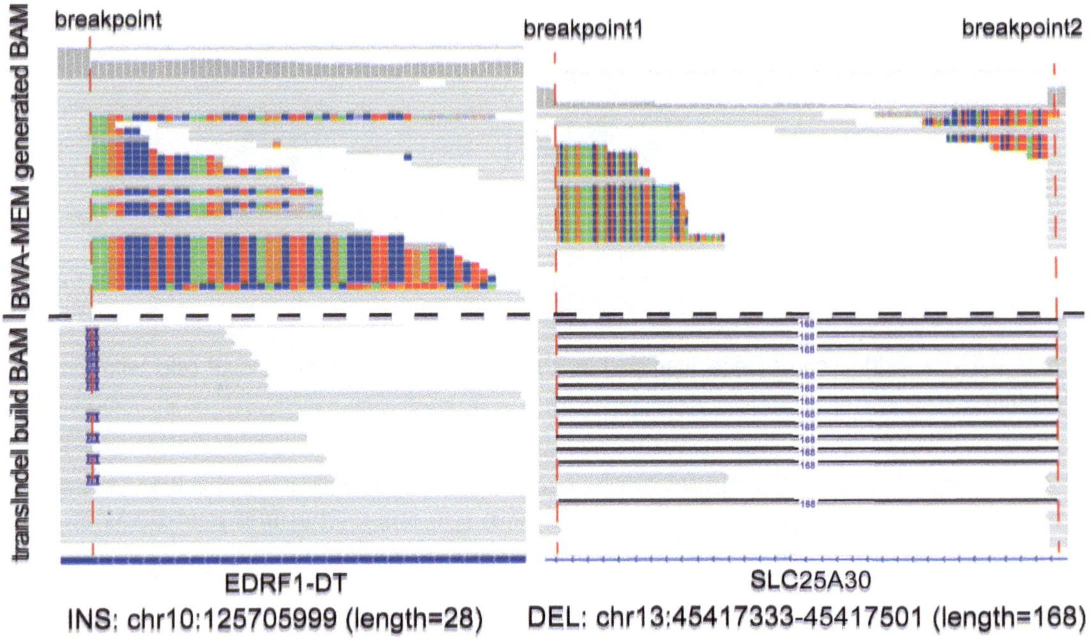

Fig. 3 Examples of medium- and large-sized indels identified in RNA-seq of prostate cancer LNCaP cells with transIndel algorithm

5. In general, a higher number of alternate allele observations (AO) with a higher number of estimated allele frequency (AB) suggest the detected indels have a higher confidence. Thus, in the step of indel calling, transIndel used AO ≥ 4 and AB ≥ 0.1 cutoffs by default.

References

1. Shlien A, Raine K, Fuligni F, Arnold R, Nik-Zainal S, Dronov S, Mamanova L, Rosic A, Ju YS, Cooke SL, Ramakrishna M, Papaemmanuil E, Davies HR, Tarpey PS, Van Loo P, Wedge DC, Jones DR, Martin S, Marshall J, Anderson E, Hardy C, Icgc Breast Cancer Working Group OBCRC, Barbashina V, Aparicio SA, Sauer T, Garred O, Vincent-Salomon A, Mariani O, Boyault S, Fatima A, Langerod A, Borg A, Thomas G, Richardson AL, Borresen-Dale AL, Polyak K, Stratton MR, Campbell PJ (2016) Direct transcriptional consequences of somatic mutation in breast cancer. Cell Rep 16(7):2032–2046. https://doi.org/10.1016/j.celrep.2016.07.028
2. Radenbaugh AJ, Ma S, Ewing A, Stuart JM, Collisson EA, Zhu J, Haussler D (2014) RADIA: RNA and DNA integrated analysis for somatic mutation detection. PLoS One 9(11):e111516. https://doi.org/10.1371/journal.pone.0111516
3. O'Brien TD, Jia P, Xia J, Saxena U, Jin H, Vuong H, Kim P, Wang Q, Aryee MJ, Mino-Kenudson M, Engelman JA, Le LP, Iafrate AJ, Heist RS, Pao W, Zhao Z (2015) Inconsistency and features of single nucleotide variants detected in whole exome sequencing versus transcriptome sequencing: a case study in lung cancer. Methods 83:118–127. https://doi.org/10.1016/j.ymeth.2015.04.016
4. Wilkerson MD, Cabanski CR, Sun W, Hoadley KA, Walter V, Mose LE, Troester MA, Hammerman PS, Parker JS, Perou CM, Hayes DN (2014) Integrated RNA and DNA sequencing improves mutation detection in low purity tumors. Nucleic Acids Res 42(13):e107. https://doi.org/10.1093/nar/gku489
5. McPherson A, Wu C, Hajirasouliha I, Hormozdiari F, Hach F, Lapuk A, Volik S,

Shah S, Collins C, Sahinalp SC (2011) Comrad: detection of expressed rearrangements by integrated analysis of RNA-Seq and low coverage genome sequence data. Bioinformatics 27(11):1481–1488. https://doi.org/10.1093/bioinformatics/btr184

6. Piskol R, Ramaswami G, Li JB (2013) Reliable identification of genomic variants from RNA-seq data. Am J Hum Genet 93(4):641–651. https://doi.org/10.1016/j.ajhg.2013.08.008

7. Zhang J, White NM, Schmidt HK, Fulton RS, Tomlinson C, Warren WC, Wilson RK, Maher CA (2016) INTEGRATE: gene fusion discovery using whole genome and transcriptome data. Genome Res 26(1):108–118. https://doi.org/10.1101/gr.186114.114

8. Sun Z, Bhagwate A, Prodduturi N, Yang P, Kocher JA (2017) Indel detection from RNA-seq data: tool evaluation and strategies for accurate detection of actionable mutations. Brief Bioinform 18(6):973–983. https://doi.org/10.1093/bib/bbw069

9. Wajnberg G, Passetti F (2016) Using high-throughput sequencing transcriptome data for INDEL detection: challenges for cancer drug discovery. Expert Opin Drug Discov 11(3):257–268. https://doi.org/10.1517/17460441.2016.1143813

10. Yang R, Van Etten JL, Dehm SM (2018) Indel detection from DNA and RNA sequencing data with transIndel. BMC Genomics 19(1). https://doi.org/10.1186/s12864-018-4671-4

11. Li H (2013) Aligning sequence reads, clone sequences and assembly contigs with BWA--MEM.arXiv:1303.3997

12. Li H, Handsaker B, Wysoker A, Fennell T, Ruan J, Homer N, Marth G, Abecasis G, Durbin R, Genome Project Data Processing S (2009) The sequence alignment/map format and SAMtools. Bioinformatics 25(16):2078–2079. https://doi.org/10.1093/bioinformatics/btp352

13. Anders S, Pyl PT, Huber W (2015) HTSeq--a Python framework to work with high-throughput sequencing data. Bioinformatics 31(2):166–169. https://doi.org/10.1093/bioinformatics/btu638

14. Kent WJ, Sugnet CW, Furey TS, Roskin KM, Pringle TH, Zahler AM, Haussler D (2002) The human genome browser at UCSC. Genome Res 12(6):996–1006. https://doi.org/10.1101/gr.229102

15. Frankish A, Diekhans M, Ferreira AM, Johnson R, Jungreis I, Loveland J, Mudge JM, Sisu C, Wright J, Armstrong J, Barnes I, Berry A, Bignell A, Carbonell Sala S, Chrast J, Cunningham F, Di Domenico T, Donaldson S, Fiddes IT, Garcia Giron C, Gonzalez JM, Grego T, Hardy M, Hourlier T, Hunt T, Izuogu OG, Lagarde J, Martin FJ, Martinez L, Mohanan S, Muir P, Navarro FCP, Parker A, Pei B, Pozo F, Ruffier M, Schmitt BM, Stapleton E, Suner MM, Sycheva I, Uszczynska-Ratajczak B, Xu J, Yates A, Zerbino D, Zhang Y, Aken B, Choudhary JS, Gerstein M, Guigo R, Hubbard TJP, Kellis M, Paten B, Reymond A, Tress ML, Flicek P (2019) GENCODE reference annotation for the human and mouse genomes. Nucleic Acids Res 47(D1):D766–D773. https://doi.org/10.1093/nar/gky955

16. Danecek P, Auton A, Abecasis G, Albers CA, Banks E, De Pristo MA, Handsaker RE, Lunter G, Marth GT, Sherry ST, McVean G, Durbin R, Genomes Project Analysis G (2011) The variant call format and VCFtools. Bioinformatics 27(15):2156–2158. https://doi.org/10.1093/bioinformatics/btr330

17. Barretina J, Caponigro G, Stransky N, Venkatesan K, Margolin AA, Kim S, Wilson CJ, Lehar J, Kryukov GV, Sonkin D, Reddy A, Liu M, Murray L, Berger MF, Monahan JE, Morais P, Meltzer J, Korejwa A, Jane-Valbuena J, Mapa FA, Thibault J, Bric-Furlong E, Raman P, Shipway A, Engels IH, Cheng J, Yu GK, Yu J, Aspesi P Jr, de Silva M, Jagtap K, Jones MD, Wang L, Hatton C, Palescandolo E, Gupta S, Mahan S, Sougnez C, Onofrio RC, Liefeld T, MacConaill L, Winckler W, Reich M, Li N, Mesirov JP, Gabriel SB, Getz G, Ardlie K, Chan V, Myer VE, Weber BL, Porter J, Warmuth M, Finan P, Harris JL, Meyerson M, Golub TR, Morrissey MP, Sellers WR, Schlegel R, Garraway LA (2012) The cancer cell line encyclopedia enables predictive modelling of anticancer drug sensitivity. Nature 483(7391):603–607. https://doi.org/10.1038/nature11003

18. Leinonen R, Sugawara H, Shumway M, International Nucleotide Sequence Database C (2011) The sequence read archive. Nucleic Acids Res 39(Database issue):D19–D21. https://doi.org/10.1093/nar/gkq1019

Chapter 6

DECoN: A Detection and Visualization Tool for Exonic Copy Number Variants

Anna Fowler

Abstract

Detection of copy number variants from targeted sequencing, including whole-exome sequencing, can be particularly difficult since the break points of the CNV are not always captured. Here we describe DECoN, a software tool which uses changes in read depth to identify CNVs that affect whole exons. It is optimized for clinical use and allows for interactive visualization of CNVs identified.

Key words Copy number variants, Sequencing, Bioinformatics, Whole exome, Targeted panels

1 Introduction

The detection of copy number variants (CNVs) from whole exome or targeted panel sequencing data is typically more complex than the detection of CNVs from whole-genome sequencing, since the break points of the CNV are often outside of the targeted region. Therefore, the primary means of detection is based on changes in read depth, whereby an increase in the number of reads in comparison to the expected number of reads indicates a duplication and a decrease indicates a deletion. Many software tools are available which aim to detect these changes in read depth [1–10].

DECoN (Detection of Exon Copy Number) [11] is a sensitive and specific tool for the detection of exon copy number variants (exon CNVs) in targeted sequencing data. For a test sample, an optimal set of highly correlated reference samples are first selected from a larger set of potential reference samples. Then the read depth of the test sample is compared to the optimal reference samples using a statistical model. DECoN has been developed with clinical use in mind and therefore provides thorough quality checks, strict version control, and interactive visualization to enhance clinical utility. DECoN is based on ExomeDepth [1],

Charlotte K. Y. Ng and Salvatore Piscuoglio (eds.), *Variant Calling: Methods and Protocols*,
Methods in Molecular Biology, vol. 2493, https://doi.org/10.1007/978-1-0716-2293-3_6,

optimized for use on targeted panel sequencing of run batches of samples.

2 Materials

DECoN is freely available for non-commercial use, under CC BY-NC License. It can be downloaded from: https://zenodo.org/record/1254945.

2.1 Software

DECoN is available for Windows, Mac, and Linux operating systems and requires the following prerequisites:

1. R [12] the specific version used should match the version of DECoN.
2. The capacity to build packages from source. In windows, this requires the installation of Rtools. On Mac/Linux, it requires gcc and gfortran compilers (available through xtools on Mac).
3. A modern internet browser such as Firefox, Chrome, or IE v.10 or later.
4. An internet connection for installation.

2.2 Download and Install DECoN

DECoN can be downloaded from: https://zenodo.org/record/1254945. This provides a compressed file which should be unpacked to a new local directory (*see* **Note 1**). It contains both a Windows version and a Mac/Linux version, you only need the version which corresponds to your operating system.

In a Windows environment, double click the setup.bat file in the Windows sub-directory. The file path to the correct R version (e.g., C:\Program Files\R\R-3.1.2\bin) must be in the PATH variable.

In a Mac/Linux environment, use the command line to navigate to the Linux sub-directory, then use the command:

```
setup.sh
```

to run the setup script.

3 Methods

Once DECoN has been downloaded with dependencies correctly installed (*see* Subheading 2), running DECoN requires four sequential steps:

1. Reading the BAM files to generate coverage metrics.
2. Running quality checks.

3. Calling exon CNVs.
4. Visualizing the calls.

A summary .RData file is created after **steps 1** and **3** which is required for **steps 2** and **4**, respectively. This approach allows the user to perform individual steps without having to re-run preceding steps. For instance, a user can make calls with a different set of parameters or change quality thresholds for failing samples without having to re-process the BAM files (which is the most time-consuming step).

In Mac/Linux, each step is launched via the command line. In Windows, there are executable files for each step which are clicked to launch. The user is then prompted to specify the required inputs (*see* **Note 2**). Each step is detailed in the following sections.

3.1 Step 1: Reading BAM Files to Generate Coverage Metrics

DECoN uses a list of BAM files and a BED file of exons to calculate a coverage metric called the fragment per kilobase and million base pairs (FPKM) for each exon specified in the BED file in each sample's BAM file. DECoN uses this metric to call exon CNVs. FPKM is calculated according to the below formula:

$$\text{FPKM} = C/(N \times L)$$

where C is the number of read pairs mapping to the exon, N is the total number of mapped read pairs divided by one million, and L is the length of the exon in kilobases.

For example, consider a sample with a total of 20 million mapped read pairs of which 200 map to an exon which is 100 bases long:

$$\text{FPKM} = 200/(20 \times 0.1)$$

Thus, FPKM for this exon in this sample is 100.

3.1.1 Inputs

There are four required inputs:

1. BAM files—these can be specified in either a text file or a path to a directory containing all the BAM files can be supplied. The text file must contain a list of the BAM files to be read in, with each file name on a separate line and each file name ending in . bam. If a path is supplied, all BAM files in the directory will be read. DECoN expects each BAM file to have a .bai file in the same location as the .bam file, with a .bai extension instead of . bam as the file name, e.g., *directory/sample.bam* and *directory/sample.bai*.
2. BED file—the targeted 1-based BED file to be used for analysis. This file must be sorted in chromosome order (*see* **Note 3**). This file does not have a header and must have four tab-separated columns corresponding to:

   1. Chromosome.
   2. Start position.
   3. End position.
   4. Gene.
3. FASTA file—the reference genome FASTA file to be used with the data.
4. Output prefix—the prefix for the summary output .RData file. If none is supplied, the default value is DECoN.

3.1.2 Running ReadInBams

In Mac/Linux, run the following command from the directory containing the DECoN scripts:

```
Rscript ReadInBams.R --bams bams.file --bed bed.file --fasta
fasta.file --out output.prefix
```

In Windows, click the *ReadInBams.bat* executable. You will be prompted to enter inputs.

3.1.3 Output

`ReadInBams` outputs a summary .RData file prefixed with the output prefix specified in the input which contains sample coverage values and sample names taken from the BAM files.

3.2 Step 2: Running Quality Checks

The summary .RData file outputted by `ReadInBams` (Subheading 3.1—Step 1) can be used to flag any samples or exons where exon CNV calling may be suboptimal. Both exons and samples are evaluated based on their median coverage level. When coverage is low, accuracy of detection will be compromised, and caution should be exercised when interpreting results. Samples are also evaluated based on their correlation with other samples. Samples which do not have a high correlation with other samples in the set are likely to have suboptimal detection across the entire target. Recommended default thresholds underpinning this quality flag are given below (*see* **Note 4**).

3.2.1 Inputs

There is one required file input and a number of threshold inputs which can be set by the user:

- Summary RData file (required)—A summary RData file containing the FPKM for each exon of an analyzed BED file, created by `ReadInBams`.
- Minimum correlation threshold—the minimum correlation between a test sample and any other sample for the test sample to be considered well-correlated. The default value is 0.98.
- Minimum coverage threshold—the minimum median coverage for any sample (measured across all exons in the target) or exon

(measured across all samples) to be considered well-covered. The default value is 100.

- Exon numbering (optional)—a file containing exon numberings with custom annotation for at least one value in the analyzed BED file. This is a tab-separated file with four columns labeled with headings:
  1. **Chromosome.**
  2. **Start**—start position from the analyzed BED file.
  3. **End**—end position from the analyzed BED file.
  4. **Custom.Exon**—custom exon name.
- Custom reporting—Boolean value indicating whether an output file containing samples or exons which affect genes with custom annotation should be generated. The default value is FALSE.
- Output prefix—the prefix for the output files. If none is supplied, the default value is DECoN.

3.2.2 Running IdentifyFailures

In Mac/Linux, run the following command from the directory containing the DECoN scripts:

```
Rscript IdentifyFailures.R --Rdata summary.file --mincorr .98
--mincov 100 --exons customNumbers.file --custom FALSE --out
output.prefix
```

In Windows, click the *IdentifyFailures.bat* executable. You will be prompted to enter inputs.

3.2.3 Outputs

If all samples and exons are above the user-defined thresholds, no output is created. If any suboptimal samples and/or exons are identified, a tab-separated text file ending in *_Failures.txt* is created with six columns:

1. **Sample**—the name of the sample. If an exon is suboptimal this column has value "All."
2. **Exon**—the number of the suboptimal exon, in the order of the analyzed BED file. If the sample has suboptimal correlation or suboptimal median coverage, this column has value "All."
3. **Type**—the type of failure, either "Whole sample" if the sample is below the correlation or coverage threshold or "Whole exon" if the exon is below the coverage threshold.
4. **Gene**—the name of the gene, from the Gene column of the analyzed BED file.
5. **Custom.numbering**—the number of the suboptimal exon from the Custom.Exon column of the custom numbering file.
6. **Info**—the underlying metric information.

If the custom option is TRUE and exons are identified which affect a gene with custom annotation, an additional file ending in *_custom_Failures.txt* is created. This is the same format as the *_Failures.txt* file and contains the subset of information pertaining to the relevant genes.

3.3 Step 3: Calling Exon CNVs

This step calls exon CNVs in each sample by selecting reference samples from all other samples contained in the input summary . RData file. The correlation between samples and the number of samples used as a reference are thus calculated and outputted to aid interpretation of call quality (*see* **Notes 5–7**).

3.3.1 Inputs

There is one required file input and a number of parameter inputs which can be set by the user:

- Summary RData file (required); this contains the FPKM for each exon of an analyzed BED file, created by `ReadInBams` (Subheading 3.1—step 1).
- Transition probability; this is the transition probability between normal copy number state and either deletion or duplication state in the hidden Markov model. The default value is set to 0.01. Use a high threshold value to increase sensitivity.
- Exon numbering (optional); a file containing exon numberings with custom annotation for at least one value in the analyzed BED file. This is a tab-separated file with four columns labeled with headings:

    1. **Chromosome.**
    2. **Start**—start position from the analyzed BED file.
    3. **End**—end position from the analyzed BED file.
    4. **Custom.Exon**—custom exon name.

- Custom reporting—Boolean value indicating whether an output file containing only calls in genes with custom annotation should be generated. The default value is FALSE.
- Output prefix—the prefix for the output files. If none is supplied, the default value is DECoN.
- Plotting of variants—takes one of "All," "Custom," or "None," will create an automated visualization (plot) for all variants, variants in exons which have custom annotation, or no variants, respectively. The default value is "All."
- Plot folder—the folder in which plots are saved, created if it does not exist already. Defaults to "DECoNPlots."

3.3.2 Running makeCNVcalls

In Mac/Linux, run the following command from the directory containing the DECoN scripts:

```
Rscript makeCNVcalls.R --Rdata summary.file --transProb tran-
sition.probability --exons customNumbers.file --custom FALSE
--out output.prefix --plot All --plotFolder DECoNPlots
```

In Windows, click the *makeCNVcalls.bat* executable. You will be prompted to enter inputs.

3.3.3 Output

If no calls are generated, no output is created. If any calls are generated, `makeCNVcalls` outputs two files:

- Summary RData file—a summary .RData file containing the FPKM for each sample and exon, all CNV calls, and quality control information.
- Table of all calls—a tab-separated text file ending in *_all.txt* detailing all exon CNV calls. This file has 19 columns (Table 1) and includes details of the sample in which the CNV is identified, the position of the CNV, and various measures of confidence in the CNV (*see* **Notes 8** and **9**).

If the plotting option is either "All" or "Custom," then a simple visualization of each variant is automatically created in the plot folder. An example plot is shown in Fig. 1. The plots show the variant compared to the reference samples and the ratio of the observed to expected coverage for the sample with the variant. A 95% confidence interval for the ratio of observed to expected coverage is also displayed. All exons in the gene are shown. If the variant is within five exons of the gene endpoints an additional five exons from a neighboring gene (as specified by the BED file) are also shown. Custom numbering is used if supplied by the user, otherwise exons are numbered in increasing chromosome order within each gene.

3.4 Step 4: Visualizing Calls

The DECoN call visualization step provides interactive exploration and visualization of the exon deletion/duplication calls and the underlying data (*see* **Note 10**).

3.4.1 Input

This step only requires a summary.RData file containing the FPKM for each sample and exon, all CNV calls, and quality control information, this is generated by the `makeCNVcalls` (Subheading 3.3—step 3).

3.4.2 Running DECoN Call Visualization

In Mac/Linux, run the following command from the directory containing the DECoN scripts:

```
Rscript runShiny.R --Rdata summary.file
```

Table 1
Details of the column headings and their interpretation for the table of CNV results generated by the makeCNV program

Column heading	Interpretation
CNV.IDS	Unique ID for the CNV
Sample	Name of the sample containing the CNV
Correlation	Correlation between the sample containing the CNV and the most correlated sample in the reference set
N.comp	Number of samples used in the reference set
Start.b	Index of the first entry of the bed file affected by the CNV
End.b	Index of the last entry of the bed file affected by the CNV
CNV.type	Deletion or duplication
N.exons	Number of exons affected by the CNV
Start	Genomic coordinates of the first base of the CNV
End	Genomic coordinates of the last base of the CNV
Chr	Chromosome
Genomic.ID	Genomic ID
BF	Bayes factor for a model containing the CNV
Reads. expected	Number of expected reads in the CNV region
Reads. observed	Number of observed reads in the CNV region
Reads.ratio	Reads.observed/Reads.expected
Gene	Gene containing the CNV
Custom.first	Custom numbering of first affected exon (if custom numbering is provided)
Custom.last	Custom numbering of last affected exon (if custom numbering is provided)

In Windows, click the *gui.bat* executable. You will be prompted to enter inputs.

3.4.3 Output

This step automatically launches a web browser with an interactive GUI. CNV calls can be visualized using the CNV Calls tab. The user can select the CNV ID to be plotted, the number of exons displayed in the plot, the other samples to be plotted (either its reference set or the full set), and the scale of the plot. The affected exon(s) is shown in red, and a 95% confidence interval is also shown

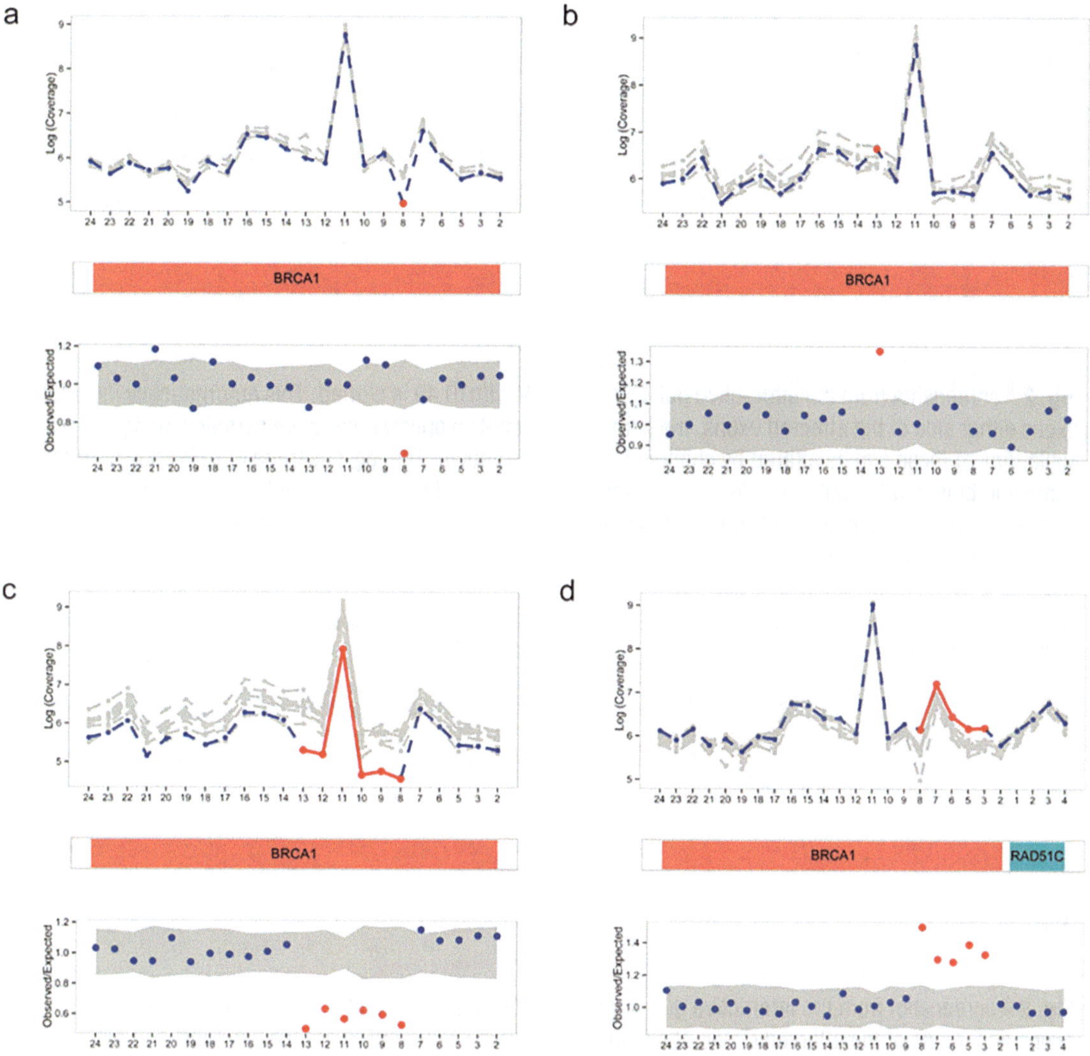

Fig. 1 Automatically generated plots of a CNV in *BRCA1*. In all panels, the top plot shows the log-normalized coverage of the sample of interest (blue) relative to reference samples (gray), and the bottom plot shows the ratio of observed to expected coverage with a 95% confidence interval in gray. The relevant gene(s) are shown between the plots in red. Deleted or duplicated exons are in red. (**a**) A single exon deletion; (**b**) A single exon duplication; (**c**) A multi-exon deletion; (**d**) A multi-exon duplication. (Reproduced from Fowler, 2016 [11])

in gray. Screenshots of CNV IDs 49 and 53 are shown in Figs. 2 and 3, respectively; these are both deletions in *BRCA1* (*see* **Note 11**).

4 Notes

1. DECoN implements strict version control over all packages and dependencies used by changing the local default R settings. Any R session launched from the same directory as the DECoN

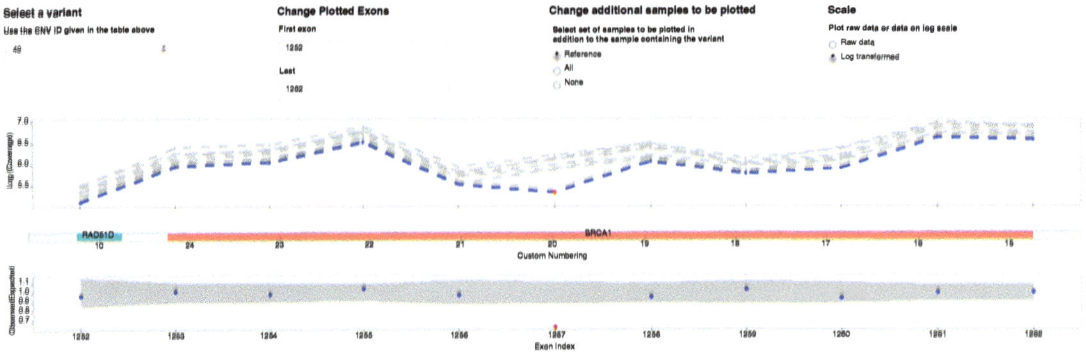

Fig. 2 A screenshot from the interactive gui, in which CNV with ID 49 is plotted. This automatically includes five exons either side of the affected exons, the samples selected as optimal reference samples, and plots on a log scale. Any of these can be altered by changing the options shown. The log-coverage (top) is shown for the test sample in blue and the optimal reference samples in grey, with deleted exons highlighted in red. The number of observed/expected reads (bottom) is shown for the test sample in blue, affected exons in red and a 95% confidence interval in gray

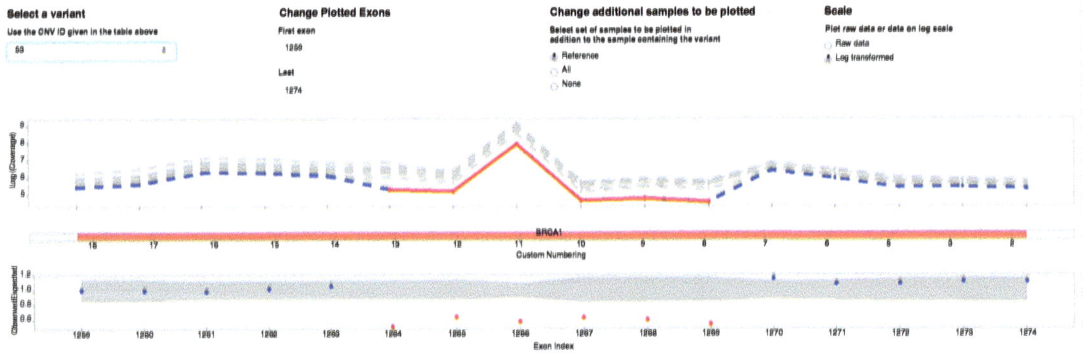

Fig. 3 A screenshot from the interactive gui, in which CNV with ID 53 is plotted. This automatically includes five exons either side of the affected exons, the samples selected as optimal reference samples, and plots on a log scale. Any of these can be altered by changing the options shown. The log-coverage (top) is shown for the test sample in blue and the optimal reference samples in grey, with deleted exons highlighted in red. The number of observed/expected reads (bottom) is shown for the test sample in blue, affected exons in red and a 95% confidence interval in gray

scripts will have these settings; therefore, it is recommended to unpack the compressed file to a new directory containing only DECoN.

2. All scripts automatically creates a log file. In a Mac/Linux environment, the log files have a .log extension, whereas in windows they have a .Rout extension and are located in the scripts folder is called setup.log. These can be essential in diagnosing any problems and checking that scripts have run successfully.

3. It is assumed that CNVs affect targeted whole exons only, and that each targeted exon is specified in the bed file. If you want to consider other regions, for instance partial exons or untargeted regions covered by off-target reads, then altering the bed file to include these regions will allow CNVs in these regions to be detected.
4. The default values for quality control used by DECoN are set fairly stringently for clinical use. There are many applications where lower values may be more suitable.
5. Calls are made relative to a reference set of samples. These reference samples should be as well correlated as possible to the test sample. Preferably they would be sequenced at the same time on the same sequencing lane as small experimental differences can result in poor correlation. It is better to have a small set of well-correlated samples than a large set of samples which may contain systematic differences.
6. Increasing the size of the potential reference set will only improve CNV detection if the new samples are well correlated with the test sample. If new samples, which are not well correlated with the test sample, are added to the potential reference set, they will simply be excluded at the CNV detection stage; therefore, this would only increase the run time without changing the results.
7. The reference samples are automatically optimized and sometimes as few as one sample can be chosen as a reference. Calls in the test sample are made relative to the reference samples; therefore, CNVs in the reference samples can result in false calls in the test sample. This is particularly a problem when there are a very small number of reference samples selected. Quality metrics include the number of samples used as reference samples and care should be taken when this is small. Inspection of plots CNVs will help identify this problem, as well as CNV calling in the reference samples and a comparison between CNVs in test and reference samples.
8. If a single call spans multiple genes on the same chromosome, the call will be represented with multiple entries (one per gene) in the output. The gene-specific information will be given in the Start.b, End.b, Gene, Custom.first and Custom.last columns. All other columns will have the same values for all entries for the call.
9. The Bayes factor and reads ratio reported in the table of CNVs (Table 1) can be used as indicators of confidence in a CNV call. High Bayes factors indicate high levels of confidence, and similarly high reads ratios indicate high level of confidence in a duplication. However, reads ratio less than one indicate a deletion, and in this case lower reads ratio indicate higher levels

of confidence. Particular care should be taken when comparing these, as direct comparisons can only be performed between CNVs in the same sample.

10. The browser launched in the visualization step can also be used for quality control, providing plots of coverage and comparison between samples and across exons.

11. If there are particular genes/exons of interest, it is recommended to visualize the read depth of those exons to examine whether they behave consistently across samples. In particular, some exons have far higher variation than others, for instance if they have pseudo genes, and this makes CNV identification challenging but would not be picked up by the quality control metric which looks at median coverage.

References

1. Plagnol V, Curtis J, Epstein M, Mok KY, Stebbings E, Grigoriadou S, Wood NW, Hambleton S, Burns SO, Thrasher AJ, Kumararatne D (2012) A robust model for read count data in exome sequencing experiments and implications for copy number variant calling. Bioinformatics 28(21):2747–2754
2. Fromer M, Moran JL, Chambert K, Banks E, Bergen SE, Ruderfer DM, Handsaker RE, McCarroll SA, O'Donovan MC, Owen MJ, Kirov G (2012) Discovery and statistical genotyping of copy-number variation from whole-exome sequencing depth. Am J Hum Genet 91(4):597–607
3. de Ligt J, Boone PM, Pfundt R, Vissers LE, Richmond T, Geoghegan J, O'Moore K, de Leeuw N, Shaw C, Brunner HG, Lupski JR (2013) Detection of clinically relevant copy number variants with whole-exome sequencing. Hum Mutat 34(10):1439–1448
4. Li J, Lupat R, Amarasinghe KC, Thompson ER, Doyle MA, Ryland GL, Tothill RW, Halgamuge SK, Campbell IG, Gorringe KL (2012) CONTRA: copy number analysis for targeted resequencing. Bioinformatics 28(10): 1307–1313
5. Krumm N, Sudmant PH, Ko A, O'Roak BJ, Malig M, Coe BP, Quinlan AR, Nickerson DA, Eichler EE, NHLBI Exome Sequencing Project (2012) Copy number variation detection and genotyping from exome sequence data. Genome Res 22(8):1525–1532
6. Amarasinghe KC, Li J, Halgamuge SK (2013) CoNVEX: copy number variation estimation in exome sequencing data using HMM. BMC Bioinformatics 14(2):1–9
7. Johansson LF, van Dijk F, de Boer EN, van Dijk-Bos KK, Jongbloed JD, van der Hout AH, Westers H, Sinke RJ, Swertz MA, Sijmons RH, Sikkema-Raddatz B (2016) CoNVaDING: single exon variation detection in targeted NGS data. Hum Mutat 37(5):457–464
8. Chiang T, Liu X, Wu TJ, Hu J, Sedlazeck FJ, White S, Schaid D, de Andrade M, Jarvik GP, Crosslin D, Stanaway I (2019) Atlas-CNV: a validated approach to call single-exon CNVs in the eMERGESeq gene panel. Genet Med 21(9):2135–2144
9. Jiang Y, Wang R, Urrutia E, Anastopoulos IN, Nathanson KL, Zhang NR (2018) CODEX2: full-spectrum copy number variation detection by high-throughput DNA sequencing. Genome Biol 19(1):1–13
10. Kang Y, Nam SH, Park KS, Kim Y, Kim JW, Lee E, Ko JM, Lee KA, Park I (2018) DeviCNV: detection and visualization of exon-level copy number variants in targeted next-generation sequencing data. BMC Bioinformatics 19(1):1–13
11. Fowler A, Mahamdallie S, Ruark E, Seal S, Ramsay E, Clarke M, Uddin I, Wylie H, Strydom A, Lunter G, Rahman N (2016) Accurate clinical detection of exon copy number variants in a targeted NGS panel using DECoN. Wellcome open research 1:20
12. Team, R.C. (2013) R: A language and environment for statistical computing

Chapter 7

FACETS: Fraction and Allele-Specific Copy Number Estimates from Tumor Sequencing

Arshi Arora, Ronglai Shen, and Venkatraman E. Seshan

Abstract

Clinical sequencing studies routinely involve molecular profiling of patients for mutations and copy number alterations. However, detection of "actionable" aberrations to guide treatment decisions require accurate, tumor purity-, ploidy-, and clonal heterogeneity-adjusted integer copy number calls. In this chapter, we describe the FACETS algorithm, an Allele-Specific Copy Number (ASCN) analysis tool with a broad application to whole-genome, whole-exome, as well as targeted panel sequencing platforms to annotate the genome for the detection of copy number alterations including homozygous/heterozygous deletions, copy-neutral loss-of-heterozygosity (LOH) events, allele-specific gains/amplifications, and cellular fraction profiles.

We will describe some methodological details on joint segmentation of total and allele-specific copy number, on the estimation of integer copy number calls adjusting for tumor purity, ploidy, and intratumor heterogeneity, along with comprehensive output and integrated visualization. We also provide a tutorial on the installation, application, and tips to run and interpret FACETS.

Key words Allele-specific copy number analysis, Integer copy number, Purity, Ploidy, High-throughput DNA sequencing

1 Introduction

Cancer genomes consist of numerous sequence alterations compared to their corresponding normal counterparts. Such alterations range from single-nucleotide variants, insertions or deletions, copy number alterations, gene fusions, and other structural variations. Various technologies including SNP arrays, whole-exome, whole-genome sequencing, and targeted panel sequencing have enabled us to capture the tumor profile toward the understanding of tumor biology, the identification of therapeutic targets, and further to inform clinical decisions under the development of precision medicine. However, analysis of the sequencing data from tumor samples is complicated by the admixture of both tumor and normal cells and clonal heterogeneity within tumor cells. The FACETS algorithm

Charlotte K. Y. Ng and Salvatore Piscuoglio (eds.), *Variant Calling: Methods and Protocols*,
Methods in Molecular Biology, vol. 2493, https://doi.org/10.1007/978-1-0716-2293-3_7,

facilitates an allele-specific copy number (ASCN) analysis accounting for these factors.

FACETS [1] (Fraction and Allele-Specific Copy Number Estimates from Tumor Sequencing) is an ASCN pipeline and clonal heterogeneity analysis tool for high-throughput DNA sequencing. It is a fully integrated stand-alone pipeline starting with a sequencing BAM file as an input and providing estimates of tumor purity, ploidy, major and minor copy number profile, and associated cellular fractions to infer clonality of the alterations. The input, paired normal–tumor pair BAM files, is summarized over allele-specific read counts for both homozygous and heterozygous SNPs, avoiding any loss of information. It should be noted here that this is an important difference between the existing methods such as FALCON [2] and TITAN [3], which may not be directly applicable for a targeted panel, as they only consider heterozygous SNP loci.

1.1 Theoretical Methods

Total copy number log ratio or logR is computed from the total read count in the tumor versus normal, considering all SNPs, as observed. Specifically, the expected value of logR can be expressed as

$$E[\log R] = \log\{(m^* + p^*)/2\} + w(.) + \lambda$$

where, $m^* = m\phi + (1 - \phi)$ and $p^* = p\phi + (1 - \phi)$ are parental copy number in the tumor samples arising from a mixed normal and aberrant (m,p) copy number with cellular fraction ϕ which is a function of tumor purity and clonal composition. Cellular fraction stands for the fraction of tumor cells carrying the genetic alteration and clonal composition is the number of different clones present in the tumor tissue. For a tumor with a single clone, the parental copy numbers m and p are integers, but for a tumor with multiple clones, they are weighted averages of parental copy numbers of the clones weighted by the relative proportions of the clones. The term w(.) denotes bias due to GC content, and λ is a location shift constant in tumor due to aneuploidy. In a diploid genome, $\log R = 0$ $(m^* = p^* = 1)$. However, in the case of aneuploidy, we first need to determine the shift in the normal diploid state (discussed in methods as diplogR).

We pointed out previously [1] that the conventional B-allele frequency based on sequencing read counts possesses inherent bias due to mapping affinity toward reference allele, and thus, we proposed log odds ratio, logOR, ratio of variant-allele count in tumor versus normal. The expected logOR can be computed as follows:

$$E[\log \mathrm{OR}] = \log\left(\frac{m^*}{p^*}\right) \text{ or } \log\left(\frac{p^*}{m^*}\right)$$

depending on which parental copy the variant allele resides on (issue of multiplicity).

1.2 Joint Segmentation and Estimation of Integer Copy Number

A bivariate genome segmentation of logR and logOR is used to identify change points such that the bivariate Hotelling T^2 statistic is maximized. These inflection points in the joint segmentation are used to identify segment cluster that belong to the same underlying genotype. The joint segmentation was made possible by extending the circular binary segmentation (CBS) algorithm by Olshen and Seshan [4].

Once the sample is segmented, regions that are allelically balanced are used to determine the normal diploid state (note that alleles are balanced in normal diploid state). For every other region, the difference between its logR and that of the diploid region along with the logOR are used to simultaneously estimate the major and minor copy numbers as well as cellular fraction. Since parental copy numbers and cellular fraction are inversely related to give the same logR and logOR, the initial fit uses the smallest copy numbers with highest cellular fraction that are consistent with the data. We finally use the EM algorithm to refine the segment level fits to obtain the segmental allele specific copy numbers and the cellular fraction (purity) and average total copy numbers (ploidy) of the sample. For more details on the theoretical methodology, refer to the FACETS paper in Nucleic Acid Research [1].

2 Software

2.1 Installation

FACETS is available as an R package (requires R version ≥3.0.0) and is computationally fast and light. The source code of FACETS is available at GitHub (https://github.com/mskcc/facets).

To install FACETS, use the install_github function from devtools package as follows:

```
devtools::install_github("mskcc/facets", build_vignettes =
TRUE)
```

FACETS require pctGCdata(>=0.3.0) package as a dependency, and it will be automatically installed the first time you try to install FACETS.

```
devtools::install_github("mskcc/pctGCdata")
```

This step is only required once. Note, that if you get an error message about pctGCdata, you can alternatively use the R CMD INSTALL option to install the data package from the source. Source files are available at https://github.com/mskcc/pctGCdata.

2.2 Input Requirements

1. As mentioned above, pctGCdata package is required. This package contains R datasets of GC percentages for human and mouse genomes.
2. Paired Normal and Tumor BAM files to be analyzed.
3. VCF files containing the exhaustive list of SNPs. It can be downloaded from NCBI (https://ftp.ncbi.nih.gov/snp/organisms/human_9606/VCF/) and selecting file 00-common_all.vcf.gz. Make sure that the VCF file is current and compatible with the genome build used for aligning the above-mentioned BAM files.
4. To get the Tumor–Normal read counts at these various SNP positions, we provide a computationally fast C++ implementation of pileup, snp-pileup. It is available along with the mskcc/facets GitHub repository: https://github.com/mskcc/facets/tree/master/inst/extcode. snp-pileup requires HTSlib to be installed. It can be downloaded here: http://www.htslib.org/download/. More details can be found in the README.txt file.

2.3 Test Dataset

We will be considering the dataset stomach.csv.gz consisting of TCGA stomach cancer samples that comes with the package. To generate these data, tumor and normal bam files were downloaded from the Genomic Data Commons Data Portal (https://portal.gdc.cancer.gov/). Reference and variant allele read counts were extracted from the BAM files for germline polymorphic sites catalogued in the dbSNP and 1000 genome database (~1.9 million polymorphic positions fall in regions targeted by whole-exome sequencing). The resulting csv file contains pileup read count on the ~1.9 million SNP positions summarized for Normal (#1) and Tumor (#2) samples over their reference allele (column File#R), alternate allele (column File#A), errors (neither ref nor alt, column File#E), and deletions (column File#D) in that position.

The test dataset can be found at https://github.com/mskcc/facets/tree/master/inst/extdata.

3 Methods

We will be using the data provided in the extdata/ folder, stomach.csv.gz to illustrate how to use FACETS and elaborate on its powerful functionalities to identify unique Copy Number events. We will also cover how to interpret FACETS output and some of the pitfalls associated with a typical FACETS run.

3.1 Getting Ready for FACETS

The data in the test dataset stomach.csv.gz is pre-processed on ~1.9 million cataloged polymorphic regions over read counts observed for:

1. Reference allele = column File1R is for normal counts and File2R is for tumor counts.
2. Alternate allele = column File1A summarizing counts form normal, File2A for tumor.
3. Errors, neither Ref nor alt = column File1E in normal, and column File2E for tumor.
4. Deletions = column File1D for deletions in normal and column File2D in tumor.

```
datafile = system.file("extdata", "stomach.csv.gz", package="-
facets")
stomach.mat = read.csv(datafile)
head(stomach.mat[,c(1:2,5:12)])
##    Chromosome Position File1R File1A File1E File1D File2R
File2A File2E File2D
## 1           1    69424    170    117      0      0    158    103
0      0
## 2           1    69515      0     76      0      0      0     77
0      0
## 3           1    69536    103      0      0      0     99      0
0      0
## 4           1   808866     96      0      0      0    133      0
0      0
## 5           1   809120     66      0      0      0    105      0
0      0
## 6           1   809176     79      0      0      0    126      0
0      0
```

In the included data, the ref and alt alleles were replaced with "N" and random noise added to the SNP positions to deidentify the samples.

If you are working with BAM files, then you need to run snp-pileup first to obtain read counts for SNPs present in your . vcf file. More details are discussed in Subheading 2.2. We recommend running snp-pileup command as follows:

```
snp-pileup -g -q15 -Q20 -P100 -r25,0 <vcffile> <outputfile>
<normalbam> <tumorbam>
```

where

-g, --gzip means: compress the output file.

-q, --min-map-quality=QUALITY: minimum threshold for mapping quality.

-Q, --min-base-quality=QUALITY: minimum threshold for base quality.

-P, --pseudo-snps=MULTIPLE: Insert a pseudo SNP every 100 positions, with the total count at that position. This is used to reduce large gaps between consecutive SNPs and still get consistent read counts across the genome.

-r, --min-read-counts=READS: minimum tumor and normal read counts for a position, in that order. For example -r25,0 means minimum depth of tumor bam is 25 and for normal bam is 0. This helps to limit the output file size by eliminating off-target reads.

After the data has been summarized in read counts over various SNP regions, we need to read in the data readSnpMatrix and apply some more filters before creating the segmentation output with the help of preProcSample and procSample.

3.2 Segmentation

We first read in the output of snp-pileup into R using readSnpMatrix. readSnpMatrix and expect the read counts to be in normal–tumor order. So, if you are using pileup output from other sources, make sure that the order of summary of ref, alt, error, and deletions is summarized first for normal and then for tumor tissue. Note, overlapping paired end reads should not be counted twice, as seen in older versions of mpileup from samtools.

You can also set the threshold read count for error and deletions in parameters err.thresh and del.thresh, respectively.

```
rcmat = readSnpMatrix(filename=datafile, err.thresh=Inf, del.
thresh=Inf)
head(rcmat)
##   Chromosome Position NOR.DP NOR.RD TUM.DP TUM.RD
## 1          1    69424    287    170    261    158
## 2          1    69515     76      0     77      0
## 3          1    69536    103    103     99     99
## 4          1   808866     96     96    133    133
## 5          1   809120     66     66    105    105
## 6          1   809176     79     79    126    126
```

where, NOR.DP and TUM.DP are total read counts of normal and tumor tissue (total read count is obtained by File#A + File#R). NOR.RD and TUM.RD are reference allele read counts.

The next step is to run joint segmentation of logR and logOR. To recap, for each position, logR is defined by the log-ratio of total read depth in the tumor versus that in the normal and logOR is defined by the log-odds ratio of the variant allele count in the tumor versus in the normal, accounting for GC bias.

Furthermore, the pileup data is processed as follows to avoid problems in the joint segmentation step, as part of preProcSample. Below we describe the default values, their explanation, and how to go about determining the parameter unique to your analysis.

```
set.seed(1234)
xx = preProcSample(rcmat, ndepth=35, het.thresh=0.25, snp.
nbhd=250, gbuild="hg19", unmatched=FALSE, ndepthmax=1000)
```

- snp.nbhd=250: To avoid hyper-segmentation, sample a random SNP from a region containing more than one SNP. snp.nbhd determines the window size to scan for more than one SNP occurrences in that interval. This causes the SNP-dense regions to space out and reduce downstream local patterns and serial correlation. Since we are sampling random SNPs, the random generator seed should be set/saved for reproducible results (*see* **Note 1**).

  The default snp.nbhd=250 is based on exome data. For whole-genome consider increasing it to snp.nbhd=500 and for target sequencing to snp.nbhd=150.
- cval=25: This determines the critical value (cval) for segmentation, used to threshold the Hoteling T^2 statistic to identify change points on the joint segmentation of logR and logOR. Lower cval value leads to a higher sensitivity for small changes. cval=[25–150] is the default based on exome data. For whole genome consider increasing to cval=[25–400] and for targeted sequencing decrease to cval=150 (*see* **Note 2**).
- ndepth=35 and ndepthmax=1000: Remove positions with total read count below 35 or exceeding 1000 (excessive coverage). Determine this by viewing the threshold of coverage as in Fig. 1. Here, the distribution of the coverage from Normal and Tumor seems quite similar. In a case where Normal or Tumor, one of them, has been sequenced with different insert-size, it might lead to waterfall effect. See more in **Note 3**.
- het.thresh=0.25: Parameter to adjust vaf threshold to call an SNP heterozygous. Default is het.thresh=0.25. In case the normal-tumor pair is unmatched, adjust it to het.thresh=0.10 or lower, and set unmatched=TRUE. An example of unmatched normal–tumor FACETS plot can be seen in Fig. 2. Although choosing a lower het.thresh is a way to circumvent the problem of a sample having an unmatched normal, the output from FACETS should be interpreted carefully. To identify further anomalies such as the Waterfall effect, *see* **Note 3**.
- gbuild=hg19: Genome build used for the alignment of the genome. FACETS support "hg38," "hg18," "mm9," "mm10." Option udef can be used to analyze other genomes. For the udef option, one needs to provide the GC percentage data for windows of 1000 bases in steps of 100, as in ugcpct.

Output from preProcSample is an intermediate step to run FACETS, and we will not go into detail of its output. Remember,

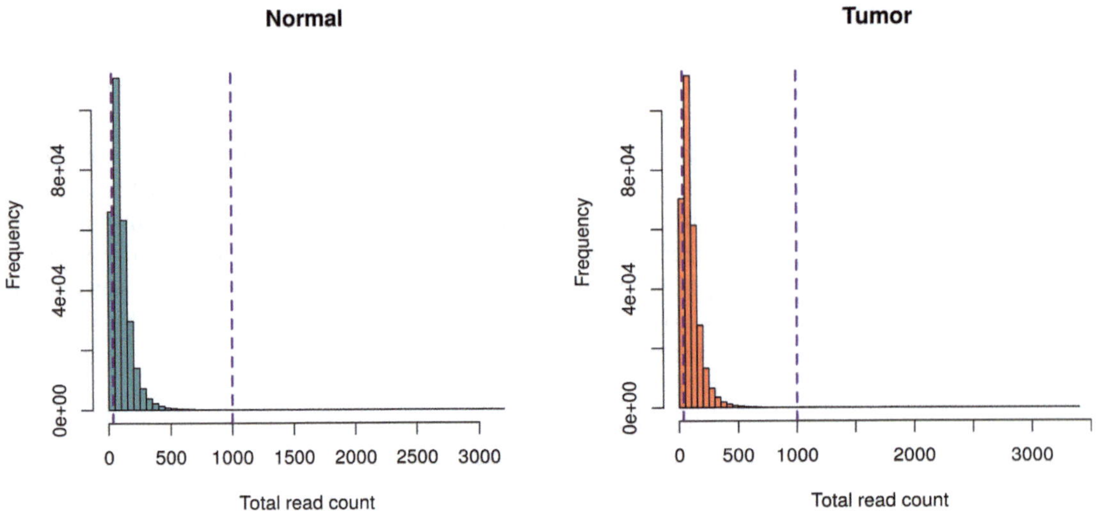

Fig. 1 Total read count in Tumor and Normal tissues. Dashed purple lining is marking read count at 35 and 1000

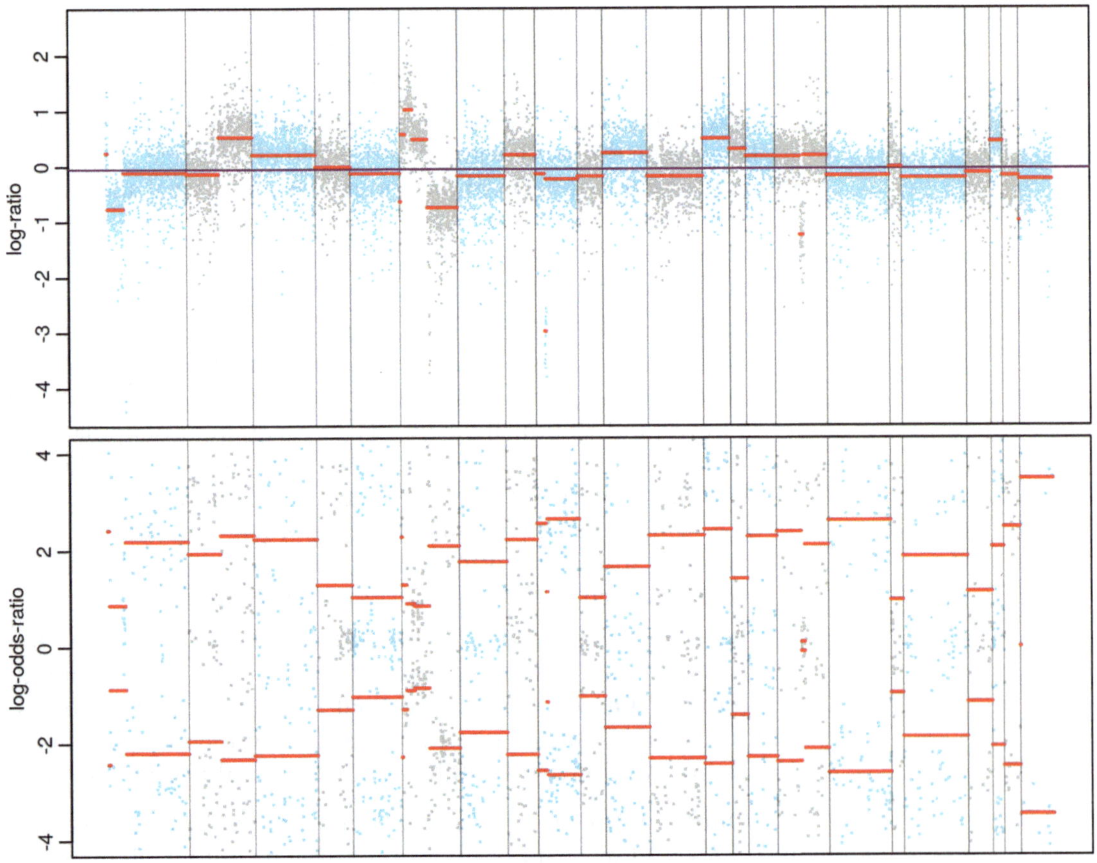

Fig. 2 An example from FACETS output when the tumor sample has an unmatched normal

Table 1
Columns of the procSample output

Column names	Meaning
chrom	Chromosome to which the segment belongs
seg	The segment number
num.mark	# of SNPs in the segment
nhet	# of SNPs that are deemed heterozygous
cnlr.median	Median logR (log-ratio) of the segment
mafR	logOR, log-odds-ratio summary for the segment
segclust	Segment cluster to which the segment belongs to
cnlr.median.clust	Median logR of the segment cluster
mafR.clust	logOR for the segment cluster
cf	Cellular fraction of the segment
tcn	Total Copy Number of the segment
lcn	Minor Copy Number of the segment

to set the set.seed() to get reproducible results when using snp. nbhd.

The joint segmented data is returned as jointseg. preProcSample essentially grows a large segmentation tree which is also used to identify focal changes, returned as seg.tree. procSample then uses its output to first identify focal changes and then prune the tree to remove all changes attributable to waves that still remain after GC correction, etc. as follows:

```
oo=procSample(xx,cval=150, min.nhet=15)
```

Note that there are two cval parameters in preProcSample and procSample. Read more about it in **Note 2**.

As explained above, procSample expects the output from preProcSample containing seg.tree and jointseg. Parameter min. nhet=15 is the number of heterozygote SNPs in a segment used for bivariate t-statistic during clustering of segments. procSample spits out object oo$out, a data frame as follows, explained in Table 1.

```
head(oo$out)
##    chrom seg num.mark nhet cnlr.median         mafR segclust
cnlr.median.clust
## 1      1   1     2631  158  0.50028101 0.305665373       12
0.49145998
```

```
## 2      1   2      4598  249 -0.03699913 0.021520755          7
-0.04444348
## 3      1   3      6342  408  0.49218666 0.404039235         12
0.49145998
## 4      2   4      7006  375 -0.03654935 0.019844538          7
-0.04444348
## 5      2   5        14    0 -0.85252562 0.000000000          6
-0.88771396
## 6      2   6      2299   96 -0.05328541 0.001226293          7
-0.04444348
##    mafR.clust        cf tcn lcn
## 1 0.38807104 0.891936   3   1
## 2 0.01655667 1.000000   2   1
## 3 0.38807104 0.891936   3   1
## 4 0.01655667 1.000000   2   1
## 5         NA 0.891936   1   0
## 6 0.01655667 1.000000   2   1

oo$dipLogR
## [1] -0.04444348
```

dipLogR—dipLogR here shows the systematic shift from the normal diploid state. In order to obtain correct genotype calls for copy number, we need to identify the location of the normal diploid state. dipLogR is used to get logR, and logOR estimates adjusting for purity, etc. *See* **Note 4**.

flags—Check the output of the flags variable to understand if there was a problem with the run of procSample in the form of helpful messages like "Noisy sample," "Low purity," etc.

3.3 Estimate Tumor Purity, Ploidy and Allele-Specific Copy Number

Once we have estimated observed Total Copy Number ooouttcn, Minor Copy Number oooutlcn, and Cellular Fraction oooutcf, we run emcncf to obtain allele-specific estimates.

To further refine these initial estimates and obtain a genome-wide optimization, we apply a genotype mixture model and maximize a joint likelihood that summarizes over all SNP loci and segment clusters across the genome. An expectation-maximization (EM) algorithm is used for the estimation procedure. It can be viewed as an estimation problem with the latent copy number states as "missing" data. In the E-step of the EM procedure, Bayes theorem is used to compute the posterior probability of the segment cluster.

In the M-step, given the imputed genotype, we update the model parameters by maximizing the complete-data likelihood. This procedure is iterated until convergence. For a segment, if the number of heterozygous SNPs is less than 15, only the total copy number is given. For male gender, the number of copy for chromosome X is 1 in the normal. The X chromosome copy number in the tumor is adjusted accordingly.

The segmentation result and the EM fit output look like as follows:

```
fit=emcncf(oo)
head(fit$cncf)
##   chrom seg num.mark nhet cnlr.median        mafR segclust
cnlr.median.clust
## 1     1   1     2631  158  0.50028101 0.305665373       12
0.49145998
## 2     1   2     4598  249 -0.03699913 0.021520755        7
-0.04444348
## 3     1   3     6342  408  0.49218666 0.404039235       12
0.49145998
## 4     2   4     7006  375 -0.03654935 0.019844538        7
-0.04444348
## 5     2   5       14    0 -0.85252562 0.000000000        6
-0.88771396
## 6     2   6     2299   96 -0.05328541 0.001226293        7
-0.04444348
##   mafR.clust     start       end     cf.em tcn.em lcn.em
## 1 0.38807104     69424  29651873 0.8767632      3      1
## 2 0.01655667  31188034 144922109 1.0000000      2      1
## 3 0.38807104 144922463 249212378 0.8767632      3      1
## 4 0.01655667     41507 197673978 1.0000000      2      1
## 5         NA 197706054 197873655 0.8919360      1      0
## 6 0.01655667 197947995 243061179 1.0000000      2      1
```

Columns 1–11 are the same as oo$out as explained above and in Table 1. Remaining columns are the EM estimates of cellular fraction, total copy number, and minor copy number, respectively (cf.em, tcn.em, lcn.em).

emcncf function also calculates purity, ploidy, and spits out useful messages in fit$emflags. *See* **Note 5** for a few recommendations if your output has NAs and **Note 6** if your purity estimate is 0.3.

```
fit$purity
## [1] 0.891936
fit$ploidy
## [1] 2.070151
```

3.4 Integrated Output

plotSample provides genome-wide view and FACETS output. *See* Fig. 3.

```
plotSample(x=oo,emfit=fit)
```

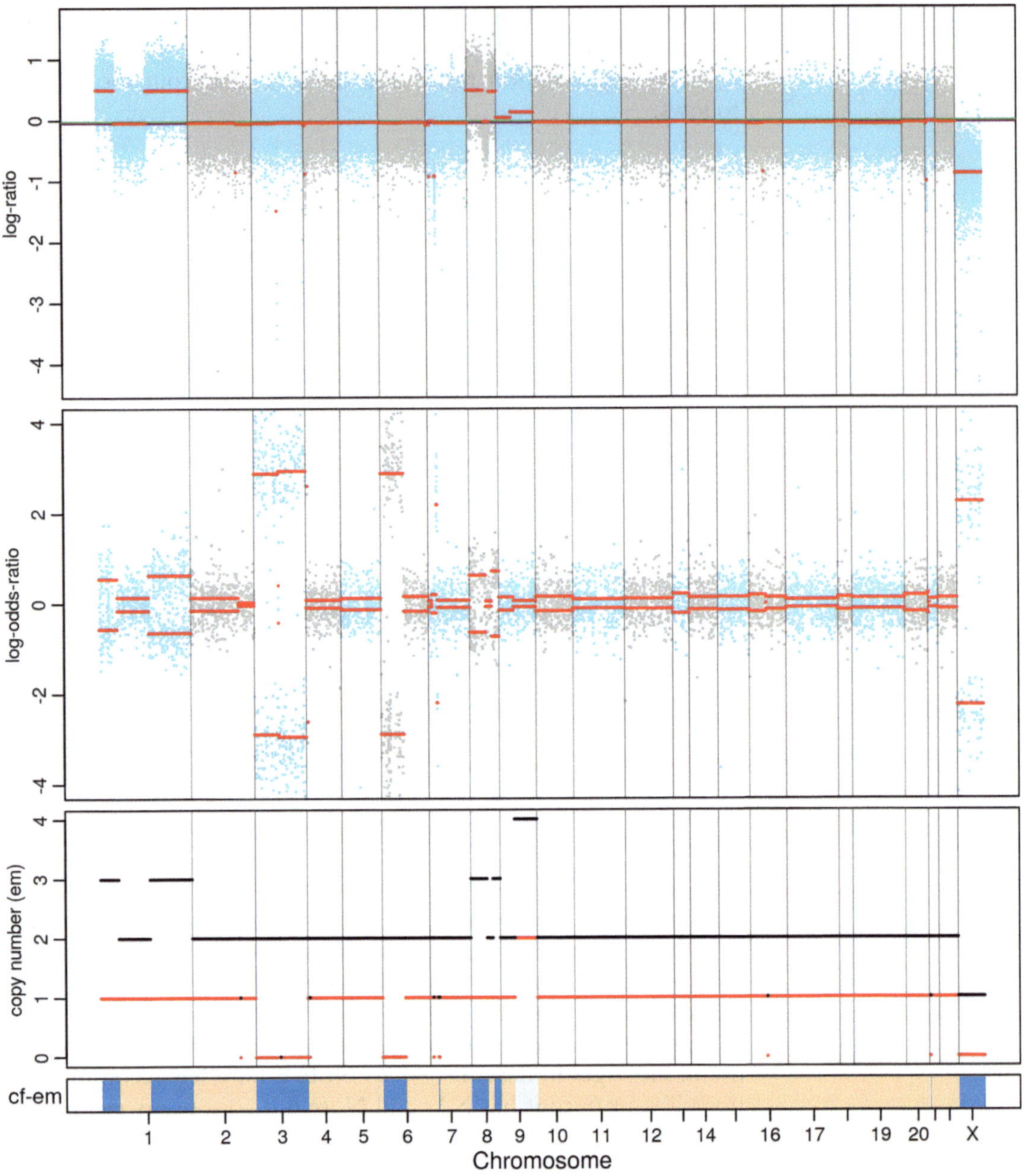

Fig. 3 Visualizing integrated output from plotSample summarizing log-ratio, log-odds-ratio and estimated Cancer Cellular Fraction

The plot above by plotSample consists of four panels summarizing FACETS output over Chromosomes 1-X, as can be seen in the bottom of the plot. The top panel displays logR with chromosomes alternating in blue and gray. The green line indicates the median logR in the sample, purple line indicates the logR of the diploid state (dipLogR).

The second panel displays logOR. Segment means are plotted in red lines.

The third panel plots the total (black) and minor (red) copy number for each segment. The bottom bar shows the associated cellular fraction (cf). Dark blue indicates high cf. Light blue indicates low cf. Beige indicates a normal segment (total = 2, minor = 1).

In this figure, one can also identify some unique Copy Number events. First, Copy Neutral LOH events were seen in chromosomes 3 and 6 (lcn.em==0 and tcn.em==2). Furthermore, sub-clonal duplication event was seen in chr 9 (lcn.em==2 and tcn.em==4) with an estimated cellular fraction of 0.89. For more details on identifying clonal and sub-clonal events, *see* **Note 7**.

3.5 Assess the Fit by Spider Plot

A diagnostic plot (Fig. 4) for the fit can be generated using the function:

```
logRlogORspider(oo$out, oo$dipLogR)
```

This is known as a spider plot. Here, we plot the trajectories for each of the CN states for expected (logR-dipLogR) vs. expected (logOR) for purity ranging from 0 to 0.95. The circles are the segment summaries as observed in the FACETS fit, with size of the circles proportional to the number of loci in the segment. For a good fit, the segment summaries (circles) should be close to one of the lines.

4 Notes

1. Hyper-segmentation occurs when we see preProcSample and procSample generating a large number of tiny fragments from the segmentation process (see an example in Fig. 5). Hyper-segmentation can happen if the sampled SNPs are not spaced out sufficiently, or an overly small cval value is used for segmentation. A higher value of cval and a higher value of snp.nbhd (e.g., 1000 for WGS) can help alleviate hyper-segmentation effects. Anytime, after joint segmentation by procSample if the number of rows in oo$out or number of segments exceed 300–400 (for gene panels), the sample is hyperfragmented.
2. The cval or critical value is a parameter in both preProcSample and procSample which controls the joint segmentation of observed logR and logOR. In preProcSample, a default value is set at cval=25 in order to capture focal changes. In procSample, a large cval value is used for segmentation. The default values are mentioned in Subheading 3.2. procSample expects

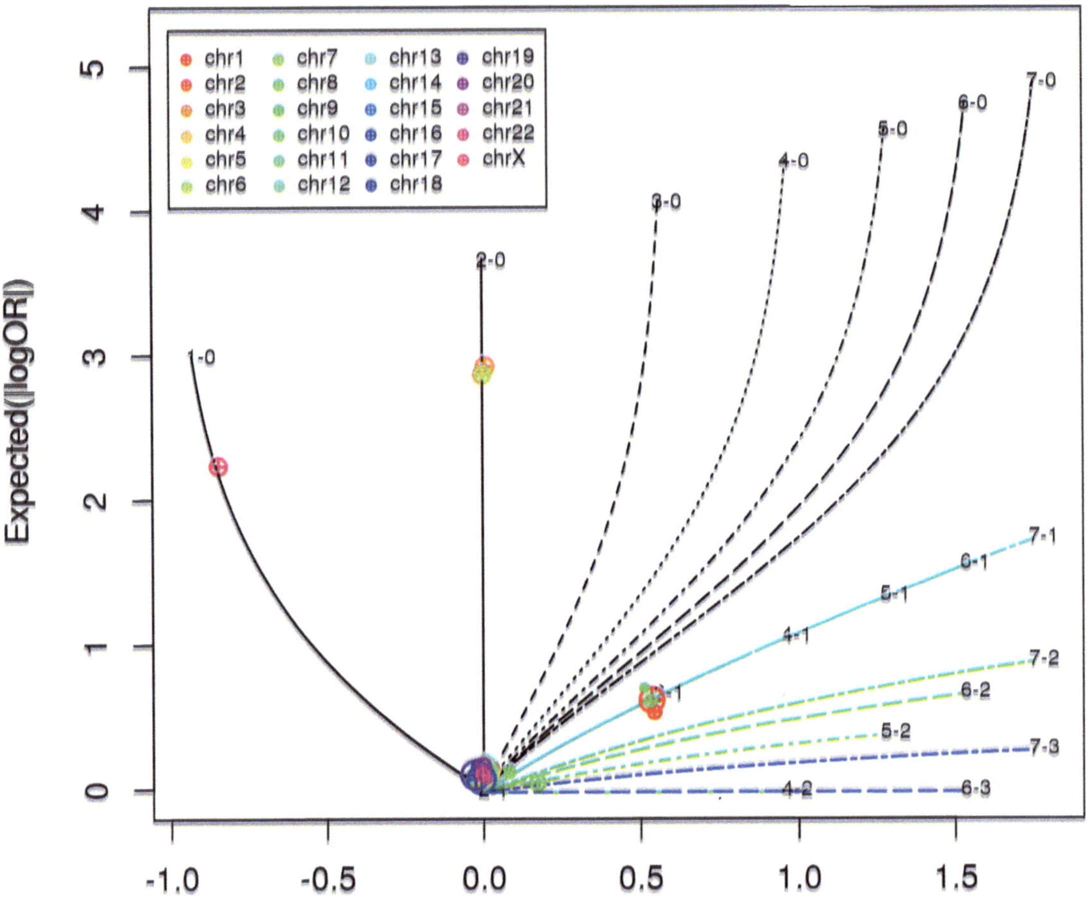

Fig. 4 A spider plot to assess the output and fit of a FACETS run

the data to be in the form of the output of preProcSample, so preProcSample should be run before. One can run procSample with any cval that is large enough to avoid hyper-segmentation but specify the dipLogR from a prior fit.

3. When insert sizes of Tumor and Normal samples do not match, it results in a waterfall effect in FACETS output (*see* Fig. 6). This is due to loss of data points as the mismatch in insert sizes do not capture all SNP positions. This is common where Tumor samples came from FFPE which had smaller insert sizes, whereas Normal samples were mostly derived from blood having larger insert sizes or vice versa.
4. cnlr.median is the segment level summary of observed data. Correcting it for dipLogR can be easily done by subtracting it from cnlr.median, which gives the log(total copy number) adjusted for ploidy. However, it is to note that this correction

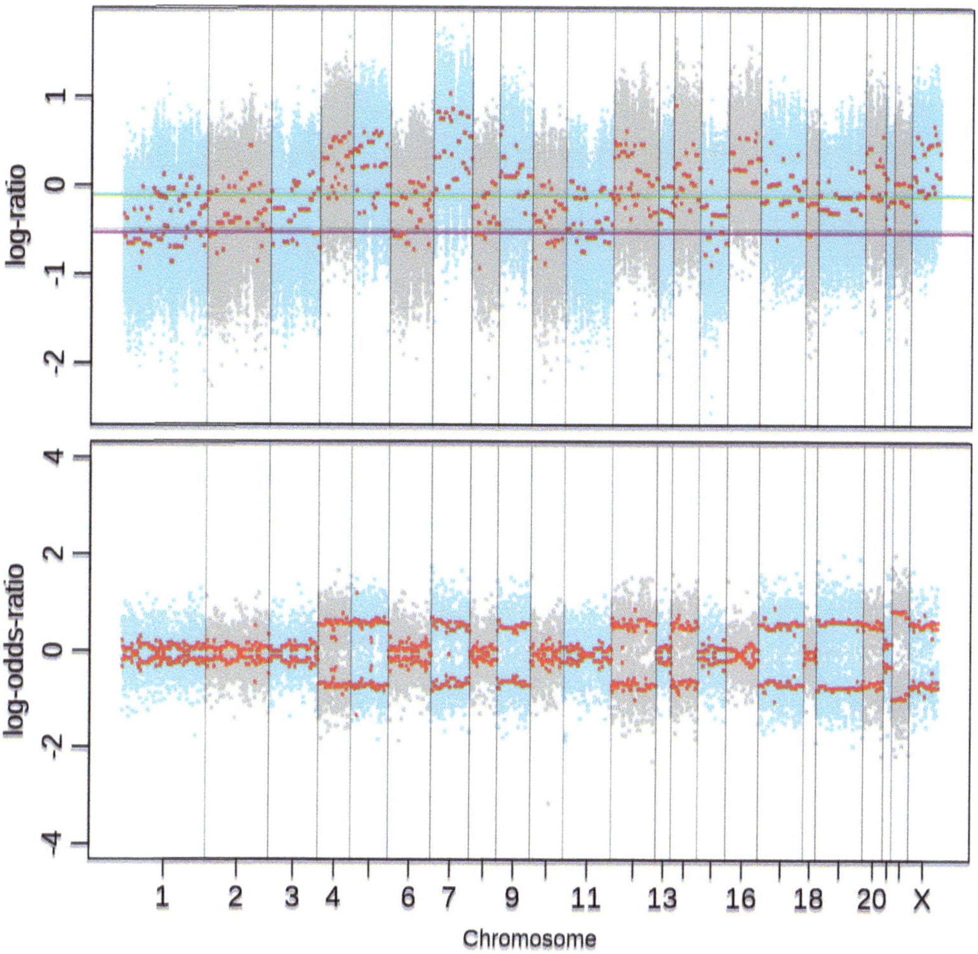

Fig. 5 Example of hyper-segmentation

does not give comparable log-ratio values for two different samples with different purity values. The final integer copy number calls are appropriate for downstream analysis.

5. NA or missing values in FACETS output, especially tcn.em, lcn.em and cf.em or for purity and ploidy estimates can arise due to a variety of reasons. One of the more common reason can be due to the fact that the segment does not possess sufficient numbers of heterozygous SNPs to facilitate the estimation of the allele specific copy number. This typically happens in focal changes (mostly amplification since deletions are mostly estimated to be zero for lcn). If the expected coverage is high, say 150× in the case of whole-exome sequencing, then one should get around 250k SNPs as a part of their analysis of which 8–10% will be heterozygous. This can be checked by looking at the numbers sum(oo$jointseg$het). Some other reasons can include low coverage in certain targeted panel design, or a genome build mismatch in the analysis.

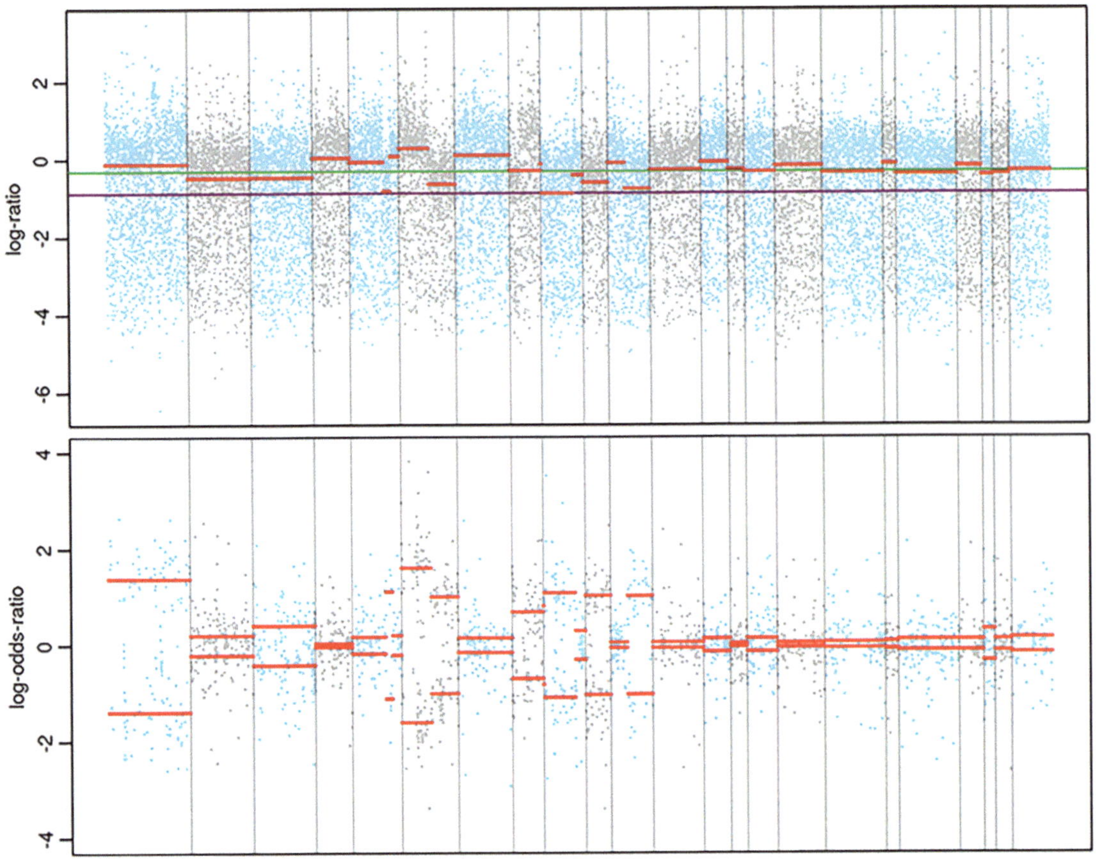

Fig. 6 An example of waterfall effect due to insert size mismatch as obtained from plotSample. See how logR is spread out for SNPs across all chromosomes looking like a "waterfall"

6. 0.3 is used as a placeholder purity estimate when a sample is mostly diploid except for some focal amplifications without many heterozygous SNPs, making it difficult to estimate the total copy number and purity as they are inversely related.

7. Given we have robust estimates of purity, ploidy, and integer number copy number calls from a typical FACETS run, one can now analyze Cellular Fraction (cf) associated at each SNP position in the tumor sample. Cf is bounded by the purity estimate and is very unlikely to be 1. An event is considered subclonal if the estimated cellular fraction is less than the purity estimate (cf. em < purity) and clonal if it's close to the purity estimate, or more than what is attributable to noise. One can also see this information from the FACETS output from plot-Sample at the bottom of the plot.

References

1. Shen R, Seshan VE (2016) FACETS: allele-specific copy number and clonal heterogeneity analysis tool for high-throughput DNA sequencing. Nucleic Acids Res 44(16): e131–e131
2. Chen H et al (2015) Allele-specific copy number profiling by next-generation DNA sequencing. Nucleic Acids Res 43(4):e23–e23
3. Ha G et al (2014) TITAN: inference of copy number architectures in clonal cell populations from tumor whole-genome sequence data. Genome Res 24(11):1881–1893
4. Venkatraman ES, Olshen AB (2007) A faster circular binary segmentation algorithm for the analysis of array CGH data. Bioinformatics 23(6):657–663

Chapter 8

Meerkat: An Algorithm to Reliably Identify Structural Variations and Predict Their Forming Mechanisms

Lixing Yang

Abstract

Next-generation sequencing technologies have been widely used to query genetic variants in normal individuals as well as in those with diseases. Large-scale structural variations are a common source of genetic diversity in human population, and some of them have significant contributions to the etiology of diseases. However, the detection of large-scale structural variations from sequencing data remains challenging. Here, we describe Meerkat—an algorithm which can reliably detect structural variations from Illumina short-read sequencing data at basepair resolution. A unique feature of Meerkat is that it can infer the variant forming mechanisms based on the DNA content and features at the breakpoints.

Key words Somatic variants, Germline variants, Genomic rearrangements, High-throughput sequencing, Bioinformatics

1 Introduction

Structure variations (SVs) refer to large-scale alterations of the genome that change the DNA structure [1]. The main difference between SVs and copy number variations (CNVs) is that SVs always have DNA breakpoints. Some CNVs have DNA dosage changes but not necessarily carry breakpoints (e.g., chromosomal gains and losses). SVs can be detected by Sanger sequencing, DNA microarray, second-generation short-read sequencing and third-generation long-read sequencing [1, 2]. In this chapter, I will describe a computational algorithm "Meerkat" which is designed to detect SVs from Illumina short-read sequencing data [3]. Meerkat can be applied to any species as long as reference genomes are available. I will use human genome as an example in this chapter. Meerkat can be used to detect different types of variants including germline, somatic, as well as de novo SVs. It is particularly suitable for somatic SV detection in cancer genomes [3].

Charlotte K. Y. Ng and Salvatore Piscuoglio (eds.), *Variant Calling: Methods and Protocols*,
Methods in Molecular Biology, vol. 2493, https://doi.org/10.1007/978-1-0716-2293-3_8,

Detecting SVs is challenging for several reasons: (1) all sequencing technologies have biases and errors; (2) repetitive sequences in the human genome (as well as other genomes) can cause problems when aligning sequencing reads, and repetitive sequences are known to induce SVs; (3) heterogeneity is common in tumor tissues and somatic SVs can be present only in a subset of the cells; and (4) some SVs can be quite complex. Various strategies have been used for SV detection using short-read sequencing data, such as read depth [4], read pair [5], split read [6], genome assembly [7], and certain combinations of them [3, 8–12]. Each strategy has its own advantages and disadvantages. Meerkat combines read pair and split read strategies to achieve optimal sensitivity and specificity of SV detection (Fig. 1). Precise breakpoints obtained from split read alignment can offer opportunity to infer the SV forming mechanisms (Fig. 1) which is typically not provided by other SV detection algorithms. Meerkat does not use read depth, because balanced SVs, such as inversions and reciprocal translocations, do not change DNA copy number or read depth. Typically, whole-genome sequencing (WGS) data are required for SV detection. However, Meerkat can be applied to any Illumina paired-end sequencing data such as whole-exome sequencing (WES) to detect SVs located in or near exons [13]. Benchmarking of SV callers have been performed by several studies and the performance of Meerkat has been consistently among the top [14–16].

2 Software

Meerkat is available for download at http://compbio.med.harvard.edu/Meerkat/. The testing files can be downloaded at https://github.com/yanglab-computationalgenomics/Meerkat.example.hg38. Academic use is free. For commercial use, please contact Boston Children's Hospital Technology and Innovation Development Office at tido@childrens.harvard.edu.

2.1 Prerequisites

The following software are required to run Meerkat:

1. Unix/Linux system.
2. CMake version 2.6.4 or above.
3. PERL 5.8.1 or above.
4. BioPERL 1.5.0 or above.
5. R 2.6.1 or above (you must be able to run Rscript directly from command line).
6. Samtools 0.1.19. Newer version 1.x will not work.

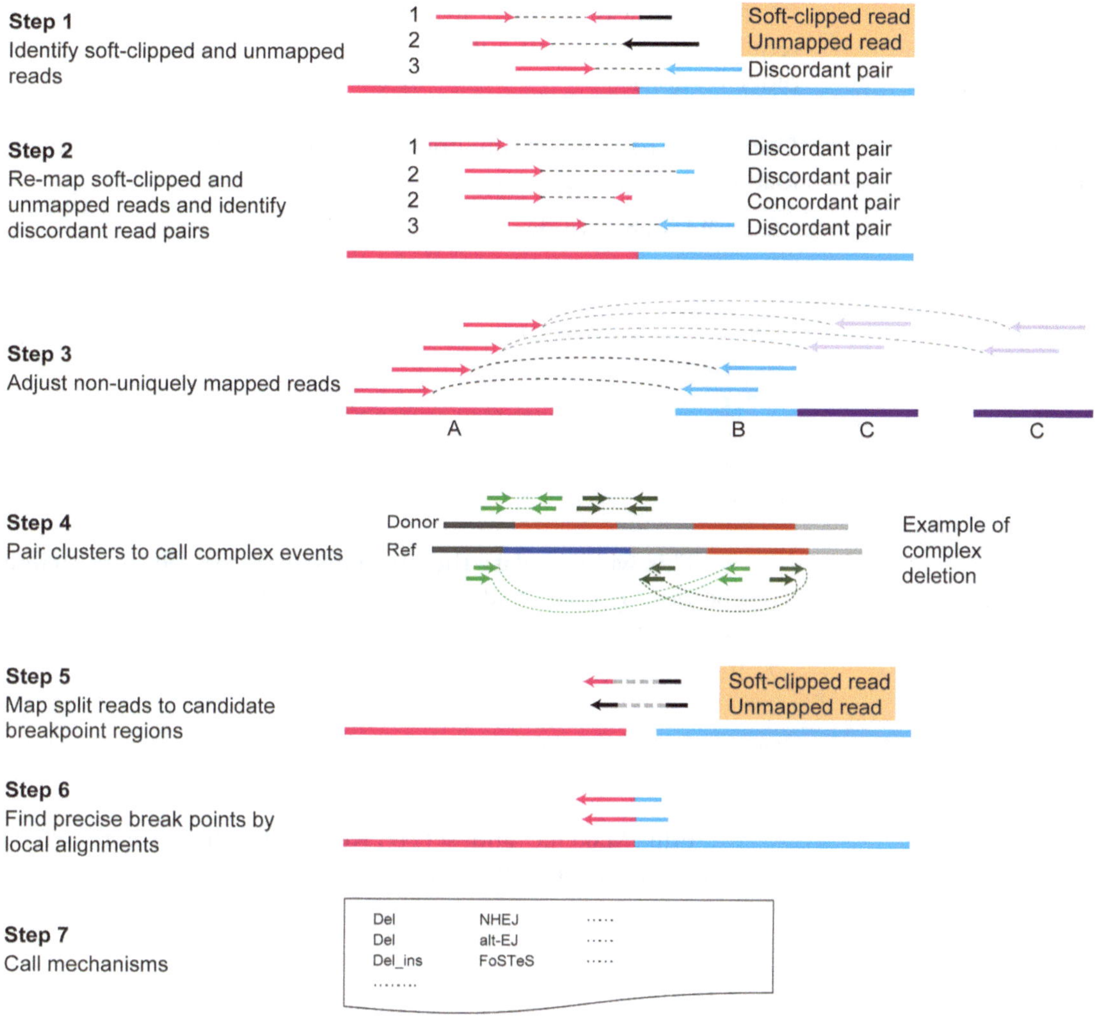

Fig. 1 Meerkat workflow. Pink and light blue lines denote two segments of the genome connected together in the formation of an SV. In Step 3, purple lines denote repetitive segments. Arrows connected by dashed lines represent paired-end reads. In Step 4, the dark blue line on the reference genome represents a deleted segment in the donor genome; the dark red line represents an inserted segment. Light and dark green arrows connected by dashed lines represent two clusters of discordant read pairs. In Step 5, the beginning and end of soft-clipped and unmapped reads identified in Step 1 are used as split reads

7. BWA 0.6.2. Users can use the latest version to align the reads and generate bam files as input, but version 0.6.2 is required in Meerkat for split read alignment.
8. NCBI legacy blast 2.2.26, can be downloaded at: ftp://ftp.ncbi.nlm.nih.gov/blast/executables/legacy.NOTSUPPORTED/2.2.26/. The blast+ suite will not work.
9. Blat 2.2.24 or above.
10. Primer3 2.2.0 or above.

Meerkat was released in 2013. The latest version of some required packages may not be compatible. If necessary, try some older versions.

2.2 Software Compilation

Users need to compile one library and three modules after downloading the Meerkat package:

1. Build mybamtools:

```
cd ./src/
tar xjvf mybamtools.tbz
cd mybamtools
mkdir build
cd build
cmake ..
make
```

   This will populate the lib folder with the requisite shared libraries. The src folder is still necessary—header files stored here are needed to compile other binaries.

2. Build bamreader (must have mybamtools compiled):

```
tar xjvf bamreader.tbz
cd bamreader
```

   Edit Makefile and set BTROOT to the path to which mybamtools was extracted.

```
vi Makefile
BTROOT = /path/to/mybamtools
```

   Continue the compilation:

```
make
mv ./bamreader ../../bin/
```

   Add path to mybamtools' compiled libraries to LD_LIBRARY_PATH.

   In bash: export LD_LIBRARY_PATH=/path/to/mybamtools/lib

   In csh: setenv LD_LIBRARY_PATH /path/to/mybamtools/lib

3. Build dre (must have mybamtools compiled):

```
tar xjvf dre.tbz
cd dre
```

Edit Makefile and set BTROOT to the path to which mybamtools was extracted.

```
vi Makefile
BTROOT = /path/to/mybamtools
```

Continue the compilation:

```
make
mv ./dre ../../bin/
```

4. Build sclus:

```
tar xjvf sclus.tbz
cd sclus
make
mv ./sclus ../../bin/
```

2.3 Reference Files

1. Reference genome sequence in fasta format and indexed by PERL module Bio::DB::Fasta. The fasta file of reference genome must be in a folder alone without any other files, see Bio::DB::Fasta manual for more details. Reference sequence in 2 bit format is required to set up blat server.
2. BWA index of reference file generated by bwa index command.
3. Fai index of reference file generated by samtools faidx command.
4. Gene annotation of reference genome which can be downloaded from UCSC. The file needs to be sorted by chromosome and coordinate by the following command:

```
sort refGene.txt -k 3,3 -k 5,5n > refGene_sorted.txt
```

5. Repeat annotation (RepeatMasker output) of reference genome which can be downloaded from UCSC.

2.4 Example of Software and Reference File Arrangements

The required packages/files are configured as follow:

1. samtools in folder /opt/samtools/.
2. BWA in folder /opt/bwa/.
3. NCBI blast in folder /opt/blast/. Binaries of blastall and formatdb are in /opt/blast/bin/.
4. Blat and gfServer in folder /opt/blat/.

5. Reference hg38.fasta in folder /db/hg38/hg38_fasta/. Perl module Bio::DB::Fasta will index the entire folder and generate a file directory.index in the same folder.
6. BWA index files hg38.fasta.* of reference in folder /db/hg38/hg38_bwa_idx/.
7. Fai index file hg38.fasta.fai of reference in folder /db/hg38/.
8. Gene annotation file refGene_hg38_sorted.txt in folder /db/hg38/.
9. RepeatMasker file rmsk-hg38.txt in folder /db/hg38/.

2.5 Input File

Meerkat is designed to analyze Illumina paired-end short-read sequencing data. It is not recommended to process long-read (>400 bp read length) or single-end data. Other short-read sequencing platforms (e.g., BGI sequencer) have not been tested. The input files need to be in BAM format. CRAM or other formats are not supported and must be converted to BAM files first. It is strongly recommended to use BWA mem [17] for read alignment and Picard [18] to mark duplicates. BAM files must be sorted and indexed.

Meerkat has a built-in function to consider non-uniquely mapped reads. This function depends on alignment performed by old version of BWA aln function with XT and XA tag to extract all alternative mapping positions. Usually, newer version of BWA mem performs better than BWA aln. If users want to use non-uniquely mapped reads, for human 75 or 100 bp reads, it is recommended to use -l 40 -k 2 for bwa aln and -N 100 for bwa sampe.

Meerkat can handle multiple read groups with different insert size and read length. Alignment for multiple read groups shall be merged into one single BAM file with RG tag properly labeled. Note, "/" is not allowed in RG string. If certain read groups are to be left out of analysis due to low quality, a read group blacklist file shall be generated with one read group ID per line. *See* Subheadings 3.1 and 3.2 for details.

Meerkat will generate a large number of intermediate files with the same prefix as the input BAM file, it is recommended to create one folder per BAM file, place the BAM file in the folder or create a soft link to the BAM file, so that all the intermediate files are generated in the same folder. Raw variants from each sample can be filtered to produce high confidence germline variants. They can also be filtered against other control samples to produce high confidence somatic and de novo variants.

3 Methods

The main SV detection and mechanism inference workflow includes four steps (pre_process.pl, meerkat.pl, mechanism.pl, and somatic_sv.pl). There are multiple sub-steps for pre_process.pl and meerkat.pl. For high coverage sequencing, users shall run sub-steps one by one, check outputs, and adjust parameters if necessary. We have suggested different sets of parameters suitable for different types of data. DO NOT use default parameters. The package also provides a few handy tools (discon.pl, meerkat2vcf.pl, fusions.pl, and primers.pl) in addition to the main workflow.

3.1 *Pre_process.pl*

This step is to calculate insert size distribution and extract soft-clipped and unmapped reads for split read alignment. The following command can be used to display all parameters:

```
perl ./scripts/pre_process.pl -h
```

For high coverage WGS (>20×) with read length between 75 and 150 bp and aligned by BWA mem, it is recommended to use -s 20 -k 1500 -q 15 -l 0. For high coverage WGS (>20×) with read length between 75 and 150 bp and aligned by BWA aln, use -s 20 -k 1500 -q 15. For low coverage WGS (<10×) with read length of 50 bp, use -s 18 -l 0 -q 0. For WES data, use -s 20 -k 10000 -q 5.

The full list of parameters for pre_process.pl is as follows:

- -b FILE: sorted and indexed bam file, required.
- -k INT: min coverage required for a nucleotide to be ignored (ignore position that covered by too many reads), 0 turn this function off, default 500.
- -r INT: range of insert size to be plotted, insert size larger than this number will not be used calculating distribution, default 1000.
- -n INT: number of reads per read group to be used calculating insert size distribution, 0 use all reads, default 10,000.
- -l INT: [0/1], extract paired soft-clipped reads and pairs with one read mapped, the other unmapped, or both unmapped, re-map mate pair, default 1.
- -q INT: Read trimming parameter. Equivalent to BWA's -q option, default 15.
- -c INT: bp to be cut off from beginning and end of unmapped and soft-clipped reads, such reads were used to find smaller events, default 35.
- -s INT: bp to be cut off from beginning and end of unmapped and soft-clipped reads, such reads were used in split reads mapping, must be the same as -s in meerkat.pl, default 20.

- -u INT: [0/1], process uu pair (both reads unmapped in a pair) into 4 pairs, default 0.
- -f INT: number of alternative mappings to print in XA tag for clipped alignments, default 100.
- -N INT: clipped reads and split reads must have <= INT Ns, default 5.
- -t INT: number of threads used in bwa alignment, default 1.
- -R STR: file name of a list of blacked read groups, one read group ID per line.
- -I STR: /path/to/reference/bwa_index for bwa alignment, required if -l 1.
- -A STR: /path/to/reference/fasta.fai for bwa alignment, required if -l 1.
- -S STR: /path/to/samtools, path only, not the command, no need to specify if samtools is in PATH.
- -W STR: /path/to/bwa, path only, not the command, no need to specify if bwa is in PATH.
- -P STR: specify step to run, [all|is|cl], default all.
  - is: extract unmapped, soft-clipped reads, calculate insert size distribution.
  - cl: map soft-clipped mate pairs to reference genome, if using large clusters to run, users may split cl step into two stages, cl1 and cl2. Cl1 can be run on multiple threads controlled by t parameter, cl2 stage can only be run on a single thread.
  - all: run all above steps.
- -h help.

Here, I describe the functions of a few important parameters and how to choose proper values.

- -k option is aimed to filter out reads mapped to telomeres, centromeres, or other highly repetitive regions by filtering out positions with unusually high coverage. In cancer genomes, it is common to observe copy gain events. Users shall use a higher value such as 1500 in order not to filter out such events.
- -l option is to control whether to re-map some unmapped reads and partially mapped (soft-clipped) reads. This function is unnecessary if the reads are aligned by BWA mem. It is only needed in BWA aln. If insert size is small, it is common to have a read spanning an SV breakpoint. Such read would be unmapped or partially mapped. For a read pair that one end is mapped, the mate is partially mapped, the unmapped part of partially mapped read is extracted and pair with mapped read to form an artificial read pair. For a read pair that one end is mapped, the mate is unmapped, a portion from the beginning and a portion

from the end of the unmapped read are extracted and form two artificial read pairs with the mapped read. The portion to be extracted is controlled by -c option. Such artificial read pairs are re-mapped to the reference genome. It can greatly increase sensitivity if the insert size is small and reads are not too short (at least 75 bp). For short read length (50 bp or shorter), do not use this option.

- -q option is to control the trimming of poor-quality bases from the end of a read similar to the -q option in BWA aln. Do not set to 0 unless you know what you are doing.
- -u option is to control generating four artificial pairs from both ends of unmapped read pairs. It is useful to identify very small events if the sequencing quality is very good and genome is not too repetitive. For human genome, it is recommended to turn this function off.
- -c option is to control how many bases to extract from unmapped or partially mapped reads to form artificial read pairs, see also -l option. It should be slightly smaller than ½ read length. The extracted part will be aligned against the original reference genome. So, users should consider mappability when selecting this parameter. Default is good for human genome with read length of at least 75 bp. Users can use smaller value if the genome is less repetitive.
- -s option is to control how much to extract from unmapped or partially mapped reads to generate split reads. Since split reads will be mapped to breakpoint regions of SVs predicted from discordant read pairs, the value can be set small without sacrificing mappability. It should be between 1/5 and 1/3 of read length.
- -R specifies the file that contains the list of read group IDs (one line per read group ID in the file) that shall be left out of analysis. Recommend not to process a read group if the uniquely mapped reads are less than 30%. If all the read groups are of high quality, users do not need to specify this option.

3.2 Meerkat.pl

This script performs the SV calling. It goes through all reads, extract discordant pairs, cluster discordant pairs, identify candidate SVs based on clusters, align clipped reads and unmapped reads to candidate SVs, detect SVs with split read support, and finally refine breakpoints to basepair resolution.

The following command can be used to display all parameters:

```
perl ./scripts/meerkat.pl -h
```

For high coverage WGS (>20×) with read length between 75 and 150 bp and aligned by BWA mem, it is recommended to use -s 20 -d 5 -p 3 -o 1 -m 0 -l 0. For all BWA mem alignment, -m 0 -l 0 are required. For high coverage WGS (>20×) with read length between 75 and 150 bp and aligned by BWA aln, use -s 20 -d 5 -p 3 -o 1. For deeper coverage, such as >60×, use -s 20 -d 5 -p 5 -o 3. If tumor genome is sequenced at 60× and matched-normal genome at 30×, use 60× parameters for both tumor and normal. For low coverage WGS (<10×) with read length of 50 bp, use -s 18 -d 5 -a 0 -l 0 -q 1. For WES data, use -s 20 -d 5 -p 3 -o 1. Always replace the *.blacklist.gz file for cancer genome by the file generated from the matched normal genome (discard the cancer. blacklist.gz file, put the normal.blacklist.gz file into the cancer folder and rename it to cancer.blacklist.gz).

The full list of parameters for meerkat.pl is as follows:

- -b FILE: sorted and indexed bam file, required.
- -k INT: [0/1], use blacklist generated in pre_process.pl, default 1.
- -d FLT: standard deviation cutoff to call discordant mate pairs, default 3.
- -c FLT: standard deviation cutoff to cluster discordant mate pairs, default equal to -d, it is recommended to use same -c as -d if –d <= 5, set -c 5 if -d > 5.
- -p INT: number of supporting read pairs required for an event to be called, default 2.
- -o INT: number of supporting full length read pairs, default 0, specify this option will decrease sensitivity on small complex events but increase specificity.
- -q INT: number of supporting split reads required for an event to be called, default 1.
- -z INT: event size cutoff, default 1,000,000,000.
- -s INT: bp to be cut off from beginning and end of unmapped reads, must be the same as -s in pre_process.pl, default 20.
- -m INT: [0/1], if set to 1, use Meerkat to remove duplicates; if set to 0, use flag d marked by Picard or other tools to remove duplicates. The bwa mem aligned bam file will not work with option 1. If the bam file is aligned by bwa mem, it has to be processed by Picard to mark duplicates and use -m 0 option. Default 1.
- -a INT: [0/1], adjust non-uniquely mapped reads, default 1.
- -u INT: [0/1], use all alignments in the BAM file, turn this option on if the BAM file is not generated by BWA, turn on this option will force turning off option a, default 0.
- -Q INT: minimum mapping quality for reads to be used, default 0.

- -g INT: number of alternative mappings to consider in main bam file, number of alternative mappings printed out in XA tag by bwa is controlled by -N, default use all in bam file.
- -f INT: number of alternative mappings to consider in clipped alignments, default use all in bam file.
- -l INT: [0/1], consider clipped alignments, default 1.
- -t INT: number of threads used in bwa alignment, default 1.
- -R STR: file name of a list of blacked read groups, one read group ID per line.
- -F STR: /path/to/reference/fasta/files, path only, not the files, required.
- -S STR: /path/to/samtools, path only, not the command, no need to specify if samtools is in PATH.
- -W STR: /path/to/bwa, path only, not the command, no need to specify if bwa is in PATH.
- -B STR: /path/to/blastall and formatdb, path only, not the command, no need to specify if blastall and formatdb are in PATH.
- -P STR: specify step to run, [dc|cl|mpd|alg|srd|rf|all], default all. Each step requires results from previous steps.
  - dc: extract discordant read pairs.
  - cl: construct clusters of discordant read pairs.
  - mpd: call events based on read pairs.
  - alg: align split reads to candidate breakpoint regions, if using large clusters to run, users may split alg step into two stages, alg1 and alg2. Alg1 can be run on multiple threads controlled by t parameter, alg2 stage can only be run on a single thread.
  - srd: confirm events based on split reads and filter results.
  - rf: refine breakpoints by local alignments.
  - all: run all above steps.
- -h help.

Here, I describe how to choose proper values for a few important parameters.

- -d option is to control how to call discordant read pairs. It is equivalent to define what is the largest fragment size (median insert size + d * sd) that could be concordant. Use default if the insert size distribution is narrow, symmetric (Fig. 2a, b), and sequencing coverage is no more than 40×. Use 5 if insert size distribution is wide or skewed with long tail (Fig. 2c, d) or sequencing coverage is higher than 40×. Do not use value less than 3.

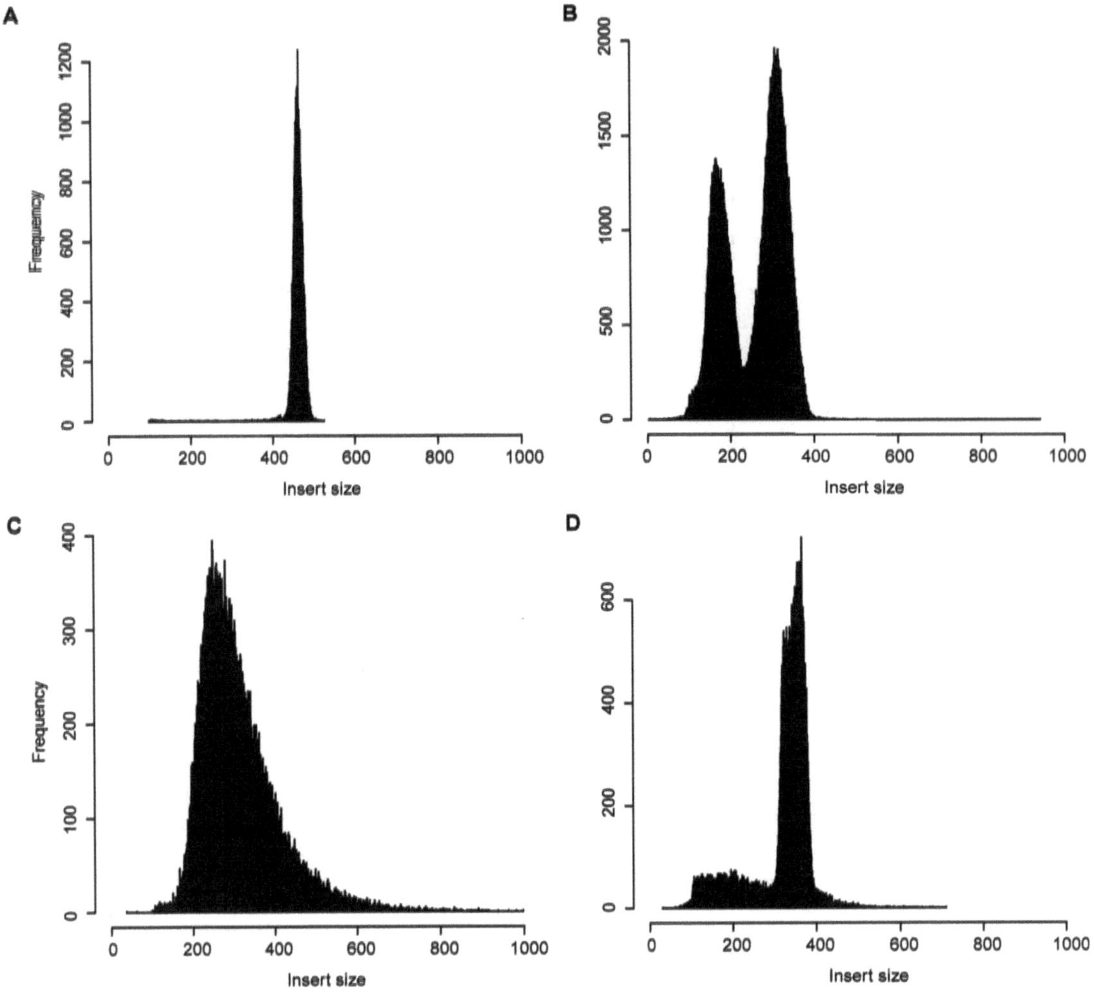

Fig. 2 Various insert size distributions

- -c option is to control how to merge discordant read pairs into clusters and construct confidence interval of breakpoints. It is equivalent to define what is the largest fragment (median insert size + c * sd) that could cover a breakpoint. Use 5 if d is set to 5 or higher values. Otherwise use the same value as d.
- -a g and f options are to control how to handle non-uniquely mapped reads. Set a to 0 to turn such function off if sequencing quality is not good or low coverage sequencing. If turning off, both reads in a pair are required to be uniquely aligned. It will depend on XT tag of uniqueness of mapping generated by BWA. If no XT tag, use option Q. Options g and f are to control number of alternative mapping positions to consider. The number of alternative mapping positions BWA provide in XA tag is controlled by -N parameter in bwa sampe command.

- -u option is to control if to use all alignments in BAM file or not. BAM files generated by BWA are recommended. For BAM files that are not aligned by BWA or aligned by BWA but there is no XT tag, users shall turn on this option. When this option is on, all reads will be used and treated as uniquely mapped, users shall use option Q to specify a minimum allowed mapping quality to get rid of poor-mapping reads. For option Q, recommend using 10. For BWA aligned BAM, users can still use option Q to remove poor-mapping reads.
- -p o and q options are to control the stringency of calling SVs. If using -a 1 -l 1, it is recommended to use -o 1 to avoid too many artifacts arisen from repeats. Specifying o option will decrease the sensitivity for very small events, especially small complex events.

3.3 Mechanism.pl

This script can predict SV forming mechanisms. It needs to be run after meerkat.pl and outputs a prefix.variants file which contains the raw SV events. Raw SVs need to be filtered (*see* Subheading 3.4 for details). The following command can be used to display all parameters:

```
perl ./scripts/mechanism.pl -h
```

Using default values should be fine for most cases. Here is an example for NA18507 run:

```
perl ./scripts/mechanism.pl -R /db/hg38/rmsk-hg38.txt -b
na18507.sorted.bam
```

The full list of parameters for mechanism.pl is as follows:

- -b FILE: sorted and indexed bam file, required.
- -o INT: [0/1], include rmsk type "Other" in TE, default 1.
- -t INT: max size of TE, default 100,000.
- -z INT: size limit of SVs to be processed, default 1,000,000,000.
- -R STR: /path/to/repeat_mask_file, required, can be downloaded from UCSC.
- -h help.

3.4 Somatic_sv.pl

This script can filter raw variants and produce high-quality calls. Even if users want to detect germline and de novo variants, running somatic_sv.pl is still necessary. Variants passing the filtering steps are final SV calls. Both input and output are variants files. The variants file format is complicated due to the presence of complex SVs. If users only need SV breakpoints, use meerkat2vcf.pl or fusions.pl to convert variants to different formats. *See* Subheadings 3.6 and 3.7 for details. The fusions format is recommended since it has gene annotations.

The following command can be used to display all parameters:

```
perl ./scripts/somatic_sv.pl -h
```

The commands below are suggested to use to produce high confidence somatic SVs from high coverage WGS (>20×) aligned by either BWA sampe or mem as well as WES.

```
perl ./scripts/somatic_sv.pl -i $variantsfile -o $somatica -F all_normal_disc/ -l 1000 -R /db/hg38/rmsk-hg38.txt
perl ./scripts/somatic_sv.pl -S /home/ly55/opt/samtools/ -i $somatica -o $somaticb -R /db/hg38/rmsk-hg38.txt -n 1 -D 5 -Q 10 -B $normal_bam -I $normal_isinfo -K $normal_blacklistrg
perl ./scripts/somatic_sv.pl -S /home/ly55/opt/samtools/ -i $somaticb -o $somaticc -R /db/hg38/rmsk-hg38.txt -u 1 -Q 10 -B $normal_bam -K $normal_blacklistrg
perl ./scripts/somatic_sv.pl -S /home/ly55/opt/samtools/ -i $somaticc -o $somaticd -R /db/hg38/rmsk-hg38.txt -f 1 -Q 10 -B $normal_bam -K $normal_blacklistrg
perl ./scripts/somatic_sv.pl -S /home/ly55/opt/samtools/ -i $somaticd -o $somatice -R /db/hg38/rmsk-hg38.txt -e 1 -D 5 -Q 10 -B $tumor_bam -I $tumor_isinfo -K $tumor_blacklistrg
perl ./scripts/somatic_sv.pl -S /home/ly55/opt/samtools/ -i $somatice -o $somaticf -R /db/hg38/rmsk-hg38.txt -z 1
perl ./scripts/somatic_sv.pl -S /home/ly55/opt/samtools/ -i $somaticf -o $somaticg -R /db/hg38/rmsk-hg38.txt -d 40 -t 20
```

The all_normal_disc folder contains all the prefix.discord files from normal genomes. Do not use -F option in somatic_sv.pl with n, u, f, e or z options, it will be extremely slow. Run those filters in separate steps. The prefix.isinfo file is generated by pre_process.pl. When using options n and e, prefix.isinfo file needs to be provided.

Extra filtering steps are needed for WES data:

1. Discard samples that have ≥100 somatic fusions.
2. Apply filter_fusions.pl as follows:

```
perl ./scripts/filter_fusions.pl -i $somaticg_fusion -o $somatich_fusion -m 4 -s 20000
```

The input file $somaticg_fusion is generated by fusions.pl from the final somatic variants (*see* Subheading 3.7 for fusions.pl usage).

The above steps are to produce somatic SVs. To detect de novo SVs, filter the raw SVs from query genomes against their parental genomes.

The following are suggested steps to produce high confidence somatic SVs from low coverage WGS (<10×) data.

Before running filtering steps, users need to set up a blat server:

```
gfServer start 10.11.240.76 17777 /reference/hg38/hg38.2bit
-stepSize=5 &
```

The above command will run gfServer in the background. When the server is ready (usually takes a few minutes), run the following steps:

```
perl ./scripts/somatic_sv.pl -i $variantsfile -o $somatica -F
all_normal_disc/ -R /db/hg38/rmsk-hg38.txt -x 1
perl ./scripts/somatic_sv.pl -i $somatica -o $somaticb -R /db/
hg38/rmsk-hg38.txt -M 1 -B $tumor_bam -C $tumor_bp_reads
perl ./scripts/somatic_sv.pl -i $somatica -o $somaticc -R /db/
hg38/rmsk-hg38.txt -L /opt/blat/ -V "10.11.240.76" -T 17777 -A
$srout_dir
perl ./scripts/merge_sv.pl $somaticd $somaticb $somaticc
```

The second and third filters are too stringent and will miss some real events, but merging the events together performs better. The $somaticd file is the final SV call set.

The following are suggested commands to produce high confidence germline SVs from high coverage WGS (>20×).

```
perl ./scripts/somatic_sv.pl -i $variantsfile -o $germa -R /
db/hg38/rmsk-hg38.txt -E 0
perl ./scripts/somatic_sv.pl -i $germa -o $germb -R /db/hg38/
rmsk-hg38.txt -E 0 -e 1 -D 5 -Q 10 -B $normal_bam -I $normal_-
isinfo -K $normal_blacklistrg
perl ./scripts/somatic_sv.pl -i $germb -o $germc -R /db/hg38/
rmsk-hg38.txt -E 0 -u 1 -Q 10 -B $normal_bam -K $normal_-
blacklistrg
perl ./scripts/somatic_sv.pl -i $germc -o $germd -R /db/hg38/
rmsk-hg38.txt -E 0 -d 40 -t 40
perl ./scripts/somatic_sv.pl -i $germd -o $germe -R /db/hg38/
rmsk-hg38.txt -E 0 -z 1 -p 5 -P 10
```

The full list of parameters for Somatic_sv.pl is as follows:

- -i FILE: input file, required.
- -o FILE: output file, required.

- -D FLT: standard deviation cutoff to call discordant read pairs, use the same value as used in d option of meerkat.pl, default 3.
- -I FILE: isinfo file produced by pre_process.pl run, required if n or e options are turned on.
- -K FILE: file name of read group to be ignored, one read group ID per line, same as R option in pre_process.pl and meerkat.pl. If all the read groups are of high quality, you do not need to specify this option. If there is any blacklist rg and any of the options n, u, f, e are enabled, K option is required.
- -F STR: name of folder contains all *.discord files from normal genomes to filter germline events, we recommend to filter against all normal genomes from one tumor type.
- -x INT: number of discordant read pairs supporting the same event to be used as filter from *.discord file, default 2.
- -l INT: distance of breakpoints to filter germline events from *. discord file, default 500.
- -s INT: minimum size of simple events, default 100.
- -E INT: filter TEI for deletions, these are typically germline events, default 1.
- -d INT: max homology allowed for deletion and intra-chr events, 0 to disable, default 100.
- -t INT: max homology allowed for inter chromosomal translocation events, 0 to disable, default 100.
- -m INT: filter events that both breakpoints fall in satellite or simple repeats, default 1.
- -n INT: filter by total number of discordant read pairs in matched normal genome. If certain number of discordant read pairs are observed in the given genome, discard the event. If enabled, B and I options are required, default 0.
- -y INT: maximum fraction of discordant pairs in normal bam, default 0.1.
- -u INT: filter by non-uniquely mapped reads in matched normal genome, determined by XT tag or mapping quality, if too many non-uniquely mapped reads are observed in the given genome, discard the event. If enabled, B option is required, default 0.
- -v INT: window size to look for non-uniquely mapped reads in normal bam file, default 100.
- -r FLT: cutoff of ratio of non-uniquely mapped reads to all mappable reads, default 0.25.
- -f INT: filter by soft-clipped reads in matched normal genome. If certain number of soft-clipped reads are observed in the given

genome, discard the event. If enabled, B option is required, default 0.

- -g INT window size to search for soft-clipped reads in normal bam file, default 10.
- -j INT: cutoff for number of soft-clipped reads in normal bam file, default 3.
- -e INT: filter by discordant read pairs in tumor genome. If certain breakpoint has no discordant read pair, it is an artifact of split read mapping, discard the event. If enabled, B and I options are required, default 0.
- -B FILE: bam file.
- -k INT: window size to search for discordant read pairs in tumor bam file, default 800.
- -z INT: to enable parameter p, q and P, default 0.
- -p INT: filter by number of supporting discordant read pairs, default 3.
- -q INT: filter by number of supporting split reads, default 1.
- -P INT: filter by sum of supporting discordant read pairs and supporting split reads, default 6.
- -M INT: filter by mate position of split reads. For reads covering breakpoint junctions, their mate should map near breakpoints. If enabled, B option is required, default 0.
- -N INT: window size to search for mate in tumor bam file, default 2000.
- -C FILE: bp_reads file from meerkat.pl output, required if M option is enabled.
- -Q INT: minimum mapping quality for reads to be used, default 0. If using the Q option in meerkat.pl, use the same value here.
- -b INT: allele frequency cutoff.
- -V STR: blat server, if enable option V and T, will filter by blat the split reads against whole genome, and require both side of breakpoints to be best hit. Users need to set up a blat server before using this filter (i.e., gfServer start 10.11.240.76 17777 /reference/hg38/hg38.2bit -stepSize=5).
- -T STR: blat port.
- -L STR: /path/to/blat, path only, not the command, no need to specify if blat is in PATH.
- -A STR: location of prefix.sr folder from meerkat.pl.
- -S STR: /path/to/samtools, path only, not the command, no need to specify if samtools is in PATH.

The remaining scripts (discon.pl, meerkat2vcf.pl, fusions.pl, primers.pl) are not part of the SV detection workflow, but they provide additional functions which can be very useful.

3.5 Discon.pl

This script can provide the numbers of discordant and concordant read pairs spanning SV breakpoint, which can be used to calculate the SV allele fractions. It takes the variants file produced by somatic_sv.pl. The output is also a variants file with an extra RP tag.

Example command:

```
perl ./scripts/discon.pl -d 5 -Q 10 -i $somaticg -o $soma-
ticg_rp -B $cancer_bam -C $cancer_cluster -I $cancer_isinfo -K
$cancer_blacklistrg -S /home/ly55/opt/samtools/
```

The full list of parameters for discon.pl is as follows:

- -i FILE: input variants file, required.
- -o FILE: output file, required.
- -D INT: count number of discordant pairs from bam, [0/1], turn this function on for genotyping, default 0.
- -B FILE: bam file, required.
- -C FILE: cluster file generated by meerkat.pl, required.
- -I FILE: isinfo file from pre_process.pl, required.
- -K FILE: file name of read group to be ignored, one read group ID per line, same as R option in meerkat.pl.
- -S STR: /path/to/samtools, path only, not the command, no need to specify if samtools is in PATH.
- -d FLT: standard deviation cutoff to call discordant read pairs, default 3.
- -Q INT: minimum mapping quality for reads to be used, default 0.
- -h help.

3.6 Meerkat2vcf.pl

The format of variants file produced by somatic_sv.pl and discon.pl is quite complex (*see* Subheading 4.2 for details). The script meerkat2vcf.pl can convert variants file into VCF format.

Example command:

```
perl ./scripts/meerkat2vcf.pl -i variantfile -o vcffile -H
headerfile -F /db/hg38/hg38_fasta/
```

Users need to generate a header file prior to running this script and the -F option is the same as -F in meerkat.pl. An example header file can be found in Meerkat.example package which can be downloaded from https://github.com/yanglab-computationalgenomics/Meerkat.example.hg38.

3.7 Fusions.pl

This script can annotate the SV breakpoints, predict functional consequences of fusion transcripts, and convert the variants file into a tab delimited text file which is easier to work with at the breakpoint level. The output file contains one breakpoint junction for each line.

Gene annotation of the reference genome is required and can be downloaded from UCSC. The file needs to be sorted by chromosome and coordinate first as follows:

```
sort refGene.txt -k 3,3 -k 5,5n > refGene_sorted.txt
```

Example command to run fusions.pl:

```
perl ./scripts/fusions.pl -i variantfile -G /db/hg38/refGe-
ne_hg38_sorted.txt
```

3.8 Primers.pl

This script can design primers for PCR validation of the predicted SVs. The fusions file generated by the fusions.pl shall be used as input. Example command to run primers.pl:

```
perl ./scripts/primers.pl -i test.fusions -o primers.txt -F /
db/hg38/hg38_fasta/ -P /opt/primer3/src/ -L /opt/blat/ -V
10.11.240.76 -T 17777 -p LY20131231_
```

The full list of parameters for primers.pl is as follows:

- -i FILE: input file, required, fusions file generated by fusions.pl.
- -o FILE: output file, required.
- -p STR: prefix of primers.
- -c INT: column offset, default 0, if you have sample name in column 1 for fusion events, set to 1.
- -f INT: flanking region, default 500, the region to design primers in.
- -s STR: primer sizes separated by ",", default 20,23,25,27.
- -m STR: primer min, opt, max Tm separated by ",", default 50,60,65.
- -g STR: primer min, max GC content separated by ",", default 40,60.
- -n INT: number of primers to design for each primer size, default 5.
- -r INT: mask repeats, default 0.

- -q INT: print out flanking sequences to design primers.
- -F STR: /path/to/reference/fasta/files, path only, not the files, required, same as option F in meerkat.pl.
- -P STR: /path/to/primer3_core, path only, not the command, no need to specify if primer3_core is in PATH.
- -L STR: /path/to/blat, path only, not the command, no need to specify if blat is in PATH.
- -V STR: blat server (e.g., run server as: gfServer start 10.11.240.76 17777 /reference/hg38/hg38.2bit -stepSize=5).
- -T STR: blat port (17777 in above example).
- -h help.

4 Output Formats

4.1 List of Output Files

1. prefix/*_1/2.fq.gz: artificial read pairs per read group.
2. prefix/*.is: sample insert size per read group (used to generate prefix.isinfo and prefix.pdf).
3. prefix.sr/*.srout: split read mapping of each breakpoint, file name is discordant cluster ID. The top line is reference seq on one side of the breakpoint, the bottom line is reference seq on the other side of the breakpoint. The lines in between are reads spanning the breakpoint junction.
4. prefix.blacklist.gz: positions that shall not be processed.
5. prefix.bp.fasta: sequences of confidence intervals of breakpoints.
6. prefix.bp.info: table of cluster id and corresponding breakpoints.
7. prefix.bp_reads: table of breakpoints and supporting read names.
8. prefix.cl.disc.sorted.bam: discordant reads in re-mapped artificial read pairs.
9. prefix.cl.dup.bam: duplicates in re-mapped artificial read pairs.
10. prefix.cl.sorted.bam: BAM file for alignments of re-mapped artificial read pairs.
11. prefix.clusters: all discordant clusters sorted by weight (one discordant read pair per line).
12. prefix.disc.sorted.bam: discordant reads in original BAM file.
13. prefix.discord: summary of discordant clusters (one discordant cluster per line).
14. prefix.dre.log: log file to extract discordant read pairs.

15. prefix.dup.bam: duplicates in original BAM file.
16. prefix.inter/intra.refined.typ.sorted: variants after refine breakpoints by local alignments, sorted by event type.
17. prefix.isinfo: read length and insert size statistics per read group.
18. prefix.mapping.raw: clusters with all possible mapping positions, sorted by coordinates.
19. prefix.mp.inter/intra.out: variants predicted by discordant read pair clusters.
20. prefix.pdf: plots of insert size distribution and quality of reads.
21. prefix.pre.log: log file for extracting unmapped, partially mapped reads, and generating artificial read pairs and split reads.
22. prefix.softclips.fq.gz: partially mapped reads after trimming poor quality bases from end.
23. prefix.softclips.rdist: base quality distribution of partially mapped reads (used to generate prefix.pdf).
24. prefix.sr.1/2.fq.gz: split read sequences in fastq format.
25. prefix.sr.inter/intra.filtered: filtered variants of prefix.sr.inter/intra.out (remove variants that use same cluster).
26. prefix.sr.inter/intra.out: variants have split reads support.
27. prefix.sr.sorted.bam: split reads alignments.
28. prefix.unmapped.fq.gz: unmapped reads.
29. prefix.unmapped.rdist: base quality distribution of unmapped reads (used to generate prefix.pdf).
30. prefix.variants: final variants with mechanism prediction. **This file contains raw SV calls.**

List of event types in column 1 of prefix.variants file:

1. del: simple deletion, no insertion at breakpoint.
2. del_ins: deletion with insertion of unknown source at the breakpoint.
3. del_inssd: deletion with insertion at the breakpoint, insertion comes from the same chromosome, same orientation, and downstream of deletion.
4. del_inssu: deletion with insertion at the breakpoint, insertion comes from the same chromosome, same orientation, and upstream of deletion.
5. del_insod: deletion with insertion at the breakpoint, insertion comes from the same chromosome, opposite orientation, and downstream of deletion.

6. del_insou: deletion with insertion at the breakpoint, insertion comes from the same chromosome, opposite orientation, and upstream of deletion.
7. del_inss: deletion with insertion at the breakpoint, insertion comes from a different chromosome, same orientation.
8. del_inso: deletion with insertion at the breakpoint, insertion comes from a different chromosome, opposite orientation.
9. del_invers: deletion with inversion at the breakpoint, inversion comes from deleted part.
10. inssd: insertion, insertion comes from the same chromosome, same orientation, and downstream of deletion.
11. inssu: insertion, insertion comes from the same chromosome, same orientation, and upstream of deletion.
12. insod: insertion, insertion comes from the same chromosome, opposite orientation, and downstream of deletion.
13. insou: insertion, insertion comes from the same chromosome, opposite orientation, and upstream of deletion.
14. inss: insertion, insertion comes from a different chromosome, same orientation.
15. inso: insertion, insertion comes from a different chromosome, opposite orientation.
16. invers: inversion with reciprocal discordant read pair cluster support.
17. invers_f: unpaired cluster, both end of read pairs mapped to same chromosome, both on forward strand.
18. invers_r: unpaired cluster, both end of read pairs mapped to same chromosome, both on reverse strand.
19. tandem_dup: tandem duplication.
20. transl_inter: inter-chromosomal translocation.

Note that del, del_ins, invers_f, inver_r, tandem_dup are all predicted from unpaired clusters. They can arise from other complex events while only one cluster is called, the other cluster is not. They can also arise from intra-chromosomal translocations rather than true deletions or tandem duplications.

List of mechanisms:

1. TEI: transposable element insertion, single or multiple TE insertion.
2. TEA: alternative TE, usually a deletion with insertion event is called, deletion is a TE in reference genome and insertion is the same type of TE in reference genome. It probably arises from sequence divergence of TE.

3. VNTR: variable number of tandem repeat, deletion or insertion of satellite repeat, simple repeat or low complexity repeat.
4. NAHR: non-allelic homologous recombination, >100 bp homology.
5. alt-EJ: alternative end joining, 3–100 bp homology.
6. NHEJ: non-homologous end joining, 0–2 bp homology, or 1–10 bp insertion at deletion breakpoint.
7. FoSTeS: fork stalling and template switching, >10 bp insertion at deletion breakpoints.
8. NA: unclassified.

4.2 Format of prefix.variants File

The variants files produced by somatic_sv.pl have the same format as prefix.variants files. Note that in these files, different events have different formats due to the presence of both simple and complex SVs. It is easier to work with vcf and fusions files. The format of variants files is as follow:

1. del, mechanism, cluster id, number of supporting read pairs, number of supporting split reads, chr, range of deletion (2 col), deletion size, homology sizes, annotation of breakpoints.
2. del_ins, mechanism, cluster id, number of supporting read pairs, number of supporting split reads, chr, range of deletion (2 col), deletion size, chr (donor), range of insertion (2 col), insert size, annotation of breakpoints.
3. del_inss/o*, mechanism, cluster id, number of supporting read pairs, number of supporting split reads, chr, range of deletion (2 col), deletion size, chr (donor), range of insertion (2 col), insert size, distance of deletion and insertion, homology at breakpoints, annotation of breakpoints.
4. del_invers, mechanism, cluster id, number of supporting read pairs, number of supporting split reads, chr, range of deletion (2 col), deletion size, chr (donor), range of inversion (2 col), inversion size, distance of inversion and deletion (2 col), homology at breakpoints, annotation of breakpoints.
5. inss/o*, mechanism, cluster id, number of supporting read pairs, number of supporting split reads, chr, range of deletion (2 col), deletion size, chr (donor), range of insertion (2 col), insert size, distance of deletion and insertion, homology at breakpoints, annotation of breakpoints.
6. invers, mechanism, cluster id, number of supporting read pairs, number of supporting split reads, chr, inversion left boundary, inversion right boundary, inversion size, homology at breakpoints, annotation of breakpoints.
7. invers_*, mechanism, cluster id, number of supporting read pairs, number of supporting split reads, chr, inversion left

boundary, inversion right boundary, inversion size, homology at breakpoints, annotation of breakpoints.

8. tandem_dup, mechanism, cluster id, number of supporting read pairs, number of supporting split reads, chr, tandem duplication boundary 1, tandem duplication boundary 2, tandem duplication size, homology at breakpoints, annotation of breakpoints.
9. del_inss/o, mechanism, cluster id, number of supporting read pairs, number of supporting split reads, chr, range of deletion (2 col), deletion size, chr of insertion donor, range of insertion (2 col), insert size, homology at breakpoints, annotation of breakpoints.
10. inss/o, mechanism, cluster id, number of supporting read pairs, number of supporting split reads, chr, range of insert site, insert site size, chr of insertion donor, range of insertion (2 col), insert size, homology at breakpoints, annotation of breakpoints.
11. transl_inter, mechanism, cluster id, number of supporting read pairs, number of supporting split reads, chr of first cluster, boundary of first cluster, orientation of first cluster, chr of second cluster, boundary of second cluster, orientation of second cluster, homology at breakpoints, annotation of breakpoints.

Regarding homology at breakpoints, positive number represents homology, negative number stands for additional bases at breakpoint. In the last column of breakpoint annotation, breakpoints are annotated sequentially by repeat contents separated by "_". SR: satellite repeat, simple repeat or low complexity repeat. TE: transposable elements. For example, BP:_TE__SR means the second breakpoint is TE and fourth breakpoint is SR, while first and third breakpoints are neither TE nor SR.

For complex events, such as deletion with insertion, deletion with inversion, typically two clusters are paired to call one event, thus two cluster IDs are given. However, it is possible that there is a third discordant read pair cluster spanning the deletion junction depending on event size and insert size of the sequecing library. In such case, three cluster IDs are given (see the 4th SV example below).

Example of SVs:

```
del TEI_SINE_AluY 2790 33 12 chr1 17578190 17578503 312 15 BP:
TE_TE
del_ins NHEJ 621 84 48 chr4 173225214 173229507 4292 - - -
1 BP:TE_TE
del_inssu FoSTeS 1434_0/1005_0 52/62 6/6 chr6 56012024
56017028 5003 chr6 55938323 55939086 764 72938 2/3 BP:_TE__TE
```

```
del_inssd FoSTeS 416_0/662_0/2774 99/82/33 9/18 chr3 147867879
147873095 5215 chr3 147877553 147877855 303 4458 2/-9 BP:
TE_TE_TE_TE
insod NA 1146_0/2090_0 58/42 10/20 chr1 104382016 104382013 -4
chr1 104383501 104384051 551 1488 6/-4 BP:___
inss NA 5424_2/6914_0 42/22 20/12 chr9 67959791 67959796 4 chr3
84522750 84522975 226 -5/0 BP:___TE
invers NA 941_0/2146_0 65/41 12/6 chr6 130889878 130893988
4110 1 BP:_TE
tandem_dup NA 1171 57 5 chr9 118553273 118555818 2545 1 BP:TE_
invers_r NA 1727 24 4 chr17 7528656 12039589 4510933 2 BP:TE_SR
transl_inter NA 3649 10 1 chr17 7528046 -1 chr5 138794678
1 0 BP:TE_
```

All nucleotides of given coordinates in the final output are inclusive in the donor genome. For example, a deletion of chrA:1234-5678 will be as follows in the donor:

```
1234| |5678
---------------| |----------------
```

The deleted part is 1235-5677. On the other hand, a tandem duplication of chrA:1234-5678 will be:

```
1234 5678|1234 5678
------------==========|===========-------------
```

4.3 Variants File Produced by discon.pl

An extra RP tag will be added to each SV with 4 numbers as RP: A_B_C_D.

A: number of full-length discordant read pairs.

B: number of discordant read pairs from partially mapped reads (clipped reads).

C: number of concordant read pairs at the first breakpoint.

D: number of concordant read pairs at the second breakpoint.

The numbers of $A/(A + C)$ and $A/(A + D)$ will be equivalent to allele fractions of the SV at the two breakpoints. The allele fractions may be different at the two breakpoints because the copy numbers may differ in these two loci.

Discon.pl can be used to count the number of concordant read pairs of any given genomic location, not necessarily SVs. For example, if there is a non-reference L1 insertion in the genome, users can format the event similar to a del event as follows:

```
tei L1 NA 16 3 chr8 136006375 136006390 14 BP:NA-NA
```

Run the following command:

```
perl discon.pl -i $inputfile -o $outputfile -B $bam -I $isinfo
-S /home/ly55/opt/samtools/
```

The output will be:

```
tei L1 NA 16 3 chr8 136006375 136006390 14 BP:NA-NA-
RP:16__29_30
```

The L1 insertion is at chr8:136006375-136006390 with 14 bp target site duplication and 16 supporting discordant read pairs. There are 29 concordant read pairs spanning chr8:136006375 and 30 concordant pairs spanning chr8:136006390. Therefore, the allele fraction for this event is about 0.35. To use this function, give a non-Meerkat event type (e.g., tei) in column 1, give genomic locations in col 6,7,8, the contents of other fields do not matter.

4.4 VCF File Produced by meerkat2vcf.pl

VCF format for SVs can be found here: https://samtools.github.io/hts-specs/VCFv4.2.pdf.

4.5 Fusions File Produced by fusions.pl

The fusions file generated by fusions.pl is the preferred file to use for studying SV breakpoints with a uniform format for all types of SVs. Each line represents one pair of breakpoints. Complex SVs are broken down into two lines. The format is as follow:

```
type1 type2 type3 chrA posA oriA geneA exon_intronA chrB posB
oriB geneB exon_intronB event_type mechanism event_id dis-
c_pair split_read homology partners
```

An example:

```
gene-gene head-tail in_frame chr21 38778477 1 ERG I1 chr21
41797322 -1 TMPRSS2 I1 del NHEJ 2284_0 16 6 0 3-5
```

The fusion types predicted should be self-explainable. When the fusion is annotated as "no_impact," it means it is an intronic deletion that would not alter gene function. In the above example, the 5′ of TMPRSS2 and 3′ of ERG are fused together and form an in-frame fusion. The 5′ and 3′ partners information is encoded in the partners column. The "in_frame" and "out_of_frame" predictions are based on exons immediately adjacent to SV breakpoints. A predicted out-of-frame fusion can be an in-frame fusion if spliced differently [13].

4.6 Output by primers.pl

Output format: ID, primer seq, primer position, primer size, tm, gc content, number of blat hit in whole genome, distance to breakpoint.

All primers are designed by Primer3. For each event, users shall pick one primer from both 1 and 2. The orientation has been considered for different types of events, so when ordering primers, just copy and paste the sequence, no need to reverse compliment. If the seq is in lower case, that means the primer is in a repeat region. Ideally, users should pick upper case primers with 1 blat hit. The number of blat hit being 0 means there are too many hits (do not pick such primer). Sometimes, even the primer is in repeat region (lower case), it is still unique in the genome (1 blat hit) because it is a quite diverged repeat element. It is OK to pick such primer. Typically, a primer with less than 20 blat hits will work fine if there is no unique one available. As some primers may not work well, users can pick two forward primers and two reverse primers to perform four PCRs at the same time. General rules of primer design still apply. For example, users should pick primers that Tms are not too different and GC contents are not too extreme.

For the following event (an *EML4-ALK* gene fusion from a lung adeno carcinoma, coordinates are based on hg19):

```
gene-gene head-tail in_frame chr2 29446548 1 ALK I19 chr2
42507907 1 EML4 I5 invers_f NA 1511021_0 3 1 4 3-5
```

After running the example command in Subheading 3.8, the output will be:

```
LY20131231_1.1 GAGTCTGCGGTGCTGTGATA 84,20 60.016 55.000 1 416
LY20131231_1.1 TGAAGCACTACACAGGCCAC 393,20 59.905 55.000 1 107
LY20131231_1.1 ATTCAGCCCCTACACTGCAC 106,20 60.142 55.000 1 394
LY20131231_1.1 ACATTCAGCCCCTACACTGC 104,20 60.142 55.000 1 396
LY20131231_1.1 TGCGGTGCTGTGATAACATT 89,20 60.142 45.000 1 411
...
LY20131231_1.2 ATCCACAGATTCAGCCAACC 160,20 59.934 50.000 4 340
LY20131231_1.2 CCCGTGGATACAGGACAACT 416,20 59.844 55.000 1 84
LY20131231_1.2 TGGGTTGAAAATATTTGGGG 181,20 59.497 40.000 1 319
LY20131231_1.2 TCCTAAAATCAGTCCCCCGT 401,20 60.683 50.000 1 99
LY20131231_1.2 CATCCACAGATTCAGCCAAC 159,20 59.090 50.000 4 341
...
```

The following two primers can be used and the expected size of PCR product will be (394 + 84) bp.

```
LY20131231_1.1 ATTCAGCCCCTACACTGCAC 106,20 60.142 55.000 1 394
LY20131231_1.2 CCCGTGGATACAGGACAACT 416,20 59.844 55.000 1 84
```

5 Notes

1. For HapMap sample NA18507 (42× coverage, 100 bp read length, 500 bp insert size), Meerkat took about 1.5 days to run and 10 GB memory when using 10 threads for BWA alignment. For 30× coverage cancer genome, it can take more than 2 days and more than 30 GB of memory. If the sequencing quality is not very good, such as many chimeric reads or many non-uniquely mapped reads, it may take a lot longer and a lot more memory.
2. There is an example bam file at example/example.sorted.bam, and three output files. After running pre_process.pl, meerkat.pl and mechanism.pl, users should get an example.variants file.
3. When running Meerkat on cancer genomes, it would be better to use the blacklist.gz files generated from their matched-normal genomes rather than use the blacklist.gz file generated from the cancer genomes. The blacklist.gz files are generated by pre_process.pl. An error message "differing read lengths" in pre_process.pl can be ignored. Eventually, users might consider to manually inspect the discordant read pair alignments with IGV [19] and inspect the split read alignments from prefix.sr folder to make sure the variants are convincing. Users may need to manually edit the split read alignments. For reads spanning the breakpoint junction, part of the reads should align to the left-hand side of reference sequence on the top, the remaining of the reads should align to the right-hand side of reference sequence at the bottom (*see* Fig. 1c of ref. 3 for an example).

References

1. Ho SS, Urban AE, Mills RE (2020) Structural variation in the sequencing era. Nat Rev Genet 21:171–189
2. Alkan C, Coe BP, Eichler EE (2011) Genome structural variation discovery and genotyping. Nat Rev Genet 12:363–376
3. Yang L, Luquette LJ, Gehlenborg N et al (2013) Diverse mechanisms of somatic structural variations in human cancer genomes. Cell 153:919–929
4. Xi R, Hadjipanayis AG, Luquette LJ et al (2011) Copy number variation detection in whole-genome sequencing data using the Bayesian information criterion. Proc Natl Acad Sci U S A 108:E1128–E1136
5. Chen K, Wallis JW, McLellan MD et al (2009) BreakDancer: an algorithm for high-resolution mapping of genomic structural variation. Nat Methods 6:677–681
6. Ye K, Schulz MH, Long Q et al (2009) Pindel: a pattern growth approach to detect break points of large deletions and medium sized insertions from paired-end short reads. Bioinformatics 25:2865–2871
7. Rizk G, Gouin A, Chikhi R et al (2014) MindTheGap: integrated detection and assembly of short and long insertions. Bioinformatics 30: 3451–3457
8. Rausch T, Zichner T, Schlattl A et al (2012) DELLY: structural variant discovery by integrated paired-end and split-read analysis. Bioinformatics 28:i333–i339
9. Wang J, Mullighan CG, Easton J et al (2011) CREST maps somatic structural variation in cancer genomes with base-pair resolution. Nat Methods 8:652–654
10. Handsaker RE, Korn JM, Nemesh J et al (2011) Discovery and genotyping of genome

structural polymorphism by sequencing on a population scale. Nat Genet 43:269–276

11. Chen X, Schulz-Trieglaff O, Shaw R et al (2016) Manta: rapid detection of structural variants and indels for germline and cancer sequencing applications. Bioinformatics 32: 1220–1222
12. Layer RM, Chiang C, Quinlan AR et al (2014) LUMPY: a probabilistic framework for structural variant discovery. Genome Biol 15:R84
13. Yang L, Lee M-S, Lu H et al (2016) Analyzing somatic genome rearrangements in human cancers by using whole-exome sequencing. Am J Hum Genet 98:843–856
14. Alaei-Mahabadi B, Bhadury J, Karlsson JW et al (2016) Global analysis of somatic structural genomic alterations and their impact on gene expression in diverse human cancers. Proc Natl Acad Sci 113:13768–13773
15. Lee AY, Ewing AD, Ellrott K et al (2018) Combining accurate tumor genome simulation with crowdsourcing to benchmark somatic structural variant detection. Genome Biol 19: 188
16. Kosugi S, Momozawa Y, Liu X et al (2019) Comprehensive evaluation of structural variation detection algorithms for whole genome sequencing. Genome Biol 20:117
17. Li H (2013) Aligning sequence reads, clone sequences and assembly contigs with BWA-MEM. http://arxiv.org/abs/1303.3997
18. Broad Institute (n.d.) Picard tools. http://broadinstitute.github.io/picard/
19. Robinson JT, Thorvaldsdóttir H, Winckler W et al (2011) Integrative genomics viewer. Nat Biotechnol 29:24–26

Chapter 9

Structural Variant Detection from Long-Read Sequencing Data with cuteSV

Tao Jiang, Shiqi Liu, Shuqi Cao, and Yadong Wang

Abstract

Structural Variation (SV) represents genomic rearrangements and is strongly associated with human health and disease. Recently, long-read sequencing technologies provide the opportunity to more comprehensive identification of SVs at an ever-high resolution. However, under the circumstance of high sequencing errors and the complexity of SVs, there remains lots of technical issues to be settled. Hence, we propose cuteSV, a sensitive, fast, and scalable alignment-based SV detection approach to complete comprehensive discovery of diverse SVs. The benchmarking results indicate cuteSV is suitable for large-scale genome project since its excellent SV yields and ultra-fast speed. Here, we explain the overall framework for providing a detailed outline for users to apply cuteSV correctly and comprehensively. More details are available at https://github.com/tjiangHIT/cuteSV.

Key words Bioinformatics, Structural variants detection, Long-read sequencing, Alignment-based calling, Germline mutation calling, Population-based calling, Scaling performance

1 Introduction

Structural Variation (SV) is the largest divergence of nuclear acids across human genomes [1], including deletion, insertion, inversion, duplication, and translocation whose sizes are larger than 50 bp [2]. SV appears widely in human genome and affects molecular and cellular processes [3], regulatory functions [4], and is associated with complex phenotypes [5], and has been implicated in genomic diseases (e.g., Mendelian disorders and cancers) [6]. There are high demands for revealing the full spectrum of genetic variation, improving understanding of mutation and evolutionary processes, resolving some of the missing heritability, which are helpful for leading to more novel biological insights [7].

With the rapid development of long-read sequencing technologies such as Pacific Biosciences (PacBio) [8] and Oxford Nanopore Technologies (ONT) [9], a new effective routine of SV

Charlotte K. Y. Ng and Salvatore Piscuoglio (eds.), *Variant Calling: Methods and Protocols*,
Methods in Molecular Biology, vol. 2493, https://doi.org/10.1007/978-1-0716-2293-3_9,

detection is established for more comprehensively discovering SVs at a higher resolution through the long-range spanning information from read alignments [10]. The widely used state-of-the-art alignment-based SV calling methods usually apply a long-read aligner to map reads against a reference genome assembly at first, and then integrate the extracted candidate SV signatures into one specific cluster to recognize final SVs [11]. Furthermore, assembly-based SV callers have been well developed to achieve haplotype-resolved SV discovering [12], though suffering from the low sensitivity in detecting heterozygous variants and intensive occupation of the computational resources.

However, there are still remaining challenges even for the state-of-the-art SV calling methods. With the sequencing errors, the SV breakpoints and sizes on the read alignments are chimeric and heterogeneous. It is high demand for supplying enough coverage of sequencing datasets in order to sensitively uncover the complex and low-frequency and rare SVs. The simple rules of SV signature analysis module will also miss plenty of reality SVs and even generate false-positive predictions. All the above issues will decrease the sensitivity, accuracy, and efficiency of SV calling.

To break these bottlenecks, we present cuteSV, a versatile long-read alignment-based SV detection approach which mainly contains three steps (refer to Fig. 1):

1. cuteSV uses multiple signature extraction methods to collect the signatures of various types of SVs comprehensively. Furthermore, it heuristically combines short insertions and deletions from fragile alignments to recover the evidence of real SVs.
2. A clustering-and-refinement approach is specifically designed to integrate the chimerical aligned reads in local regions and further refine the clusters to precisely distinguish the SV signatures from heterozygous SVs.
3. cuteSV uses several tailored rules to implement SV calling and genotyping based on the refined clusters of SV signatures.

The major contribution of cuteSV includes three beneficial features:

1. Compared with other state-of-the-art SV callers, cuteSV has excellent SV yields while maintaining higher sensitivity and accuracy, even at low coverage sequencing data (*see* **Note 1**).
2. It supports all kinds of datasets produced by mainstream long-read sequencing platforms with various error rates, and can identify various types of SVs (i.e., deletions, insertions, duplications, inversions, and translocations).

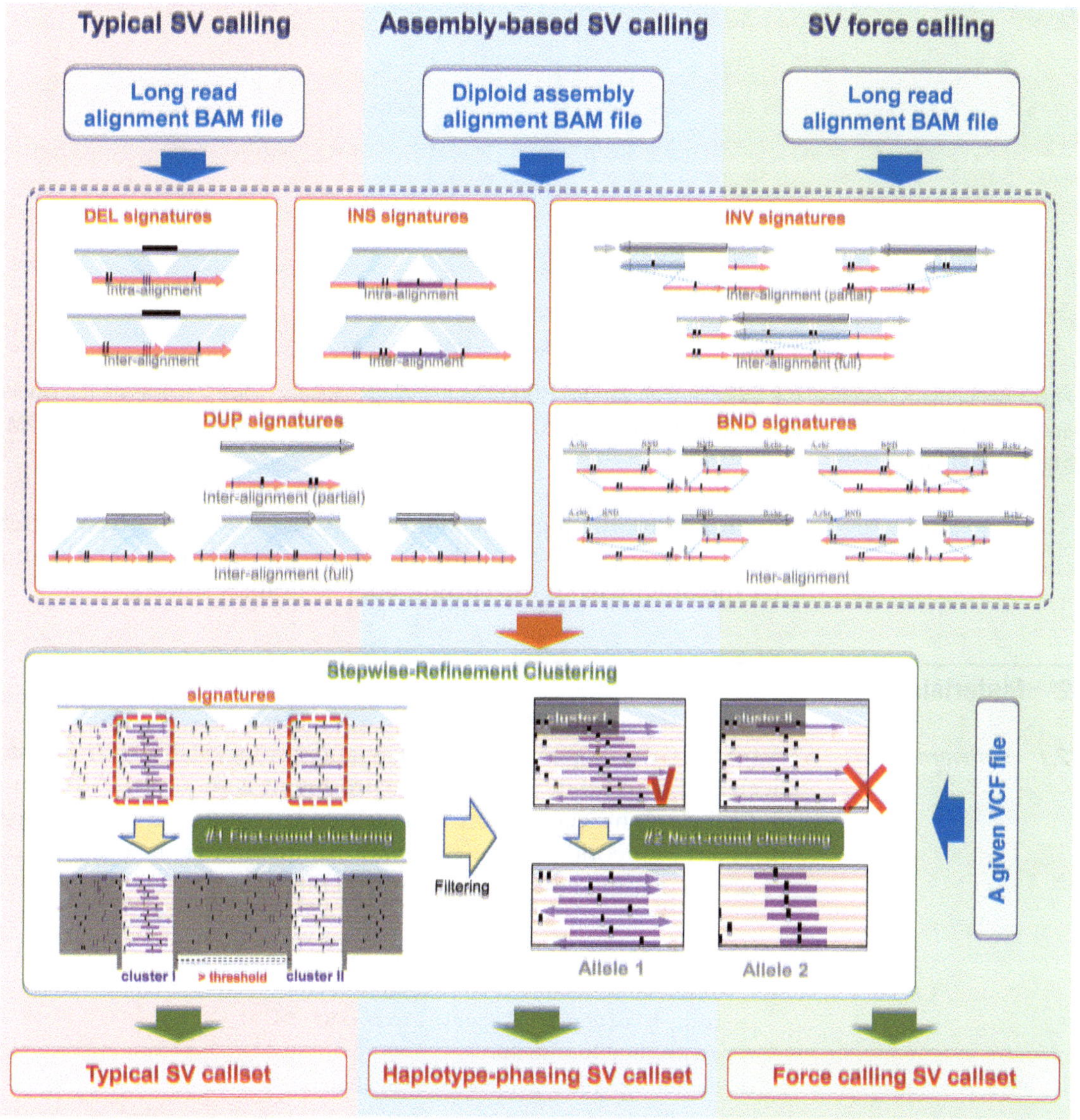

Fig. 1 The schematic illustration of cuteSV

3. cuteSV achieves almost linear speedup with multiple CPU threads on SV detection through its outstanding scalability, while maintaining lower RAM usage (*see* **Note 2**).

A series of benchmarking results (refer to Fig. 2) on various sequencing datasets from different platforms show that cuteSV obtained better F1 scores comparing to three state-of-the-art tools (i.e., Sniffles [13], SVIM [14], and PBSV (https://github.com/PacificBiosciences/pbsv)). It is able to sensitively detect various types of SVs while maintaining high accuracy, and the genotypes can be correctly assigned as well (refer to Fig. 2).

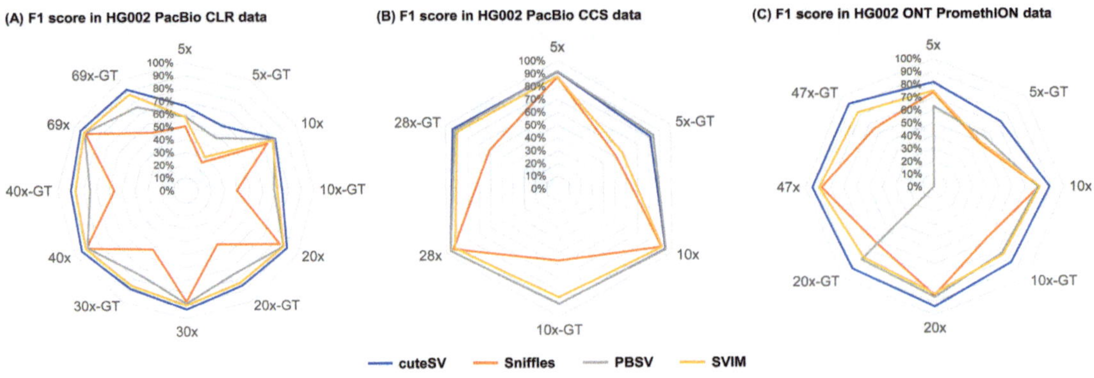

Fig. 2 Benchmark results of the SV callers on various HG002 sequencing datasets. The F1 scores on the whole and down-sampled HG002 (**a**) PacBio CLR, (**b**) PacBio CCS, and (**c**) ONT PromethION datasets of cuteSV, SVIM, Sniffles, and PBSV

Furthermore, the high scalability on SV detection of cuteSV (refer to Fig. 3) suggests it is suitable to large-scale data analysis tasks and has the potentials to the cutting-edge genomics studies.

2 Materials

2.1 Software

cuteSV is developed in Python 3 and under MIT open source license. To install it, a working environment is required. A Python 3 environment can be created using Anaconda or its minimal installer Miniconda. Here we apply Miniconda as an example to create a customized environment for cuteSV:

```
> wget https://repo.anaconda.com/miniconda/Miniconda3-latest-Linux-x86_64.sh
> bash Miniconda3-latest-Linux-x86_64.sh
> conda update conda -y
> conda create -n cute_env python=3 -y
> conda install -n cute_env -c bioconda minimap2 samtools survivor -y
```

After that, a customized environment for installation of cuteSV has been created, and three genomic analysis tools (i.e., Minimap2 [15], Samtools [16], and SURVIVOR [17]) to be used later are installed in advance. In the current version of cuteSV, the packages of Pysam (≥0.16.0), Biopython (≥1.70), Cigar (≥0.1.3), Numpy (≥1.19.4), and Pyvcf (≥0.6.8) are required dependences for processing various format files and computing several critical values. To automatically install cuteSV, there are three optional installation methods:

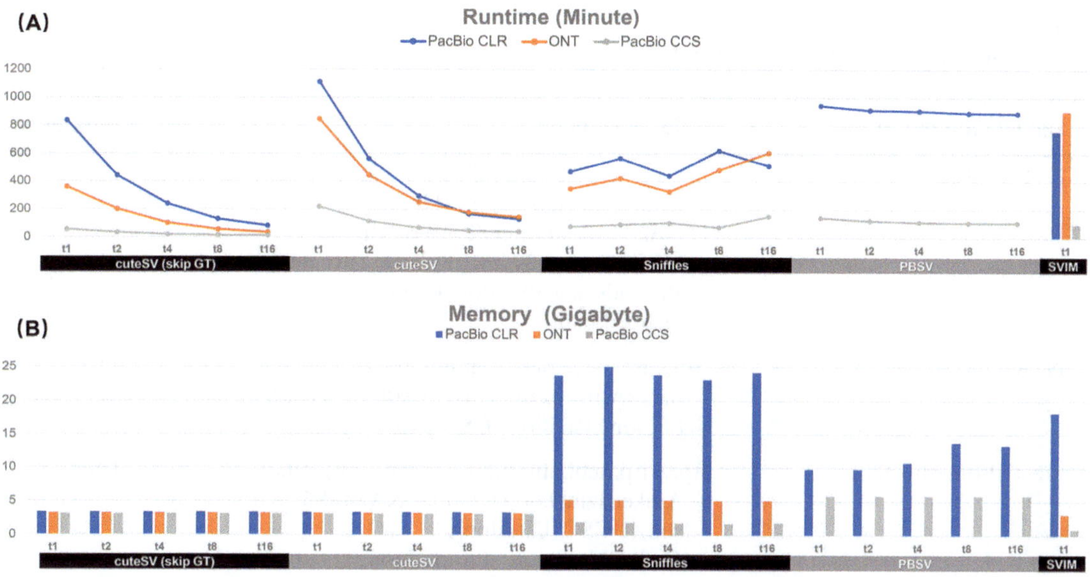

Fig. 3 Performance of the benchmarked SV callers. The (**a**) runtimes and (**b**) memory footprints of cuteSV, SVIM, Sniffles, and PBSV with 1, 2, 4, 8, and 16 CPU threads. "Skip GT" indicates the statistics without genotyping

1. **Installation via conda:**

```
> conda activate cute_env
> conda install cuteSV -c bioconda -y
> cuteSV -h
```

2. **Installation via pip:**

```
> conda activate cute_env
> pip install cuteSV
> cuteSV -h
```

3. **Installation via git:**

```
> conda activate cute_env
> git clone https://github.com/tjiangHIT/cuteSV.git
> cd cuteSV
> python setup.py install
> cuteSV -h
```

2.2 Hardware

cuteSV has been tested on a computing workstation operating on Ubuntu 18.04 LTS with the following system specifications:

1. Processor: Intel(R) Core (TM) i7-6700K CPU @ 4.00 GHz.
2. Installed memory (RAM): 64.0 GB.
3. Storage space (ROM): 100 GB free disk space at least is recommended.

Table 1
The availability of sequencing datasets

Sequencing platform	Data availability	Reference
PacBio CLR	https://ftp.ncbi.nlm.nih.gov/giab/ftp/data/AshkenazimTrio/HG004_NA24143_mother/PacBio_MtSinai_NIST/Baylor_NGMLR_bam_GRCh37/all_reads.fa.giab_h004_ngmlr-0.2.3_mapped.bam	[18]
PacBio CCS	ftp://ftp-trace.ncbi.nlm.nih.gov/giab/ftp/data/AshkenazimTrio/HG002_NA24385_son/PacBio_CCS_15kb/	[19]
ONT PromethION	ftp://ftp.ncbi.nlm.nih.gov/giab/ftp/data/AshkenazimTrio/HG002_NA24385_son/UCSC_Ultralong_OxfordNanopore_Promethion/	[20]

2.3 Datasets

To evaluate the performance of cuteSV, we applied several recently available long-read sequencing data (refer to Table 1) from the Ashkenazim son HG002/NA24385 produced by Pacific Biosciences (PacBio) and Oxford Nanopore Technologies (ONT). Below, we will demonstrate the use of cuteSV based on analysis of these datasets.

3 Methods

cuteSV requires sorted alignment file and reference genome as input and identifies different types of SVs from the alignments with high precision and sensitivity. The position parameters are essential for running cuteSV, and users can adjust the optional parameters properly to achieve better performance in variant calling. The detailed parameters instructions are shown in Table 2, and users can see them under cuteSV --help or cuteSV -h in practice.

On the basis of the specifically designed modules, there are three different application scenarios for cuteSV (*see* **Note 3**):

1. Discovering SVs using long-read alignments.
2. Discovering SVs using diploid-assembly alignments.
3. Cohort-based SVs calling.

The following will serve as a tutorial, detailing the usages and corresponding parameters utilized in generating the SV callsets as presented in various application scenarios.

Table 2
The parameters used in cuteSV

Parameter	Description	Default
Position parameter		
[BAM]	Sorted bam file of the calling sample	--
Reference	The reference genome in fasta format	--
Output	Output VCF format file	--
Work directory	Work directory for distributed jobs	--
Optional parameter		
-t / --threads	Number of threads to use	16
-b / --batches	Batch of segmentation interval to extract signatures	10000000
-S / --sample	Sample name/id	NULL
--retain_work_dir	Enable to retain temporary folder and files	False
--report_readid	Enable to report supporting read IDs for each SV, which is shown as "RNAMES" in INFO in output vcf	False
-p / --max_split_parts	Maximum number of split segments a read may be aligned before it is ignored	7
-q / --min_mapq	Minimum mapping quality value of alignment to be taken into account	20
-r /--min_read_len	Minimum alignment length reported by reads	500
--merge_del_threshold	Maximum distance of deletion signals to be merged	0
--merge_ins_threshold	Maximum distance of insertion signals to be merged	100
-s / --min_support	Minimum number of reads that support an SV to be reported	10
-l / --min_size	Minimum size of SV to be reported	30
-L / --max_size	Maximum size of SV to be reported	100,000
-sl / --min_siglength	Minimum length of SV signal to be extracted	10
--genotype	Enable to generate genotypes, otherwise cuteSV will output genotypes as "./."	False

(continued)

Table 2
(continued)

Parameter	Description	Default
--gt_round	Maximum round of iteration for alignments searching if perform genotyping	500
-Ivcf	Cohort-level vcf. Enable to perform force calling	NULL
--max_cluster_bias_INS	Maximum distance to cluster reads together for insertion	100
--diff_ratio_merging_INS	Minimum bases ratio of merged insertion breakpoints	0.3
--max_cluster_bias_DEL	Maximum distance to cluster reads together for deletion	200
--diff_ratio_merging_DEL	Minimum bases ratio of merged deletion breakpoints	0.5
--max_cluster_bias_INV	Maximum distance to cluster reads together for inversion	500
--max_cluster_bias_DUP	Maximum distance to cluster reads together for duplication	500
--max_cluster_bias_TRA	Maximum distance to cluster reads together for translocation	50
--diff_ratio_filtering_TRA	Minimum bases ratio of merged translocation breakpoints	0.6

3.1 Discovering SVs Using Long-Read Alignments

It is acceptable for the long-read alignment generated from the currently well-applicated aligners (e.g., Minimap2, LRA [21]). Samtools as a post-processing toolkit is performed for sorting and indexing the alignment. These can be completed with the following commands:

```
> minimap2 REFERENCE.FASTA LONG_READS.FASTA/Q -a -t THREADS
-x map-pb/map-ont -o alignment.sam
> samtools sort -@ THREADS -o alignment.bam alignment.sam
> samtools index -@ THREADS alignment.bam
```

After preparation of the sorted long-read alignment BAM file, cuteSV can be applied to resolve various variant signatures and generate SV callsets with predicted genotype. Notably, it is essential to ensure the stable running of cuteSV by using the same reference genome in both read alignment and SV calling (*see* **Note 4**). Detailed running commands are shown as below:

1. For PacBio CLR long-reads:

```
> cuteSV alignment.bam REFERENCE.FASTA clr.vcf WORK_DIR-
ECTORY \
    --threads THREADS --min_support READ_SUPPORT --min_size
SV_LENGTH \
    --genotype --report_readid --sample SAMPLE_ID \
    --merge_ins_threshold 500 --merge_del_threshold 500 \
    --max_cluster_bias_INS 100 --diff_ratio_merging_INS 0.3 \
    --max_cluster_bias_DEL 200 --diff_ratio_merging_DEL 0.5
```

2. For PacBio CCS long-reads:

```
> cuteSV alignment.bam REFERENCE.FASTA ccs.vcf WORK_DIR-
ECTORY \
    --threads THREADS --min_support READ_SUPPORT --min_size
SV_LENGTH \
    --genotype --report_readid --sample SAMPLE_ID \
    --merge_ins_threshold 500 --merge_del_threshold 500 \
    --max_cluster_bias_INS 1000 --diff_ratio_merging_INS 0.9 \
    --max_cluster_bias_DEL 1000 --diff_ratio_merging_DEL 0.5
```

3. For ONT long-reads:

```
> cuteSV alignment.bam REFERENCE.FASTA ont.vcf WORK_DIR-
ECTORY \
    --threads THREADS --min_support READ_SUPPORT --min_size
SV_LENGTH \
    --genotype --report_readid --sample SAMPLE_ID \
    --merge_ins_threshold 500 --merge_del_threshold 500 \
    --max_cluster_bias_INS 100 --diff_ratio_merging_INS 0.3 \
    --max_cluster_bias_DEL 100 --diff_ratio_merging_DEL 0.3
```

See **Notes 5**, **6**, and **7** for recommendations on some of the options.

3.2 Discovering SVs Using Diploid-Assembly Alignments

It is available to discover haplotype-phasing SVs from diploid-assembly alignments. There are three major steps to achieve this, the running commands are shown below:

1. Mark haplotype-unique tags:

```
> cat < (cat MATERNAL.FASTA | sed 's/>/>cutesvh1_/g') \
  < (cat PATERNAL.FASTA | sed 's/>/>cutesvh2_/g') \
  > assemblies.fasta
```

2. Perform alignment of assemblies and post-processing:

```
> minimap2 --paf-no-hit -a -x asm5 --cs -r 2k -t THREADS \
    REFERENCE.FASTA assemblies.fasta -o assemblies.sam
> samtools sort -@ THREADS -o assemblies.bam assemblies.sam
> samtools index -@ THREADS assemblies.bam
```

3. Diploid-assembly-based SV calling:

```
> cuteSV assemblies.bam REFERENCE.FASTA pre_assemblies.vcf
WORK_DIRECTORY \
    --min_support 1 --genotype --report_readid \
    --merge_ins_threshold 500 --merge_del_threshold 500 \
        --max_cluster_bias_INS 1000 --diff_ratio_merging_INS
0.9 \
        --max_cluster_bias_DEL 1000 --diff_ratio_merging_DEL
0.5
  > python3 cuteSV/src/cuteSV/diploid_calling.py pre_assem-
blies.vcf assemblies.vcf
```

The detailed parameters in cuteSV/src/cuteSV/diploid_calling.py are shown in Table 3. Users can type python3 cuteSV/src/cuteSV/diploid_calling.py --help to see them.

3.3 Cohort-Based SV Calling

cuteSV also adapts to comprehensively resolve full-spectrum SVs and corresponding allele frequencies in a multiple-samples combined cohort. The detailed descriptions of how to generate cohort-based SV callsets are shown below.

1. Generation of SV callsets on each sample:

```
> cuteSV alignment.bam REFERENCE.FASTA sample.vcf WORK_-
DIRECTORY \
  --min_support READ_SUPPORT --sample SAMPLE_ID --genotype
```

2. Merging sample-level SV callsets:

```
> ls *sample.vcf > sample_level_calls.txt
> SURVIVOR merge sample_level_calls.txt 1000 1 1 -1 -1 -1
cohort_level.vcf
```

3. Force calling cohort-level SVs across all samples:

```
> cuteSV alignment.bam REFERENCE.FASTA sample_fc.vcf WORK_-
DIRECTORY \
  -Ivcf cohort_level.vcf --min_support READ_SUPPORT --sample
SAMPLE_ID
```

Table 3
The parameters used in cuteSV/src/cuteSV/diploid_calling.py

Parameter Position parameter	Description	Default
invcf	The vcf format file generated from cuteSV	--
outvcf	The output diploid based SV callsets	--

4. Merging force calling SVs to generate final cohort callsets:

```
> ls *sample_fc.vcf > fc_calls.txt
> SURVIVOR merge fc_calls.txt 1000 -1 1 -1 -1 -1 final_co-
hort.vcf
```

In any application scenario, cuteSV outputs a file in Variant Call Format (VCF), which is a text file format for storing marker and genotype data, for storage of SV callsets. The VCF file is composed of two main parts: the header and the variant call records. The header contains information about the dataset and relevant reference sources, as well as definitions of all the annotations used to qualify and quantify the properties of the variant calls contained in the VCF file. There are abundant variant call records below the header. For each site record, the information is structured into columns (also called fields) as shown in Table 4. The first eight columns of the VCF records represent the properties observed at the level of the variant site. Sample-specific information such as genotype and individual sample-level annotation values are contained in the FORMAT column and in the sample data columns.

4 Notes

1. cuteSV has been evaluated on the Genome in a Bottle (GIAB) HG002 sequencing data with various coverage from diverse sequencing platforms (refer to Fig. 2). It respectively achieves over 0.9 F1 scores on 5× PacBio CCS, 15× ONT PromethION, and 20× PacBio CLR datasets, which indicates its great power to uncover the full spectrum of genomic structural variants even on the relatively low coverage data.
2. The runtime of cuteSV using 16 CPU-thread for a typical 69× whole-genome long-read sequencing human sample is approximately 2 h, and the memory footprints are only about 3 GB.

Table 4
The description of output format in cuteSV

Column number	Column name		Description
1	CHROM		Chromosome name
2	POS		The SV occurrence position on reference
3	ID		The identifier of SV
4	REF		The reference allele that is observed
5	ALT		The alternate allele that is observed
6	QUAL		Phred-scaled probability of SV
7	FILTER		Filter status of SV
8	INFO	PRECISE/IMPRECISE	Precise SV or imprecise SV
		SVTYPE	Type of SV (i.e., INS, DEL, INV, DUP, and TRA)
		SVLEN	Difference in length between REF and ALT alleles
		CHR2	Chromosome for END coordinate in case of a translocation
		END	End position of the variant described in this record
		CIPOS	Confidence interval around POS for imprecise variants
		CILEN	Confidence interval around inserted/deleted material between breakends
		RE	Number of reads support this SV
		RNAMES	Supporting read names of SVs (comma separated)
		STRAND	Strand orientation of the adjacency in BEDPE format (DEL:+- , DUP:-+, INV:++/--)
9	FORMAT	GT	Genotype
		DR	The number of high-quality reference reads
		DV	The number of high-quality variant reads
		PL	Phred-scaled genotype likelihoods rounded to the closest integer
		GQ	Genotype quality
10	Sample data		The specific information of sample

This also suggests a great potential for its scaling performance than the state-of-the-art long read alignment-based SV callers.

3. With the version of cuteSV ($\geq$1.0.10), it can perform force calling on given SV callsets. Besides, cuteSV also supports diploid-assembly-based SV calling to generate haplotype-phasing callsets. Both of the new functions are able to further improve its capability and application in cutting-edge studies.
4. It is necessary to use the same reference genome both on the stage of read mapping and variant calling. Otherwise, cuteSV will get confusion and terminate its processing since the discordance of the chromosome ID from different versions of reference genome.
5. It is worth mentioning that --min_support is a critical parameter that determines the minimum number of reads that support an SV to be reported. cuteSV can detect more SVs when setting a smaller number for --min_support; however, much more false-positive SV calls might be produced at the same time, and vice versa. It is advisable to use a number that equals about one-sixth of the corresponding sequencing coverage as the threshold of --min_support, which has been assessed that can obtain the better performance on various sequencing datasets.
6. Besides, --min_size is another crucial parameter which indicates the minimum size of SV signatures considered in clustering. When decreasing the value of –min_size, cuteSV can extract much shorter SV signatures from the alignments, thus much more short SVs will be detected through the variant calling. However, the smaller value of –min_size will also bring a few false SV signatures like sequencing errors, which might decrease the precision of detection.
7. In addition, --merge_ins_threshold and --merge_del_threshold are two vital parameters as well, which decide the minimum interval for recognizing a whole SV signature. When extracting different signatures from alignments, the fragile signatures are merged into one if the gap between them is less than the threshold. We recommend to set them as 500 in order to discover more SVs that might be divided into several parts mistakenly by read mapping. Users can also adjust these two parameters considering their specific conditions.

References

1. Alkan C, Coe BP, Eichler EE (2011) Genome structural variation discovery and genotyping. Nat Rev Genet 12(5):363–376. https://doi.org/10.1038/nrg2958
2. Sudmant PH, Rausch T, Gardner EJ, Handsaker RE, Abyzov A, Huddleston J, Zhang Y, Ye K, Jun G, Hsi-Yang Fritz M, Konkel MK, Malhotra A, Stütz AM, Shi X, Paolo Casale F, Chen J, Hormozdiari F, Dayama G, Chen K,

Malig M, Chaisson MJP, Walter K, Meiers S, Kashin S, Garrison E, Auton A, Lam HYK, Jasmine Mu X, Alkan C, Antaki D, Bae T, Cerveira E, Chines P, Chong Z, Clarke L, Dal E, Ding L, Emery S, Fan X, Gujral M, Kahveci F, Kidd JM, Kong Y, Lameijer E-W, McCarthy S, Flicek P, Gibbs RA, Marth G, Mason CE, Menelaou A, Muzny DM, Nelson BJ, Noor A, Parrish NF, Pendleton M, Quitadamo A, Raeder B, Schadt EE, Romanovitch M, Schlattl A, Sebra R, Shabalin AA, Untergasser A, Walker JA, Wang M, Yu F, Zhang C, Zhang J, Zheng-Bradley X, Zhou W, Zichner T, Sebat J, Batzer MA, McCarroll SA, Mills RE, Gerstein MB, Bashir A, Stegle O, Devine SE, Lee C, Eichler EE, Korbel JO, The Genomes Project C (2015) An integrated map of structural variation in 2,504 human genomes. Nature 526(7571):75–81. https://doi.org/10.1038/nature15394

3. Sudmant PH, Rausch T, Gardner EJ, Handsaker RE, Abyzov A, Huddleston J, Zhang Y, Ye K, Jun G, Fritz MH-Y (2015) An integrated map of structural variation in 2,504 human genomes. Nature 526(7571):75–81

4. Weischenfeldt J, Symmons O, Spitz F, Korbel JO (2013) Phenotypic impact of genomic structural variation: insights from and for human disease. Nat Rev Genet 14(2):125–138

5. Zichner T, Garfield DA, Rausch T, Stütz AM, Cannavó E, Braun M, Furlong EE, Korbel JO (2013) Impact of genomic structural variation in Drosophila melanogaster based on population-scale sequencing. Genome Res 23(3):568–579. https://doi.org/10.1101/gr.142646.112

6. Macintyre G, Ylstra B, Brenton JD (2016) Sequencing structural variants in cancer for precision therapeutics. Trends Genet 32(9):530–542. https://doi.org/10.1016/j.tig.2016.07.002

7. Logsdon GA, Vollger MR, Eichler EE (2020) Long-read human genome sequencing and its applications. Nat Rev Genet 21(10):597–614. https://doi.org/10.1038/s41576-020-0236-x

8. Roberts RJ, Carneiro MO, Schatz MC (2013) The advantages of SMRT sequencing. Genome Biol 14(6):405. https://doi.org/10.1186/gb-2013-14-6-405

9. Jain M, Olsen HE, Paten B, Akeson M (2016) The Oxford Nanopore MinION: delivery of nanopore sequencing to the genomics community. Genome Biol 17(1):239. https://doi.org/10.1186/s13059-016-1103-0

10. Sedlazeck FJ, Lee H, Darby CA, Schatz MC (2018) Piercing the dark matter: bioinformatics of long-range sequencing and mapping. Nat Rev Genet 19(6):329–346. https://doi.org/10.1038/s41576-018-0003-4

11. Ho SS, Urban AE, Mills RE (2020) Structural variation in the sequencing era. Nat Rev Genet 21(3):171–189. https://doi.org/10.1038/s41576-019-0180-9

12. Mahmoud M, Gobet N, Cruz-Dávalos DI, Mounier N, Dessimoz C, Sedlazeck FJ (2019) Structural variant calling: the long and the short of it. Genome Biol 20(1):246. https://doi.org/10.1186/s13059-019-1828-7

13. Sedlazeck FJ, Rescheneder P, Smolka M, Fang H, Nattestad M, von Haeseler A, Schatz MC (2018) Accurate detection of complex structural variations using single-molecule sequencing. Nat Methods 15(6):461–468. https://doi.org/10.1038/s41592-018-0001-7

14. Heller D, Vingron M (2019) SVIM: structural variant identification using mapped long reads. Bioinformatics 35(17):2907–2915. https://doi.org/10.1093/bioinformatics/btz041

15. Li H (2018) Minimap2: pairwise alignment for nucleotide sequences. Bioinformatics 34(18):3094–3100. https://doi.org/10.1093/bioinformatics/bty191

16. Li H, Handsaker B, Wysoker A, Fennell T, Ruan J, Homer N, Marth G, Abecasis G, Durbin R (2009) The sequence alignment/map format and SAMtools. Bioinformatics 25(16):2078–2079

17. Jeffares DC, Jolly C, Hoti M, Speed D, Shaw L, Rallis C, Balloux F, Dessimoz C, Bähler J, Sedlazeck FJ (2017) Transient structural variations have strong effects on quantitative traits and reproductive isolation in fission yeast. Nat Commun 8(1):14061. https://doi.org/10.1038/ncomms14061

18. Zook JM, Catoe D, McDaniel J, Vang L, Spies N, Sidow A, Weng Z, Liu Y, Mason CE, Alexander N, Henaff E, McIntyre ABR, Chandramohan D, Chen F, Jaeger E, Moshrefi A, Pham K, Stedman W, Liang T, Saghbini M, Dzakula Z, Hastie A, Cao H, Deikus G, Schadt E, Sebra R, Bashir A, Truty RM, Chang CC, Gulbahce N, Zhao K, Ghosh S, Hyland F, Fu Y, Chaisson M, Xiao C, Trow J, Sherry ST, Zaranek AW, Ball M, Bobe J, Estep P, Church GM, Marks P, Kyriazopoulou-Panagiotopoulou S, Zheng GXY, Schnall-Levin M, Ordonez HS, Mudivarti PA, Giorda K, Sheng Y, Rypdal KB, Salit M (2016) Extensive sequencing of seven human genomes to characterize benchmark reference materials. Sci Data 3(1):160025. https://doi.org/10.1038/sdata.2016.25

19. Wenger AM, Peluso P, Rowell WJ, Chang P-C, Hall RJ, Concepcion GT, Ebler J, Fungtammasan A, Kolesnikov A, Olson ND, Töpfer A, Alonge M, Mahmoud M, Qian Y, Chin C-S, Phillippy AM, Schatz MC, Myers G, DePristo MA, Ruan J, Marschall T, Sedlazeck FJ, Zook JM, Li H, Koren S, Carroll A, Rank DR, Hunkapiller MW (2019) Accurate circular consensus long-read sequencing improves variant detection and assembly of a human genome. Nat Biotechnol 37(10):1155–1162. https://doi.org/10.1038/s41587-019-0217-9
20. Shafin K, Pesout T, Lorig-Roach R, Haukness M, Olsen HE, Bosworth C, Armstrong J, Tigyi K, Maurer N, Koren S, Sedlazeck FJ, Marschall T, Mayes S, Costa V, Zook JM, Liu KJ, Kilburn D, Sorensen M, Munson KM, Vollger MR, Monlong J, Garrison E, Eichler EE, Salama S, Haussler D, Green RE, Akeson M, Phillippy A, Miga KH, Carnevali P, Jain M, Paten B (2020) Nanopore sequencing and the Shasta toolkit enable efficient de novo assembly of eleven human genomes. Nat Biotechnol 38(9):1044–1053. https://doi.org/10.1038/s41587-020-0503-6
21. Ren J, Chaisson MJP (2020) lra: the long read aligner for sequences and contigs. bioRxiv:383273. https://doi.org/10.1101/2020.11.15.383273

Chapter 10

Identifying Somatic Mitochondrial DNA Mutations

Jisong An, Kyoung Il Min, and Young Seok Ju

Abstract

Mitochondria are cellular organelles that play an essential role in eukaryotes, producing the energy needed for a cell to survive. Beyond the ~3.2 Gb of nuclear genomic DNA, each human cell has hundreds of mitochondria which carry one or a few copies of the 16.5 kb circular mitochondria DNA (mtDNA). Despite its small size, the circular genome encodes 37 genes, including 13 proteins that generate respiratory chain complexes together with other proteins of nuclear origin. Similar to nuclear genome, mtDNA in cancer cells frequently harbor somatically acquired alterations. Whole-genome or whole-exome sequencing of the tumor and its matched normal tissues (frequently blood or adjacent non-tumor tissues) enables sensitive and efficient detection of somatic mtDNA mutations. Because each cancer cell commonly carries hundreds to thousands of mtDNA copies, detection of mtDNA mutations is dependent on the heteroplasmic level of each mutation. Here, we describe strategies to accurately identify somatic mtDNA mutations in cancer genome studies.

Key words Mitochondria, Somatic mutations, Bioinformatics, Genomics, Next-generation sequencing, Genome sequencing, Heteroplasmy

1 Introduction

Given the abundance of mtDNA per cell, a standard whole-genome sequencing provides ultra-high depth sequences on the mitochondrial genome, typically ranging from 3000× to 100,000×. Although common enrichment designs for whole-exome sequencing lack baits for mtDNA, a significant amount of mtDNA is captured as off-target reads in whole-exome sequencing. Typical sequencing coverage for mitochondrial genome in whole-exome sequencing ranges from 30× to 300× [1]. Although detection sensitivity is compromised, particularly for low heteroplasmic variants (<3%), datasets from whole-exome sequencing can also be used to explore somatic mtDNA mutations in cancer. Although specific considerations are sometimes necessary, somatic mutations in mtDNA can be identified through a combination of bioinformatics tools which will be described in this chapter (Fig. 1a).

Charlotte K. Y. Ng and Salvatore Piscuoglio (eds.), *Variant Calling: Methods and Protocols*,
Methods in Molecular Biology, vol. 2493, https://doi.org/10.1007/978-1-0716-2293-3_10,

2 Materials

2.1 Input Data

To discover somatic mtDNA mutations of cancers, analyzing DNA sequences from the tumor and its matched normal tissue is necessary. The normal genome sequences act as a control to remove germline variants (Fig. 1b). For following analyses, these sequencing reads should be aligned on the human reference genome as an indexed, position-sorted BAM file format. The read-aligned BAM files can be generated using typical read-alignment tools such as BWA-MEM [2] (Fig. 1a), as used in the 1000 Genomes Project [3], and the Pan-Cancer Analysis of Whole-Genome (PCAWG) Consortium of the International Cancer Genome Consortium (ICGC) and The Cancer Genome Atlas (TCGA) [4]. The BAM file must contain "Read Group" information and, if not, Picard AddOrReplaceReadGroups can be used to add the information (available at http://broadinstitute.github.io/picard/). In addition, the process of PCR duplicate removal is helpful to eliminate artifacts caused by multiple PCR products amplified from a single spurious DNA molecule. To this end, Picard MarkDuplicates can be used, which identifies duplicate reads based on the start and stop positions of paired-end reads. Due to ultra-high depth on the mitochondrial genome, ~1% of non-duplicate reads may have same positions by chance and can be sacrificed, but its negative impact in the mtDNA variant calling is usually negligible. The indel realignment step, provided in the previous version of GATK (https://gatk.broadinstitute.org/hc/en-us), can also be considered, but it is optional given the longer reads in recent whole-genome sequencing (i.e., 2 × 151 bp) and ultra-high depth of the mitochondrial genome in typical datasets.

The standard reference sequence in the mitochondria field is the revised Cambridge Reference Sequence (rCRS; GenBank number: NC_012920.1) [5], which is not always the sequence in the human reference genomes. In order to communicate using the same genomic coordinates, the use of human reference genome harboring the rCRS for mitochondrial genome sequence is recommended. The human reference genomes build 37 (GRCh37) and 38 (GRCh38) innately contain rCRS for mitochondria. In these reference genomes, the mitochondrial genomes are named "MT" and "chrM," respectively. Older human reference genomes include slightly different mitochondrial DNA sequences (such as GenBank number: NC_001807.4). The mitochondrial genome sequences in the old reference should be replaced with rCRS before they are used in read-mapping and somatic mtDNA mutation detection. For downstream processes, the human reference genome sequences should be prepared in FASTA format.

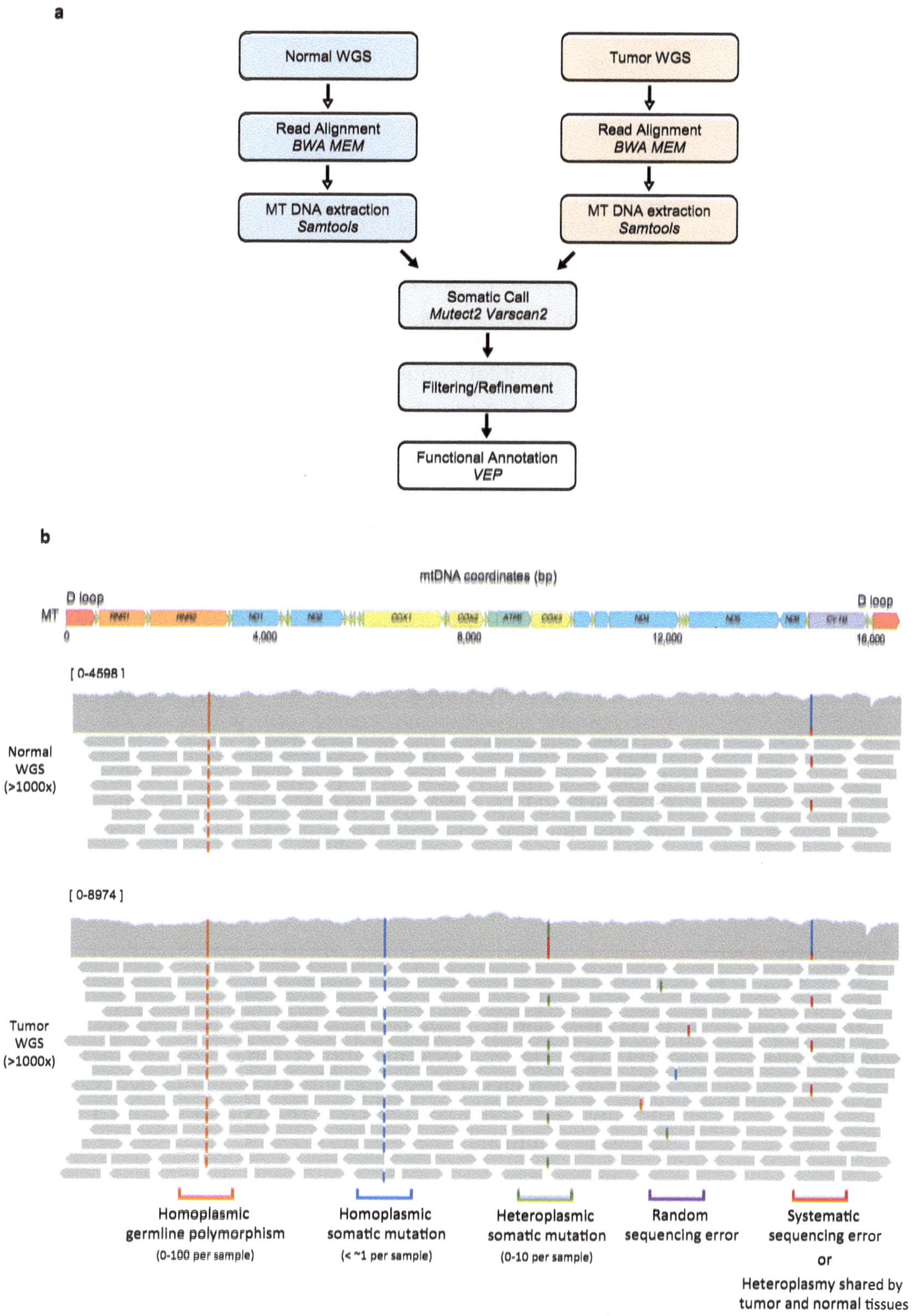

Fig. 1 Detection of somatic mtDNA mutations in cancer. (**a**) Workflow for identifying somatic mtDNA mutations is illustrated. (**b**) The genomic characteristics of mtDNA variants are illustrated according to the mutation classes (inherited polymorphisms, homoplasmic and heteroplasmic somatic mutations). The genomic features of artifacts are also shown

2.2 Tools

The tools listed below should be installed in a Linux or Unix environment for calling mtDNA somatic mutations:

1. Samtools [6]: A tool for handling BAM files. Available at https://github.com/samtools/samtools.
2. Mutect2 [7]: A tool for detecting somatic single-nucleotide variants (SNVs) and short indels in NGS data. Available at https://github.com/broadinstitute/gatk/releases (Mutect2 is a component in GATK).
3. VarScan2 [8]: A tool for detecting germline and somatic variants, such as SNVs and short indels, in NGS data. Available at http://dkoboldt.github.io/varscan/.
4. Bcftools [6]: A tool for handling VCF and BCF files. Available at https://github.com/samtools/bcftools.
5. Integrative Genomics Viewer (IGV) [9]: An interactive tool for the visual exploration of genomic data (indexed BAM files). Available at http://software.broadinstitute.org/software/igv/ .
6. Ensembl Variant Effect Predictor (VEP) [10]: A tool for computing functional effects of genomic variants. Available at https://www.ensembl.org/vep.

After the initial calling of mtDNA somatic mutations, filtering false-positive calls can be carried out in the R environment [11]. For efficient refinement of mtDNA variant call set, known germline variants deposited in public databases such as dbSNP [12], mtDB [13], and Mitomap [14], can be used. These datasets can be downloaded from https://ftp.ncbi.nih.gov/snp/latest_release/VCF/, http://www.mtdb.igp.uu.se/, and https://www.mitomap.org/MITOMAP, respectively.

3 Methods

3.1 Mitochondrial Region Extraction

BAM files of whole-genome sequences are huge in file size, typically >100 Gb. The mtDNA mutation calling process can be more efficiently conducted by slicing out mtDNA sequence reads from the entire BAM files. To this end, Samtools can be used with the following command:

```
samtools view -hbo output.MT.bam input.bam chrM
```

If the human reference genome build 37 is used for read-mapping, "chrM" should be replaced with "MT" where the target region is specified. After the extraction of mtDNA sequence reads, the new BAM file (output.MT.bam) should be indexed again:

```
samtools index output.MT.bam
```

3.2 Somatic Variant Calling

Many mutation calling tools, including Mutect2 and VarScan2, can be used for initial somatic variant calling for the mitochondrial genome. Mutect2 provides mitochondrial mode, which automatically optimizes parameters for mtDNA variant calling. After initial calling using Mutect2, tools in GATK, such as LearnReadOrientationModel, and FilterMutectCalls, can be applied subsequently to the output VCF to remove false-like calls. In order to use LearnReadOrientationModel, orientation bias metric should be saved to a file using --f1r2-tar-gz option of Mutect2. FilterMutectCalls also provides mitochondria mode for specialized refinement of unfiltered calls produced by Mutect2 mitochondria mode. Adjusting parameters such as the lower cutoff of variant allele fraction (VAF) is possible. If DNA minor-contamination is suspected, estimated contamination levels could be given to FilterMutectCalls with --contamination-estimate. Mutect2 mtDNA variant calling can be conducted using commands shown below:

```
[running Mutect2]
gatk Mutect2 \
 --mitochondria-mode \
-R reference.fasta \
 -L chrM \
 -I tumor.MT.bam \
 -I normal.MT.bam \
 -normal normal.samplename \
--f1r2-tar-gz orientation_bias_metric \
 -O output.vcf.gz

gatk LearnReadOrientationModel \
-I orientation_bias_metric \
-O readorientation_file

gatk FilterMutectCalls \
--mitochondria-mode \
-V output.vcf.gz \
-O output.filtered.vcf.gz \
-R reference.fasta \
--orientation-bias-artifact-priors readorientation_file
```

For VarScan2, samtools should first be used to “mpileup” the tumor and normal BAM files. To conduct samtools-mpileup, thresholds for read-mapping quality and base quality need to be set. Conventionally, a threshold of 20 (in Phred-like quality score; the score of 20 allows <1% error) has been used for both read-mapping and base quality. Sequence reads or bases with poor quality (below the thresholds) are not to be counted in variant calling. In VarScan2, several parameters can be adjusted for more accurate and/or sensitive results, for example, --min-var-freq (minimum VAF cutoff), --min-avg-qual (minimum base quality), --min-coverage (minimum read depth), --min-reads2 (minimum number of variant supporting read), --somatic-p-value (*p*-value threshold of calling somatic variants), and --strand-filter (remove variants with >90% strand bias):

```
[mpileup tumor and normal BAM files]
samtools mpileup -B -Q 20 -q 20 \
   -f reference.fasta normal.MT.bam > normal.MT.mpileup

samtools mpileup -B -Q 20 -q 20 \
-f reference.fasta tumor.MT.bam > tumor.MT.mpileup

[running VarScan2]
java -jar VarScan.jar somatic normal.MT.mpileup tumor.MT.
mpileup \
    --output-snp output.snp.vcf \
    --output-indel output.indel.vcf \
    --output-vcf 1 \
    --min-var-freq 0.01 \
    --min-avg-qual 20 \
    --min-coverage 100 \
    --min-reads2 5 \
    --somatic-p-value 0.1 \
    --strand-filter 1
```

Here, the minimum heteroplasmic level, minimum coverage, and minimum number of variant supporting reads are set to 1%, 100, and 5, respectively. These parameter settings can be adjusted according to the aim and/or condition of the study. For a more precise analysis, these thresholds can be enhanced.

Initial SNV and indel calls from VarScan2 are written in output.snp.vcf and output.indel.vcf, respectively. To merge SNV and indel calls, each output VCF file should be compressed and indexed, then be integrated into a single VCF file using Bcftools:

```
[compressing and indexing VCF files]
bgzip output.snp.vcf && tabix -p vcf output.snp.vcf.gz
```

```
bgzip output.indel.vcf && tabix -p vcf output.indel.vcf.gz

   [merging SNV and indel VCF files]
bcftools concat output.snp.vcf.gz output.indel.vcf.gz \
   -o output.combined.vcf -O v -a
```

Bgzip and tabix functions are available to use after installing samtools and bcftools. In the VarScan2 output file, variants with the "SOMATIC" flag can be considered as somatic mutations.

Each tool has *pros* and *cons* in detecting mtDNA mutations. Mutect2 performs local realignment for indel-rich homopolymer regions internally, so it is, in theory, able to more accurately call the indels of homopolymer regions. The mitochondria mode in Mutect2 endogenously considers conditions that are particular to the mitochondrial genome, such as ultra-high depth, heteroplasmy and the nuclear DNA of mitochondrial origin (NuMT). Although Mutect2 can identify mtDNA somatic mutations with ~0.2% VAF level at a higher precision, its recall is relatively lower than VarScan2. In addition, more customized calling is possible with VarScan2, because researchers can freely control many parameters in VarScan2. Taken together, Mutect2 has a higher positive predictive value and VarScan2 has a higher sensitivity in mtDNA mutation calling. By using both Mutect2 and VarScan2 simultaneously, we may yield sensitive and specific mtDNA mutation calls. For example, in projects aiming for more sensitive detection, the union set of the variants from the two callers can be considered for downstream analyses. In such a scenario, an extensive refinement step may be necessary to remove false-positive calls. Instead, in projects aiming for more specific detection, the intersection set of the variants can be taken.

3.3 mtDNA-Specific Consideration and Filtering

After the initial variant calling using Mutect2 and VarScan2, downstream steps for filtering false-positive variants are usually required. The filtering criteria may vary according to sequencing characteristics. For example, high-quality whole-genomes with sufficient read depth in the mitochondrial genome allow detection of very low heteroplasmic variants (i.e., ~0.5% VAF). However, applying similar conditions on mitochondrial sequences of poor quality and/or insufficient read depth can generate many false positives. To set optimal filter conditions interactively, visual inspection of mtDNA mutation calls through IGV may be helpful. Features illustrated in **Notes 1–11** can be considered for more accurate mutation calling in the mitochondrial genome.

3.4 Functional Annotation of mtDNA Variants

Not all somatic mutations have a significant impact on the tumor sample. In order to analyze the impact of somatic mutations, functional annotation must follow. Functional annotation is a computational prediction of genomic functions using genes affected by variants, location of variants, how it affects the amino acids, etc. Although most missense mutations are selectively neutral [1], knowing the functional impact of the mutation can hint at finding somatic mutations that have a critical impact on the tumor. Functional impacts of each variant can be annotated using VEP [10]. Functional annotation by VEP can be conducted using commands shown below:

```
[running VEP]
vep -i input.filtered.vcf \
-o output.filtered.vep.vcf \
--vcf \
--species homo_sapiens \
--database \
--everything \
--assembly GRCh37
```

Here, an input file and an output file with variants are in VCF formats so an option of --vcf is given. Also, VEP annotates every information of the public Ensembl database in the commands above. After functional annotation, commonly mutated genes and variants with high impact, such as frameshift and nonsense mutation, can be checked.

4 Notes

1. With a VAF cutoff >1%, typical cancer whole-genomes harbor less than 10 mtDNA somatic mutations, most frequently 1–2 mutations. The vast majority of the mtDNA somatic mutations are heteroplasmic in cancer tissues (median and average VAFs ~5% and 20%, respectively) [15]. Counterpart normal tissues show much lower VAF at the corresponding locus, close to 0% VAF. At error-prone mtDNA loci (see consideration **Note 8** in this section; Fig. 1b), counterpart normal tissues also show a considerable VAF (~0.5–1%).
2. Although sequencing coverage of the mitochondrial genome is usually high (i.e., ~10,000×) in cancer whole-genome sequences, detection of ultra-low VAF somatic mutations (i.e., <0.5%) is challenging. For example, in case of very minor DNA cross-contamination (i.e., 0.5% contamination which is pervasive in the population-scale tumor genome sequencing), all the mitochondrial germline polymorphisms of the contaminant DNA will appear as low heteroplasmic, somatic-like mutations (Fig. 2a) [16]. Minor DNA cross-contamination is highly likely if (1) a huge number of mtDNA mutations (>~20) is detected from a cancer genome

sequence, (2) the vast majority of the mutations are previously known germline polymorphisms, and (3) most of the mutations show more-or-less similar heteroplasmic levels. Of note, such low-level contamination does not impact the mutation calling of the nuclear genome because sequencing coverage is moderate for the nuclear genome sequences.

3. Sequence reads produced from the nuclear DNA of mitochondrial origin (NuMTs) can confound mtDNA mutation analysis, particularly when sequence homology is high and its read-alignment is not specific (Fig. 2b) [17]. Because mtDNA copy numbers are 100–1000 times higher than a specific NuMT copy, the NuMT issue does not practically impact mutation calls at 1% VAF cutoff. However, it may generate false positives under a more lenient VAF cutoff.
4. Identifying somatic mtDNA mutations is also prone to artificial mutations caused by the poor quality of DNA library, such as oxidative DNA damage during library preparation (Fig. 2c) [18]. Most of the artifacts are C:G>A:T substitutions at a low heteroplasmic level (1% VAF or less). This can be controlled by applying a moderate VAF cutoff (~1%).
5. The strand bias of variant supporting reads and position bias in supporting reads can also be considered. For most true variant calls, the ratio between forward and reverse supporting reads ranges from 1:10 to 10:1. Similarly, the variant-allele supporting bases should overall be randomly positioned in the sequencing reads. False calls are likely if most of the bases are enriched at the end of variant supporting reads (i.e., in the first 10 bases, and/or last 10 bases of the variant supporting reads).
6. Generally, the mapping quality cutoff (typically 20) ensures specific mutation calling in the mitochondrial region. However, due to the NuMTs of high sequence homology in the human reference genome, several regions of the mitochondrial genome show a systematically lower fraction of sequencing reads in mapping quality (rCRS:4400-5725, rCRS:8760-8900). If mapping quality cutoff is increased to 30, the detection sensitivity of somatic mtDNA mutations may be compromised in these regions. Specific considerations, such as realigning reads to reference containing only mitochondrial sequence or initially aligning reads to NuMT masked FASTA, may be required for a more sensitive calling [19].
7. Even though mtDNA is circular, a linearized reference genome (i.e., rCRS) is used in most studies investigating the mtDNA variation. In rCRS, for example, an artificial breakpoint is introduced near the origin of the mtDNA control region with the start position at chrM:1 and the end position at chrM:16,569 [20]. As a result, sequence reads that

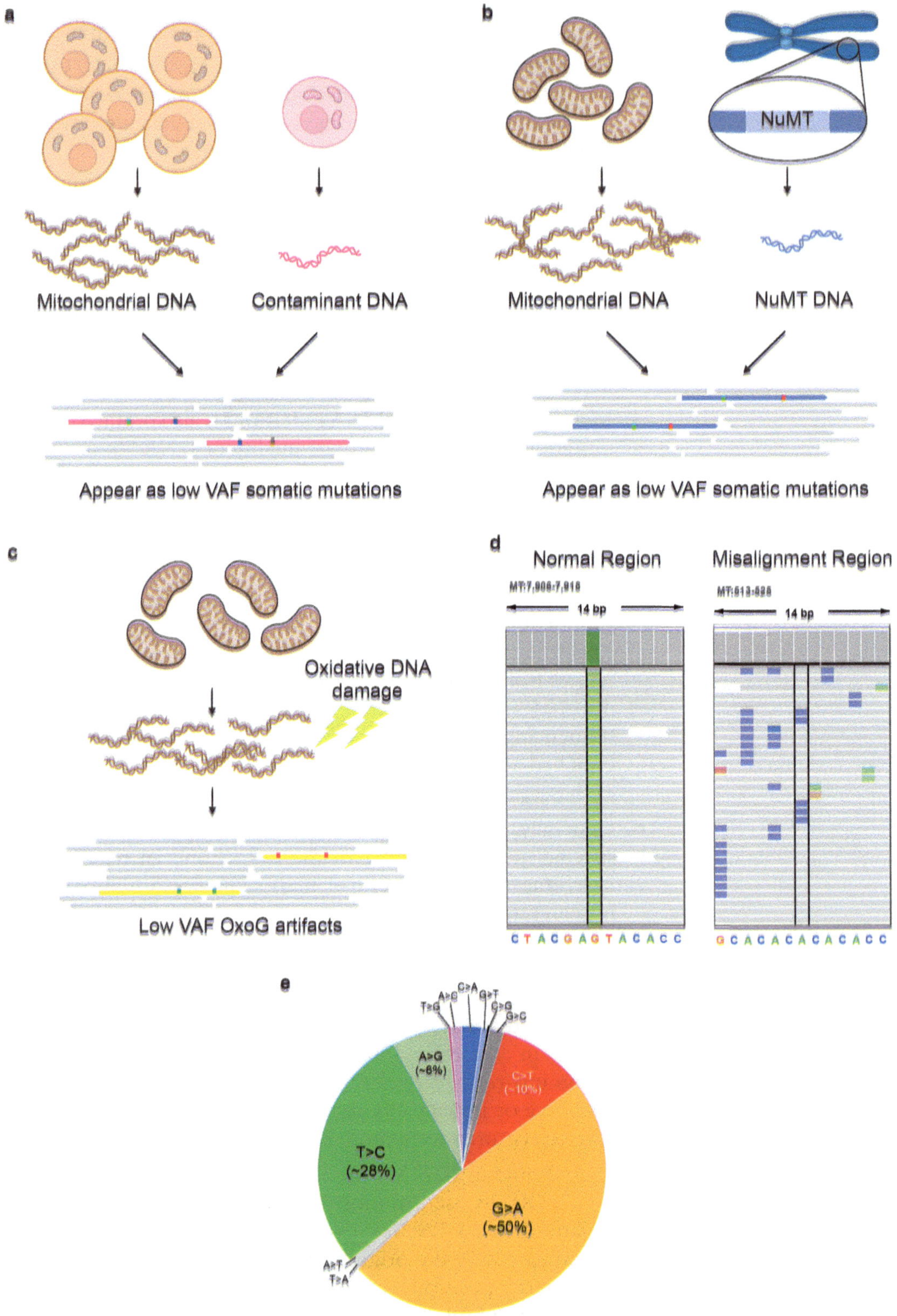

Fig. 2 Specific considerations in somatic mtDNA mutation calling procedures. (**a**) Impact of the minor cross-contamination of foreign DNA. (**b**) Impact of the sequencing reads produced from NuMTs. (**c**) Impact of the

encompass the artificial breakpoint are not properly aligned and are frequently discarded. Consequently, the detection sensitivity of mtDNA variants in the vicinity of the breakpoint is compromised. To overcome the limitation, Ding and colleagues proposed a "double alignment" strategy, using two linear reference genomes in sequence read alignment [20]. In such a case, the conventional rCRS becomes the reference for the first sequence alignment followed by variant detection in the mitochondrial genome, particularly in the internal region of the genome. Then, sequencing reads are re-aligned on the shifted rCRS, which starts at chrM:8000 and ends at chrM:7999. Variant calling can then be conducted in the end regions of the original reference sequence.

8. In the mitochondrial genome, the following regions are homopolymeric and/or include a gap (base "N"). Because these regions show an extensive level of sequencing errors or misalignment (Fig. 2d) [1, 21], hard-filtering can be considered for low VAF mtDNA mutation calls identified in the region:
   (a) Misalignment due to ACCCCCCCTCCCCC (rCRS:302-315).
   (b) Misalignment due to GCACACACACACC (rCRS:513-525).
   (c) Misalignment due to CCCCCC (rCRS 568-573).
   (d) Misalignment due to CCCCCTCCCC (rCRS:956-965).
   (e) Misalignment due to 3107N in rCRS (rCRS:3105-3109).
   (f) Misalignment due to CCCCC (rCRS 5895-5899).
   (g) Misalignment due to CCCCCTCTACCCCCTCTA (rCRS:8272-8289).
   (h) Misalignment due to AAACCCCCTCCCCAT (rCRS:16180-16195).
9. As mentioned in feature **Note 8** above, background noise rate is variable across loci in the mitochondrial genome. The noise level can be measured by a pool of mtDNA sequences collected from many normal samples. The noise matrix obtained from the normal sequences can be considered for filtering out false-positive calls in tumor genome sequences.
10. Because mitochondrial genome is short and its innate mutation rate is high, identical mtDNA mutations can be recurrently acquired in unrelated tumors without apparent selective pressure. Unlike somatic mutations in the nuclear genome, a

Fig. 2 (continued) oxidative DNA damage during DNA library preparation. (**d**) Examples of error-prone regions in the mitochondrial genome. (**e**) The mutational signature of somatic mtDNA base substitutions observed from 2658 cancer genomes [15]

substantial fraction of mtDNA somatic mutations can overlap known germline polymorphisms. In large-scale cancer genome analyses [1, 15], ~35% of the apparent somatic mutations were registered in the database of germline mtDNA polymorphisms [13, 14].

11. Previous studies suggest that the signature of somatic mtDNA mutations is more-or-less uniform, with C:G>T:A (~60%) and T:A>C:G (~35%) substitutions being the most and second most frequent mutation classes, respectively (Fig. 2e) [1, 15]. Surprisingly, G>A substitutions are approximately five-fold more frequent than C>T substitutions, although guanine bases are 2.4-fold less common than cytosine bases on the positive strand of rCRS (the light strand of the mitochondrial genome). Therefore, the rate of G>A substitutions is ~12-fold higher than that of C>T substitutions on the light strand of the mitochondrial genome. Moreover, the impact of well-known carcinogens (i.e., C:G>A:T substitutions for tobacco smoking-mediated mutations; CC:GG>TT:AA double-base substitutions for ultraviolet light-mediated mutations) is generally minimal on the mitochondrial genome. Given the mtDNA mutational signature, the accuracy of mutation calling can be collectively assessed by illustrating the overall spectrum of mtDNA somatic mutations identified from a sequencing study.

References

1. Ju YS, Alexandrov LB, Gerstung M et al (2014) Origins and functional consequences of somatic mitochondrial DNA mutations in human cancer. elife 3:e20395
2. Li H (2013) Aligning sequence reads, clone sequences and assembly contigs with BWA-MEM. arXiv: Genomics
3. Genomes Project C, Auton A, Brooks LD et al (2015) A global reference for human genetic variation. Nature 526:68–74
4. The I, Of Whole TP-CA, Consortium G (2020) Pan-cancer analysis of whole genomes. Nature 578:82
5. Andrews RM, Kubacka I, Chinnery PF et al (1999) Reanalysis and revision of the Cambridge reference sequence for human mitochondrial DNA. Nat Genet 23:147
6. Danecek P, Bonfield JK, Liddle J et al (2021) Twelve years of SAMtools and BCFtools. Gigascience 10:giab008
7. Cibulskis K, Lawrence MS, Carter SL et al (2013) Sensitive detection of somatic point mutations in impure and heterogeneous cancer samples. Nat Biotechnol 31:213–219
8. Koboldt DC, Zhang Q, Larson DE et al (2012) VarScan 2: somatic mutation and copy number alteration discovery in cancer by exome sequencing. Genome Res 22:568–576
9. Robinson JT, Thorvaldsdottir H, Winckler W et al (2011) Integrative genomics viewer. Nat Biotechnol 29:24–26
10. Mclaren W, Gil L, Hunt SE et al (2016) The Ensembl variant effect predictor. Genome Biol 17:122
11. Team RC (2013) R: a language and environment for statistical computing
12. Sherry ST, Ward MH, Kholodov M et al (2001) dbSNP: the NCBI database of genetic variation. Nucleic Acids Res 29:308–311
13. Ingman M, Gyllensten U (2006) mtDB: human mitochondrial genome database, a resource for population genetics and medical sciences. Nucleic Acids Res 34:D749–D751
14. Ruiz-Pesini E, Lott MT, Procaccio V et al (2007) An enhanced MITOMAP with a global mtDNA mutational phylogeny. Nucleic Acids Res 35:D823–D828
15. Yuan Y, Ju YS, Kim Y et al (2020) Comprehensive molecular characterization of

mitochondrial genomes in human cancers. Nat Genet 52:342–352
16. Weissensteiner H, Forer L, Fendt L et al (2021) Contamination detection in sequencing studies using the mitochondrial phylogeny. Genome Res 31:309–316
17. Maude H, Davidson M, Charitakis N et al (2019) NUMT confounding biases mitochondrial heteroplasmy calls in favor of the reference allele. Front Cell Dev Biol 7:201
18. Costello M, Pugh TJ, Fennell TJ et al (2013) Discovery and characterization of artifactual mutations in deep coverage targeted capture sequencing data due to oxidative DNA damage during sample preparation. Nucleic Acids Res 41:e67–e67
19. Lareau CA, Ludwig LS, Muus C et al (2021) Massively parallel single-cell mitochondrial DNA genotyping and chromatin profiling. Nat Biotechnol 39:451–461
20. Ding J, Sidore C, Butler TJ et al (2015) Assessing mitochondrial DNA variation and copy number in lymphocytes of~ 2,000 Sardinians using tailored sequencing analysis tools. PLoS Genet 11:e1005306
21. Tanaka M, Cabrera VM, Gonzalez AM et al (2004) Mitochondrial genome variation in eastern Asia and the peopling of Japan. Genome Res 14:1832–1850

Chapter 11

Identification, Quantification, and Testing of Alternative Splicing Events from RNA-Seq Data Using SplAdder

Philipp Markolin, Gunnar Rätsch, and André Kahles

Abstract

Alternative splicing (AS) is a regulatory process during mRNA maturation that shapes higher eukaryotes' complex transcriptomes. High-throughput sequencing of RNA (RNA-Seq) allows for measurements of AS transcripts at an unprecedented depth and diversity. The ever-expanding catalog of known AS events provides biological insights into gene regulation, population genetics, or in the context of disease. Here, we present an overview on the usage of SplAdder, a graph-based alternative splicing toolbox, which can integrate an arbitrarily large number of RNA-Seq alignments and a given annotation file to augment the shared annotation based on RNA-Seq evidence. The shared augmented annotation graph is then used to identify, quantify, and confirm alternative splicing events based on the RNA-Seq data. Splice graphs for individual alignments can also be tested for significant quantitative differences between other samples or groups of samples.

Key words Bioinformatics, Alternative splicing, Transcriptomics, Sequence analysis, RNA-sequencing, Data analysis

1 Introduction

Depending on the organism, up to 95% of expressed genes are transcribed into multiple transcript variants [1, 2]; furthermore, the same gene locus can give rise to various transcripts by utilizing differing exon compositions. A common limitation of most RNA-seq based splicing tools, including SplAdder, is the lack of identification of complete transcripts. However, focusing on local splicing variation and representing the full variability of splicing connections in a graph allows for capturing the complexity of the whole transcriptome. In SplAdder, we leverage evidence from RNA-Seq data to compute a more complete representation of the splicing diversity within a sample and tackle the complexity with a reduction to alternative splicing events instead of full transcripts. For many applications, the identification and quantification of full transcript structures is not necessary and focusing on local splicing

Charlotte K. Y. Ng and Salvatore Piscuoglio (eds.), *Variant Calling: Methods and Protocols*,
Methods in Molecular Biology, vol. 2493, https://doi.org/10.1007/978-1-0716-2293-3_11,

events is already sufficient. This is the use case addressed by SplAdder [3], which has the following key features:

1. Augmentation of a given gene annotation with additional intron- and exon-structures derived from given RNA-Seq evidence, generating a comprehensive splicing graph for each gene.
2. Splicing graphs can be derived per input sample and then integrated across a whole cohort.
3. Extraction of six different event-type from the splicing graph: exon skips, intron retention, alternative 3′ splice site, alternative 5′ splice site, mutually exclusive exons, multiple exon skip.
4. Quantification of all extracted events across all given input samples; derivation of percent-spliced-in (PSI) values for each event.
5. Testing for differential event usage between groups of input samples.

2 Software

The SplAdder tool is open source and can be freely downloaded and accessed at https://github.com/ratschlab/spladder. It can be directly installed via the Python Package Index (PyPI) or can be installed from the source using the above code repository.

2.1 Software Requirements

To run SplAdder, the computing environment must include the following tools and packages:

1. Python (3.6 or newer).
2. numpy (v1.14.6 or newer).
3. pysam (v0.15.3 or newer).
4. matplotlib (v2.2 or newer).
5. scipy (v1.3 or newer).
6. intervaltree (v3.0.0 or newer).
7. h5py (v2.2 or newer).
8. statsmodels (v0.9.0 or newer).

When installing SplAdder from PyPi or from source (*see* Subheading 2.2), the installation routine will automatically provide all required dependencies. Currently, SplAdders' latest release version is v2.4.3, and the protocols in this chapter refer to that version of SplAdder. Since it was first published in 2016, the SplAdder toolbox has been continuously improved and extended. We recommend that readers interested in the latest changes refer to the

SplAdder documentation available at https://spladder.readthedocs.io/en/latest/.

2.2 Download and Installation of SplAdder

2.2.1 Installation from PyPI

The installation from PyPI is straightforward. One can install the latest version of SplAdder using pip:

```
pip install spladder
```

Alternatively, for installation of a specific version, one can do the following (e.g., for version 2.4.2):

```
pip install spladder=2.4.2
```

2.2.2 Installation from Source

It is also possible to get the latest changes available on GitHub by cloning the repository:

```
git clone https://github.com/ratschlab/spladder.git
```

This will create a directory named spladder inside the current local directory. Next, one can install SplAdder by first changing into the local spladder directory and then typing:

```
python setup.py install
```

For most applications, the current release version of SplAdder will be sufficient. If one is interested in using a specific branch, e.g., development, you would need to change into that branch first before installation. You can achieve this with:

```
git checkout development
python setup.py install
```

For this, we recommend using a python virtual environment, such as venv or a conda environment.

3 Methods

The SplAdder toolbox contains several steps of a splicing analysis pipeline. Each of these steps is represented by a run mode, collecting the relevant functionality and associated parameters. We will first provide an overview of the different run modes and their capabilities and will then proceed to outline a typical analysis workflow of a small sample set.

The workspace of a SplAdder analysis is the SplAdder project directory. This directory will remain the same for all run modes that are applied throughout the analysis, allowing the pipeline to store

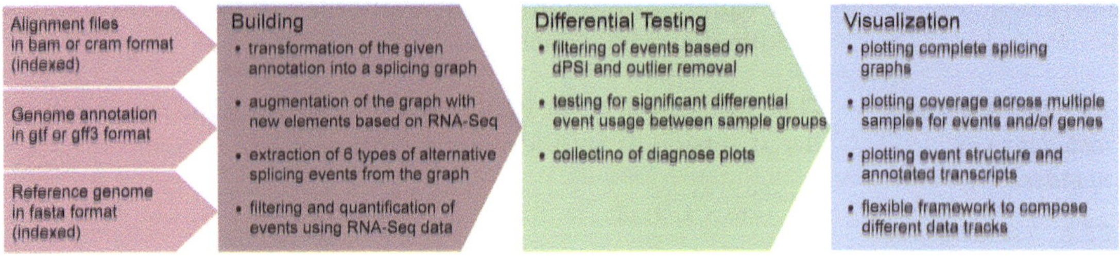

Fig. 1 SplAdder analysis flowchart. The main steps of the SplAdder workflow consist of (1) building a splicing graph from annotation and RNA-Seq data and extract alternative splicing events, (2) differential analysis of splice event usage between different sample groups, (3) generating visualizations of event structure and quantification across samples

intermediate results and, most importantly, to re-use results between different run modes.

SplAdder has three different run modes that reflect the different steps of a typical analysis pipeline (Fig. 1). In the following, we will describe each of the run modes and provide usage examples.

3.1 The Build Mode

This first step in the SplAdder alternative splicing pipeline is concerned with transforming the given annotation into a splicing graph, augmenting it with additional events from RNA-Seq data, and extracting six different types of alternative splicing events.

Graph construction In this initial phase, SplAdder parses the given annotation file and summarizes all transcripts of a gene into a splicing graph. This graph contains all alternative splicing variants present in the annotation and forms the basis for all further steps in the workflow.

Graph augmentation Given at least one RNA-Seq alignment file, the splicing graph of each gene is then augmented with new introns and exon segments that were detected in the alignment file. There are several options how multiple input files can be combined to inform splicing graph augmentation: considering each file individually, merging all input files, or first consider each file individually and then merge the resulting graphs. The last option is the default and the recommended strategy in most settings. At the end of this phase, each gene is represented with an augmented graph that contains not only annotated splice connections but also all novel connections found in the data. Depending on the chosen confidence level, determining the stringency for adding novel elements to the graph, the graph will have a higher or lower density.

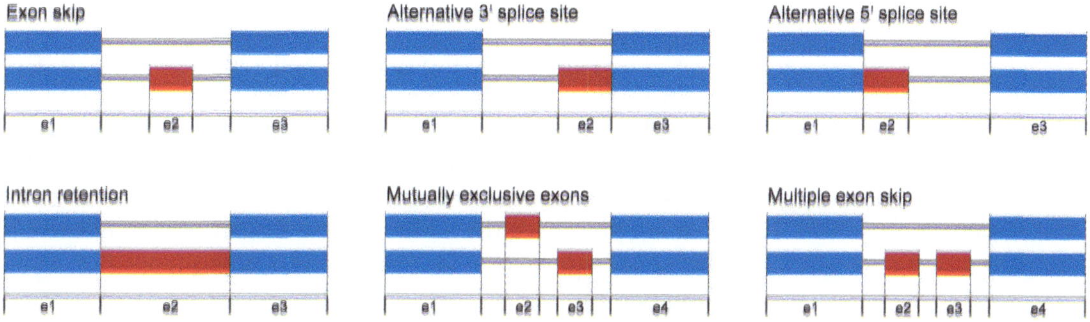

Fig. 2 Set of different alternative splicing event types supported by SplAdder. Colored boxes represent exon segments. Segments in blue do not change between the two alternative splice forms, while segments in red are alternatively used. Below each event, the numbering of segments indicates the nomenclature used for describing feature counts. Exon segments are always numbered from left to right following genomic coordinates, irrespective of the strand the event is on. The above numbering shows the positive strand. The numbering on the negative strand would be reversed

Graph quantification Once a graph is constructed for each gene, all nodes and edges (exons and introns, respectively) of the graph can be quantified using the input alignment files. The quantification values are then used subsequently to quantify splicing events and to compute percent-spliced-in (PSI) values.

Event detection Again, using the splicing graph of each gene, SplAdder can detect six different types of alternative splicing events: exon skipping, intron retention, alternative 3′ splice sites, alternative 5′ splice sites, mutually exclusive exons, and multiple (coordinated) exon skips (Fig. 2). Each event can be quantified using the graph quantifications from the previous step.

The building mode of SplAdder is invoked by:

```
spladder build
```

followed by any additional parameters. When invoked without any parameters, a detailed listing of all parameters together with their descriptions is output.

3.1.1 Understanding Key Parameters

The build mode has three mandatory inputs that have to be specified by the user and that, when provided, suffice to start the SplAdder analysis:

```
spladder build --bams alignment.bam \
               --annotation annotation.gtf \
               --outdir spladder_out
```

These parameters have the following meaning:

- --bams: The path to the input RNA-Seq alignment files in bam or cram format. Multiple files should be provided as a comma separated list (*see* **Note 1**). For large cohorts, files can also be provided as a text file, containing one path per line. This text file needs to have the .txt file ending. When alignment files in cram format are provided, the additional parameter --reference needs to be specified to provide the reference sequence for cram decompression.
- --annotation: The path to the gene annotation file in gtf or gff3 format (*see* **Note 1**).
- --outdir: The path to the output directory. This directory is the SplAdder project directory for this analysis and should be re-used throughout all phases of the project.

Although these three inputs should suffice to run SplAdder, a large number of additional optional configurations can be made. Below, we list some of the key parameters relevant for the build mode. Many more additional parameters are available. However, they often correspond to very specific use cases or cover corner cases. We refer to the software manual for a closer description of those.

Parameters that further specify the inputs:

- --no-primary-only: In its default settings, SplAdder will only consider a single alignment per read, namely the one labeled by the aligner as primary. With this flag, one can consider all alignments (even multi-mappers) provided in the input.
- --labels label1,label2, . . .: Per default, the names for each sample are automatically derived from the input alignment file names. These names can be overwritten with user-defined labels. The labels need to be presented as a comma-separated list, having the same order and length as the list of alignment files provided via --bams.
- --filter-consensus: When extracting splice-junction candidates from the alignment files, all split-read alignments are considered. With this option, one can require that any junction taken from the the alignment files complies with the GT/AG splice site consensus sequences. This option requires the use of --reference.
- --reference ref.fa: When alignment files are provided in cram format or the splice site consensus is enforced using --filter-consensus, the corresponding reference genome sequence needs to be provided.

Parameters that further specify the graph construction and augmentation (*see* **Note 1**):

- --confidence: Choice of predefined confidence levels, from 3 (highest) to 0 (lowest). This parameter determines how strictly input alignment data are filtered before being used for graph augmentation. Thresholds are automatically derived from confidence level and read length.
- --merge-strat STRAT: Here, STRAT is one of single, merge_-bams, or merge_graphs. The default, merge_graphs, computes a splicing graph for each gene for each input sample and then merges all single-sample graphs per gene to generate a merged graph per gene. On the contrary, single handles each input sample independently throughout the process and merge_bams merges all input alignment files, assuming they all represent the same sample. In most practical cases, the default can be used.
- --sparse-bam: This option allows SplAdder to summarize the given input alignment field into a sparse representation of the alignment counts. Setting this flag is recommended if some additional disk space is available (~10% of the total alignment file size), as it greatly speeds up later analysis steps in SplAdder. This also requires writing permissions to the location where the alignments are stored.
- --validate-sg and --validate-sg-count INT If --validate-sg is used, every edge in a merged multi-sample graph will be removed, if it was not present in at least --validate-sg-count many sample graphs before merging. This is useful especially in very large sample cohorts to increase confidence in the detected splicing variability.

Parameters that further specify the detection of alternative splicing events:

- --event-types TYPES Here, TYPES refers to one or more of the six event types that SplAdder extracts per default: exon_skip, intron_retention, alt_3prime, alt_5prime, mutex_exons, and mult_exon_skip. That is, the user can define a subset of these event types to be extracted. If multiple event types are chosen, they have to be provided as a comma-separated list.
- --no-extract-ase This parameter completely switches off event extraction. This can be useful if the analysis should be performed in several stages.
- --ase-edge-limit INT Depending on the amount of information provided in the inputs, the splicing graph of a gene can grow very complex. For overly complex graphs, it can take very long to extract all events. To skip outliers, there is a default limit of 500 edges imposed on graphs. If this threshold is surpassed, the gene is skipped. With this parameter, this threshold can be increased.

- --quantify-graph In default mode, SplAdder does not quantify the splicing graphs and alternative events. The framework only extracts the event structure. For computing PSI values per sample or for later testing event usage between groups of samples, this parameter needs to be set.
- --min-reads-psi For computing percent spliced in values for each event and sample, SplAdder relies on the number of split-alignments supporting the introns in each of the event isoforms. As the PSI is a ratio that becomes unstable for a very small number of supporting split-alignments, the default minimal threshold is 10. It can be adapted with this parameter.

Parameters that further define the output generation:

- --output-txt In default mode, SplAdder only outputs confidently called events. That is events that have passed the internal heuristic thresholds for read support defined by the confidence level. This parameter allows for the output of all events detected in the graphs, irrespective of whether they were confirmed based on the alignment data.
- --output-gff3 This parameter has the same meaning as --output-txt, outputting the events in gff3 format.
- --no-compress-text In default settings, all text outputs are gzip-compressed. This behavior can be switched off with this parameter.

3.1.2 Description of Output Files

SplAdder produces several human-readable output files that contain the structure and quantification information of the detected alternative splicing events. These files contain the structural information of the detected splicing events, including position, strand, and start/end coordinates of all involved exons and introns. The names of these files are constructed using the following naming convention:

```
<merge_strategy>_<event_type>_C<conf_level><conf_tag>.<file-
format>
```

- <merge_strategy> is one of single, merge_bams, merge_graphs.
- <event_type> is one of exon_skip, intron_retention, alt_3prime, alt_5prime, mutex_exons, mult_exon_skip.
- <conf_level> is the chosen confidence level (between 0 and 3).
- <conf_tag> will be set as .confirmed in default mode, where only confirmed events are reported (see output options above).
- <fileformat> corresponds to the chosen output file format (s) and will be one of bed, txt, gff3.

So for instance, if we look at all confirmed exon skipping events that were detected in the merge_graphs mode with confidence level 3 and reported in gff3 format, then the file would be named: merge_graphs_exon_skip_C2.confirmed.gff3.

Following, we provide a brief description of the different output file formats:

- gff3 format. This format follows the common specification provided by Ensembl[1] and contains only the event structure. Each event is represented as a mini-gene containing two isoforms, one for each alternative event path.
- txt format. This format is a custom tab-delimited columnar output format, containing both the event position and structure as well as quantification information for the input samples (if --quantify-graph was selected). The columns are defined as follows:

  1: Chromosome of the event.
  2: Strand of the event.
  3: Unique event_id.
  4: Name of gene the event is located in.
  5-5 + n: Start and stop coordinates of the event exons (n depends on the event type).
  5 + n and following: Count values for each of the samples with the following layout (features are event type-specific and are further outlined in the next paragraph):

```
<sample1>:<feature1>
<sample1>:<feature2>
<sample1>:<feature3>
...
<sample2>:<feature1>
...
```

- counts.hdf5 format. This format allows for automatic loading of different data elements using the HDF5 standard. It is hierarchically organized (like a file system) and has the following structure:
  - conf_idx 0-based index set, containing the index of the events that are confirmed in the provided samples based on the given confidence level.
  - confirmed same information as in conf_idx, given as a Boolean array over all events (this is mostly provided for convenience).

[1] http://www.ensembl.org/info/website/upload/gff.html.

- event_counts three-dimensional matrix (S × F × E) containing counts for each of the E events, F features and S samples.
- event_features list that contains the names of the counted features per event type (features are event type-specific and are further outlined in the next paragraph).
- event_pos position of all event exons encoded as start,stop pairs for each event (events are rows, coordinates are columns).
- gene_chr chromosome for each gene in the gene list.
- gene_idx index that maps each event to a gene in the gene list (0-based).
- gene_names gene name for each gene in the gene list.
- gene_pos position of each gene in the gene list encoded as start,stop pair.
- gene_strand strand for each gene in the gene list.
- iso1 counts for each isoform1 of each event used for PSI computation.
- iso2 counts for each isoform2 of each event used for PSI computation.
- num_verified For each event type, SplAdder uses different verification criteria to check for the validity of an event in a sample. This matrix provides an overview in how many samples each of the criteria was verified for each of the events. For more details on the criteria, we refer to the original SplAdder publication [3].
- psi The percent spliced in value computed for each confirmed event in each of the samples. The value is NA, if not enough spliced reads are available to compute the ratio. The default is 10 reads and can be adapted via the parameter --min-reads-psi.
- strains names of the samples counted.
- verified 3-dimensional bool matrix (S × V × E), where a cell is 1/True if event E was verified for criterion V in sample S and 0/False otherwise. (See comment on num_verified for further details.)

All output files containing event count data provide the count information for each event using a set of event count features. These features can be distinguished into intron-confirmation counts and exonic segment coverages. To describe which element of the alternative event is quantified, SplAdder uses the numbering of exon segments as outlined in Fig. 2, to unambiguously describe a feature. To describe the count that supports the junction between exon segments e1 and e2, the feature e1e2_conf is used. Similarly, to describe the coverage of exon segment e3, the feature e3_cov is used. Following this logic, every event type has a different set of output features. Which features are used can either be derived from the column headers in the txt files or using the event_features data element in the hdf5 output.

3.2 The Test Mode

This SplAdder mode is for differentially testing the usage of alternative events between two groups of samples. A prerequisite for this is that all samples that are to be subjected to testing have been jointly analyzed in build mode. However, not the full set of samples collected in the build mode has to be part of the testing, but any subsets of samples can be used instead.

It is recommended that for each condition to be tested (e.g., control against a perturbation), the number of available replicates is at least three. Further, the mean–variance relationship for intron counts is estimated on the set of tested events. If the number of events to be tested becomes too small, then this estimate becomes unstable and might result in an error.

The testing mode of SplAdder is invoked by:

```
spladder test
```

followed by any relevant parameters. When invoked without any parameters, a detailed listing of all parameters together with their descriptions is output.

3.2.1 Understanding Key Parameters

Three mandatory parameters form the minimal input of the testing mode:

```
spladder test --conditionA alignmentA1.bam,alignmentA2.
bam,... \
 --conditionB alignmentB1.bam,alignmentB2.bam,... \
 --outdir spladder_out
```

Analog to the way a list of alignments can be provided in build mode, also in test mode the comma-separated file list can be substituted with a file containing the paths to the respective files:

```
spladder test --conditionA alignmentsA_list.txt \
 --conditionB alignmentsB_list.txt \
 --outdir spladder_out
```

A number of parameters are shared with build mode and are used to select the corresponding graph for testing. They have the same meaning as in build mode and will only be shortly listed here:

- --confidence confidence level.
- --merge-strat merge strategy.
- --validate-sg splice graph validation.

A further set of parameters is concerned with further specifying the pre-filtering of events and the actual settings of the test:

- --event-types STRING contains a comma-separated list of all event types that should be subjected to testing. If not used, all event types will be tested.
- --correction selects the method for multiple testing correction. As each event is tested individually, the family-wise error rate needs to be controlled through a correction of the *p*-values. The default correction uses the method of Benjamini and Hochberg, but others such as Bonferroni's method can be chosen.
- --max-zero-frac FLOAT SplAdder tests each isoform of an event independently. With this filter, such an isoform is not subjected to testing, if the fraction of 0 counts across all samples is larger than the given threshold. The default is 0.5.
- --min-count INT Similarly as above, this provides the minimum count threshold for an event isoform to be subjected to testing. The count is the total count across all samples. Default is 10.
- --cap-outliers Splice event quantifications that are detected as outliers by SplAdder can be capped using this parameter. This can make the testing more stable.
- --no-cap-exp-outliers As gene expression is taken into account for testing, the capping of expression outliers is done by default in SplAdder. This capping can be switched off using this parameter.

Lastly, the following parameters are useful to control the output of the testing mode:

- --labelA STRING will assign the given label to the group of samples submitted using --conditionA. This can be useful to keep track of the groups in the outputs, e.g., by assigning condition A the name "permutation" and condition B the name "control."
- --labelA STRING is the analog parameter for the samples submitted with parameter --conditionB.
- --diagnose-plots generates diagnose plots for the test, such as QQ-plots or showing the quality of the fit.
- --plot-format STRING defines the output format of the diagnose plots. Available options are pdf, png, and ps. Default is png.

Further parameters are available. A complete list of all available options can be found by typing:

```
spladder test -h
```

3.2.2 Description of Output Files

The results of testing mode are collected in a specific sub-directory entitled testing. If labels have been set via --labelA and --labelB, the sub-directory name will change to testing_labelA_vs_labelB. This is useful if multiple tests on different sub-sets of the data are performed. All result files related to the test run will be collected in that sub-directory.

There are currently four different result files generated for each tested event type:

```
I.   test_results_C<confidence>_<event_type>.tsv
II.   test_results_C<confidence>_<event_type>.gene_unique.tsv
III.   test_results_extended_C<confidence>_<event_type>.tsv
IV.   test_setup_C<confidence>_<event_type>.pickle
```

The field names have the following meaning:

- <event_type> is one of exon_skip, intron_retention, alt_3prime, alt_5prime, mutex_exons, mult_exon_skip.
- <confidence> is the chosen confidence level (between 0 and 3).

Files I and II have the same structure. While result I contains the test results for all tested events of that type, result II contains the results made unique per gene, choosing the most significant event of that type for each gene. The content of the files is structured as follows:

1. event_id The id of the tested event.
2. gene_id The ID of the corresponding gene.
3. gene_name The common name of the corresponding gene (if available in the given annotation file).
4. p_val The raw *p*-value resulting from the test.
5. p_val_adj The *p*-value after multiple testing correction.
6. dPSI The delta-PSI value between the two tested groups. This is a measure of the effect size.
7. mean_event_count_A The mean event count for condition A.
8. mean_event_count_B The mean event count for condition B.
9. log2FC_event_count The log 2-fold change of event counts A over B.
10. mean_gene_exp_A The mean gene expression for condition A.
11. mean_gene_exp_B The mean gene expression for condition A.

12. log2FC_gene_exp The log 2-fold change of gene expression counts A over B.

File III provides the same content as file I. In addition, the event counts and gene expression values for each individual sample subjected to testing is provided. This file can be useful for secondary analyses or further testing.

File IV contains a snapshot of the testing parameters and is only used internally by SplAdder.

3.3 The viz Mode

SplAdder's visualization mode can generate graphical overviews of splicing graphs and events and the associated coverage available in the underlying RNA-seq samples.

In general, the plots are organized as individual tracks, which can be stacked to visualize several sources of information jointly. Thereby, the order, number, and repetition of tracks can be defined by the user. This allows for the generation of simple overview plots as well as for more complex multi-track visualizations. If more than one track is present, all tracks share the same joint coordinate system on the *x* axis.

3.3.1 Understanding Key Parameters

The content of the plots is provided using one of several instances of the --track parameter. Specific genomic regions can be selected with the --range parameter. We will first explain the definition of tracks in more detail and will then provide some information on how to define a specific range.

- --track TYPE [TYPE_INFO [TYPE_INFO ...]] is the general structure of the track parameter group, where TYPE is one of the track types splicegraph, transcript, event, coverage, or segments (and TYPE_INFO is specifically defined for each type). The individual track types are described in the following:

  splicegraph shows the structure of the splicing graph for each of the given genes. If no TYPE_INFO is provided, the gene(s) from the --range argument are used.

  transcript shows the structure of all annotated transcripts for each of the given genes. If no TYPE_INFO is provided, the gene(s) from the --range argument are used.

  event shows the structure of the given events, where TYPE_INFO is replaced with a space-separated list of events, each specified by its ID. If no specific event ID is given but only the **event type**, all events of that type for the genes given in --range are shown. If all events should be shown, one can use the special event type any.

  coverage shows the coverage information in the given range for all samples provided in TYPE_INFO. Samples are given by a space-separated list of bam files if each sample is to be plotted individually. If two samples should be shown jointly, the list must be comma-separated so they form a sample group. A sample group can be assigned a label, using the label as a colon-separated prefix to the first file of the group. For instance, --track coverage wildtype:aln1.bam,aln2bam mutant:aln3.bam, aln4.bam will show two groups of coverages, the first labeled wild-type and the second mutation.

segments shows the coverage information in the given range using the average segment quantification used by SplAdder in the splicing graph. The usage is analog to coverage.

- --range TYPE TYPE_INFO [TYPE_INFO ...] is the general structure for setting a plot range. Using this parameter, the user determines exactly which genomic range is to be considered for plotting the data tracks. This information can be provided as coordinates or using the ID information of one or several elements, defined through one of the following range TYPEs: gene, event, coordinate. The usage of --range overrules any range determined based on the elements given via --tracks. A detailed description of the individual range types follows:

  gene allows for providing at least one gene ID to be considered. If multiple genes should be used, they can be provided as a space-separated list.

  event allows for providing at least one event ID to be considered. If multiple events should be used, they can be provided as a space-separated list.

  coordinate allows for specifying a coordinate range to be used. Here, the TYPE_INFO consists of the list of coordinates to be used. As all ranges will be combined into a joint range eventually, there is little use in providing several coordinate ranges, as the union would be taken. Ranges spanning across multiple chromosomes are not allowed.

The order of the tracks is determined by the order they are provided in at the command line. This is true for both the order of keywords within a single --track parameter, as well as for the order of multiple --track parameters.

There is one additional parameter that is mandatory for visualization, which is, as for the other SplAdder modes, the usage of -o for the SplAdder output directory. All plots will be stored in a subdirectory named plots in the SplAdder output directory.

In addition to --track and --range, the special track mode --test has been introduced to visualize events based on the results of an experiment run in test mode. In principle, it works as the other tracks but follows a specific syntax of its elements. There is a default set of two tracks that is generated using this option: an event track showing the event of interest and a segment track showing the quantification of segments used for the test for each of the two groups. The general structure is:

```
--test TESTCASE EVENT_TYPE TOP_K
```

While any of the elements is optional, the order is important and elements can only be omitted at the end but not in the middle.

- TESTCASE describes the differential testing experiment that is to be used as a basis for the plots. Depending on how the sample groups are named in test mode, the output can be found in different subdirectories. So if the two groups were WT and MUT, the output name used by SplAdder would be testing_WT_vs_MUT. However, if no group names were defined, one can just use default.
- EVENT_TYPE specifies the event type whose test results should be considered. For each of the TOP_K events, a separate plot will be generated. One can comma-separate multiple event types or use any, for all event types.
- TOP_K defines the number of top events (following the ranking in the testing result file) that should be plotted. If the value is omitted, the default of 1 is used.

It can happen that the output for testing with SplAdder was written into a user-defined directory and not into the default SplAdder output directory. In this case, the directory can be specified using --testdir. This will also influence where the plots for the test are placed.

As already mentioned for test mode, several of the parameters from build mode are also valid in viz mode, so the correct event database can be used for plotting. Their meaning remains the same:

- --confidence confidence level.
- --merge-strat merge strategy.
- --validate-sg splice graph validation.

Lastly, there are a number of parameters available to influence the output of the visualization mode.

- --outbase NAME provides means to specify a filename stub for the plotting files. Irrespective of this prefix, all output files will be placed in the SplAdder plot directory.
- --mincount INT sets the minimum number of reads supporting an intron for it to be displayed in the coverage plot. Especially spurious split alignments can clutter the output. The default setting for this parameter is 0—all introns are shown.
- --format STRING sets the output format for the plotting files. Valid choices are pdf, png, or ps.
- --log uses a log-scale for all coverage tracks.

Further parameters are available. A complete list of all available options can be found by typing:

```
spladder viz -h
```

3.3.2 Description of Output Files

The user has to choose an output file name for each plot generated. This is specified as the basename of the output file, not containing the output directory or the file ending (which is chosen based on the format). The relevant parameter for this is --outbase. The default format of the plots is pdf, but any format supported by Matplotlib can be used.

Please note that when using the special plotting mode --test and providing a test directory with --testdir (see below), the plots are not placed in the SplAdder output directory but in the given test directory.

4 Example Run on Public RNA-Seq Dataset

4.1 Download Example Data from SRA

To demonstrate usage and utility of the pipeline, we use a publicly available RNA-seq dataset (https://www.ebi.ac.uk/ena/browser/view/PRJNA255099) throughout the tutorial. The dataset is based on the knockdown RNA-binding proteins Tra2α and Tra2β, which have both been implicated in the regulation of alternative splicing, in MDA-MB-231 cells. RNA-seq was performed using three biological replicates of negative control siRNA-treated MDA-MB-231. The corresponding fastq files can be directly downloaded from ENA following the above link. After downloading, you should have the following fastq read-pair files available in your local directory:

```
SRR1513329_1.fastq.gz, SRR1513329_2.fastq.gz
SRR1513330_1.fastq.gz, SRR1513330_2.fastq.gz
SRR1513331_1.fastq.gz, SRR1513331_2.fastq.gz
SRR1513332_1.fastq.gz, SRR1513332_2.fastq.gz
SRR1513333_1.fastq.gz, SRR1513333_2.fastq.gz
SRR1513334_1.fastq.gz, SRR1513334_2.fastq.gz
```

4.2 Read Alignment to the Reference Genome

Before SplAdder can be used on the RNA-seq files, the sequencing reads need to be aligned to the human reference genome. This will require a split-read aligner such as STAR [4] or HISAT2 [5] that is able to identify novel splice junctions during alignment. We recommend using the most recent genome version (hg38/GRCh38) as well as a recent genome annotation, such as GENCODE [6] or Ensemble [7]. We further recommend using the same genome and annotation versions for SplAdder later on that have already been used for alignment.

After genome alignment, the sample bams should all be collected in a directory to be further used for SplAdder:

```
genome_alignment/SRR1513329.bam
genome_alignment/SRR1513330.bam
genome_alignment/SRR1513331.bam
genome_alignment/SRR1513332.bam
genome_alignment/SRR1513333.bam
genome_alignment/SRR1513334.bam
```

For a large number of files, it is usually recommended to collect all absolute file paths for the .bam files in a single list text file, containing one file path per line:

```
genome_alignment/alignments.txt
```

SplAdder expects all input alignment files to be sorted by alignment coordinate and to be indexed. Both can be achieved, e.g., by using the samtools toolbox [8].

4.3 Construct Graphs and Detect Events with Spladder Build Mode

As SplAdder can be run with multiple alignment files as input, there are several ways on how these files can be combined into forming augmented splicing graphs. This behavior is controlled with the setting of the merging strategy using --merge-strat. In our example, we will perform a joint analysis of all samples, using the default strategy merge_graphs. According to this strategy, SplAdder will first generate a splicing graph for each gene for each input sample. These several graphs per gene are then merged across samples, and the merged graph will be used for any further analysis.

Although all steps of the building process can be called with a single command (which we will show in the end), we will progressively go through the process here to better understand the intermediate results.

4.3.1 Graph Construction

At first, each of the input samples is used to generate a sample-specific splicing graph for each gene. We provide short BASH statements describing each of the steps. Assuming that our input data is available as specified in annotation, genome, and alignments, we can write:

```
genome=genome/GRCh38.p12.genome.fa
annotation=annotation/gencode.v32.annotation.gtf
alignment_dir=genome_alignment
outdir=splicing

ls -1 ${alignment_dir}/*.bam > bamfiles.txt
```

```
spladder build --annotation $annotation --bams bamfiles.txt \
--outdir $outdir --sparse-bam --merge-strat single \
--no-extract-ase --confidence 2 \
--filter-consensus lenient --reference $genome
```

The above example takes the annotation and generates a splicing graph per gene per sample. Here, we have used --merge-strat single, so the process stops after creating the individual splicing graphs. We will use merge_graphs in later steps. Further, we decided to not yet extract alternative splicing events (--no-extract-ase), use confidence level 2 for augmenting the splice graphs (--confidence 2) and check for splice site consensus sequences (--filter-consensus lenient). We also used --sparse-bam to generate a sparse summary of the input bam files. This creates slight overhead on the initial compute time and needs some extra storage space (in $alignment_dir), but speeds up later access to the alignment data significantly. We would always recommend using it. After completion of this step, you should find the following output in splicing/spladder/:

```
genes_graph_conf2.SRR1513329.pickle
genes_graph_conf2.SRR1513330.pickle
genes_graph_conf2.SRR1513331.pickle
genes_graph_conf2.SRR1513332.pickle
genes_graph_conf2.SRR1513333.pickle
genes_graph_conf2.SRR1513334.pickle
```

Each of the files contains the splicing graphs of all genes for one of the samples. They are not meant for direct inspection through the user. Also consider that here we use confidence level 2 (*see* **Note 2**) which sets a reasonable restriction on read support for detection splicing events. Lastly, the above command can be trivially parallelized across samples. For this, one can just call the same command providing a single alignment file to the --bams parameter at a time. All results will be written to the output directory independently.

4.3.2 Graph Quantification

In the next step, you will use the single sample graphs to generate a merged splicing graph. Re-using the same variables as defined for the previous call, you can write:

```
spladder build --annotation $annotation --bams bamfiles.txt \
--outdir $outdir --sparse-bam --merge-strat merge_graphs \
--no-extract-ase --confidence 2 \
--filter-consensus lenient --reference $genome --quantify-graph
```

This command line contains only two small changes. First, you updated the merging strategy from single to merge_graphs. Second, you added the --quantify-graph option, which will quantify nodes and edges in the splicing graphs for each of the input samples. As the output directory remains the same as before, all initial steps for generating the splicing graphs will not have to be repeated. That is, SplAdder re-uses any existing results. If everything was successful, three additional output files are now present under splicing/spladder/:

```
genes_graph_conf2.merge_graphs.pickle
genes_graph_conf2.merge_graphs.count.hdf5
genes_graph_conf2.merge_graphs.gene_exp.hdf5
```

The first file contains the merged splicing graphs across all samples (one per gene). The second and third files are in HDF5 format and contain the graph quantifications per sample and the gene expression estimates per gene per sample, respectively. Although they can be accessed directly by the user, this will usually not be necessary.

4.3.3 Alternative Option: Simultaneous Quantification Across Samples

Especially for larger cohorts, it can be challenging to process through all the alignment files for quantification in a sequential manner, as it is the default when all files are provided to SplAdder at the same time. To parallelize quantification, one can use a similar approach as with graph construction, collecting single quantification files per sample using the merged graph. The command is the same as before, just applied per sample, using the additional parameter --qmode single:

```
for bamfile in ${alignment_dir}/*.bam
do
 spladder build --annotation $annotation --bams $bamfile \
 --outdir $outdir --sparse-bam --merge-strat merge_graphs \
 --no-extract-ase --confidence 2 --filter-consensus lenient \
 --reference $genome --quantify-graph --qmode single
done
```

Of course, the SplAdder build command would need to be started in parallel or on different machines to achieve a speedup.

After the individual runs are completed, all results need to be merged into a joint count representation for further processing using an additional run of SplAdder:

```
spladder build --annotation $annotation --bams bamfiles.txt \
--outdir $outdir --sparse-bam --merge-strat merge_graphs \
--no-extract-ase --confidence 2 --filter-consensus lenient \
```

```
--reference $genome --quantify-graph --qmode collect
```

Note, that now the mode for quantification is set as --qmode collect and the set of samples to be collected is determined by all files available via --bams bamfiles.txt. The expected result files are the same as in the previous section for non-simultaneous construction.

The last step consists of extracting different alternative splicing events from the now merged and quantified splicing graphs. For this, it suffices to drop the parameter --no-extract-ase. Hence, the full command consists of:

```
spladder build --annotation $annotation --bams bamfiles.txt \
--outdir $outdir --sparse-bam --merge-strat merge_graphs \
--confidence 2 --filter-consensus lenient \
--reference $genome --quantify-graph
```

When run on an empty output directory, the above command will perform all steps described up to now and create all necessary intermediate results. However, it will process all samples sequentially, and one might decide to parallelize individual steps as outlined above.

After successful completion of the command, you should see the following files in the output directory splicing/:

```
merge_graphs_alt_3prime_C2.confirmed.gff3
merge_graphs_alt_3prime_C2.confirmed.pickle
merge_graphs_alt_3prime_C2.confirmed.txt.gz
merge_graphs_alt_3prime_C2.counts.hdf5
merge_graphs_alt_3prime_C2.pickle
...
```

For space reasons, we only show the output files for alternative 3′ splice site events. The same files should also be present for the other event types, replacing alt_3prime with alt_5prime, exon_skip, intron_retention, mult_exon_skip, and mutex_exons, respectively.

Note, the files in .pickle format contains SplAdder internal information and is not expected to be used by the user. They will likely move into an internal directory in future versions.

4.4 Differential Analysis Using SplAdder Test

For this example, the input samples can be distinguished into two groups, the wild-type and the TRA2A/B knockdown group. The next step in the analysis is to test for differential event usage between samples. At first, you need create two lists that define which samples (absolute path to the .bam file) belong to which experimental group:

```
list_controls.txt
```

```
which contains:
- genome_alignment/SRR1513329.bam
- genome_alignment/SRR1513330.bam
- genome_alignment/SRR1513331.bam

list_tra2KO.txt
which contains:
- genome_alignment/SRR1513332.bam
- genome_alignment/SRR1513333.bam
- genome_alignment/SRR1513334.bam
```

Again, SplAdder is run on the same output directory, to re-use the events generated in the previous analysis steps. Here, we will only focus on testing a single event type (exon_skip), but the commands for the other event types are analogous.

```
spladder test -v -o $outdir --conditionA list_controls.txt \
--conditionB list_tra2KO.txt --labelA control --labelB KO \
--merge-strat merge_graphs --confidence 2 \
--diagnose-plots --plot-format pdf --event-types exon_skip
```

The key parameters for differential testing are the two sample lists provided via --conditionA and --conditionB. Additional labels set with --labelA and --labelB help keeping track of which tests were performed. For the above setup, the output directory would be testing_control_vs_KO/, and it would contain the following files:

```
test_results_C2_exon_skip.gene_unique.tsv
test_results_C2_exon_skip.tsv
test_results_extended_C2_exon_skip.tsv
test_setup_C2_exon_skip.pickle
plots/
 count_distribution.exon_skip.log10.pdf
 count_distribution.exon_skip.pdf
 dispersion_adjusted_exon_skip.pdf
 dispersion_fitted_exon_skip.pdf
 dispersion_raw_exon_skip.pdf
 ma_plot_exon_skip.pdf
 qq_plot_exon_skip.adj.log10.pdf
 qq_plot_exon_skip.adj.pdf
 qq_plot_exon_skip.log10.pdf
 qq_plot_exon_skip.pdf
```

The outputs per event type in .tsv format summarize for each tested event the resulting *p* value (raw and adjusted) as well as the effect size in terms of delta PSI. The result file ending in .gene_unique.tsv summarizes the test result per gene, while the file containing the tag extended provides the quantifications not only aggregated per group but also per sample. This is useful if further analyses should be performed with the events.

4.5 Visualizing Splicing Events Using SplAdder viz

The SplAdder visualization mode can generate plots that are composed of different tracks. The type of tracks to be used and how they ordered from top to bottom of the plot can be defined by the user. Here, we will use a small set of typical plots to demonstrate the use in practice.

First, let us construct a plot that contains three different tracks. At the top, a track showing the complete splice graph of a given gene, below a specific alternative splicing event and in the third row a coverage track, illustrating the use in the two sample groups that were also used for testing earlier.

```
spladder viz --outdir $outdir \
--track splicegraph ENSG00000148677.6 --track event exon_s-
kip_9029 \
--track coverage control:controls.txt KO:list_tra2KO.txt \
--format png --confidence 2 --outbase full_gene
```

The plot is generated in png format (as defined with --format png) and uses the string full_gene as its basename. After this command has been run, the output directory should contain the file:

```
plots/full_gene.png
```

The generated plot is shown in Fig. 3. As the event selected for plotting is one that also has been found to be significantly different between the two groups (in the results of the previous section), one can clearly observe the difference in coverage for the alternative exon in the bottom track.

In the above plot, the --range parameter has not been used, and the plotting range was implicitly determined by the coordinates of the gene. In general, SplAdder will use the most inclusive range across all implicit track ranges but will be overwritten by the --range parameter. To illustrate this, we will generate a plot omitting the splice graph track.

```
spladder viz --outdir $outdir --track event exon_skip_9029 \
--track coverage control:controls.txt KO:list_tra2KO.txt \
--format png --confidence 2 --outbase single_event
```

Again, no range is provided, and only the implicit range of the alternative event track is taken. We have also changed the --outbase parameter, to not overwrite the first plot file. The plot will be generated as plots/single_event.png and is shown in Fig. 4.

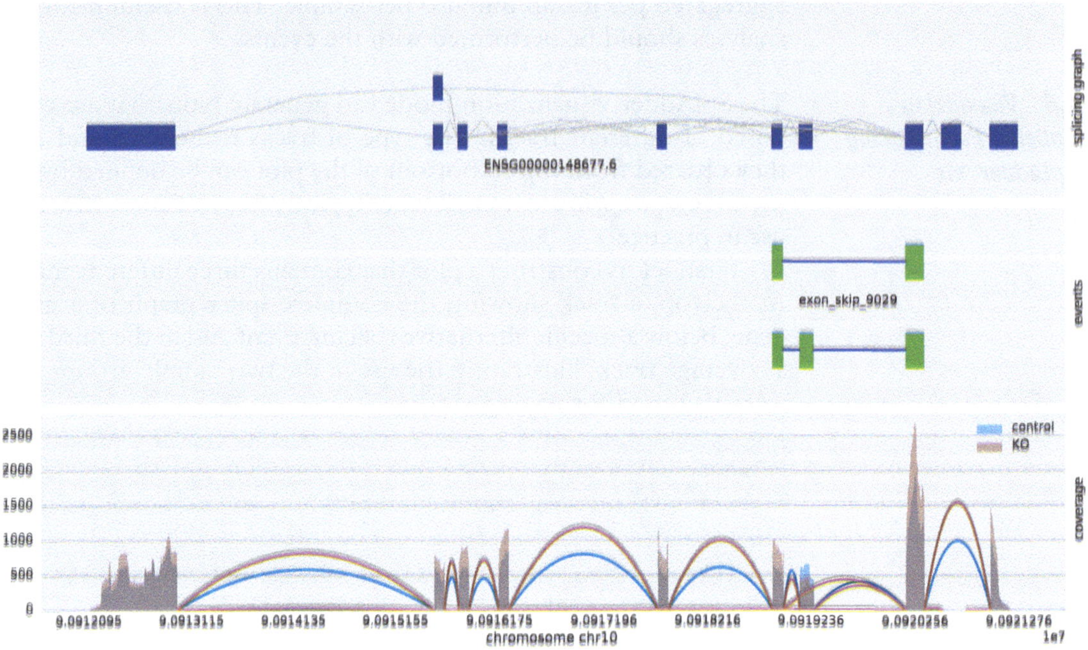

Fig. 3 Result plot written to plots/full_gene.png. The top track contains the splicing graph for gene ENSG00000148677.6, where blue boxes represent exons and gray lines represent introns/junctions. The middle track shows the exon skipping event with ID 9029, where green boxes represent exons and blue lines represent introns. The bottom track shows the average coverage in the two sample groups labeled with the given labels "control" and "KO"

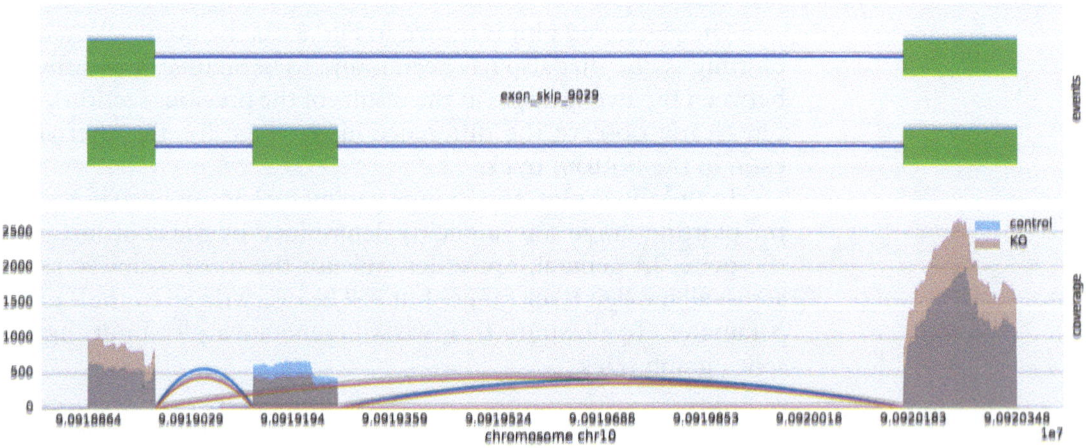

Fig. 4 Result plot written to plots/single_event.png. The top track shows the exon skipping event with ID 9029, where green boxes represent exons and blue lines represent introns. The bottom track shows the average coverage in the two sample groups labeled with the given labels "control" and "KO." The plotting range is now fully determined by the event coordinates

5 Notes

1. SplAdder is a heuristic approach that employs a system of empirical filter rules on RNA-Seq data to extend a splicing graph predefined by a given annotation. Due to this heuristic nature, there is a list of considerations one should take into account when working with the software:
   (a) Although the algorithm is deterministic, it is sensitive to the order of input data. Especially when working with many input samples, integrating their information to generate a single joint splicing graph, the order of input files might influence the outcome. However, in most cases, the generated splicing graphs are robust against changes of the order of the input.
   (b) Depending on the annotation file that is being used, some annotated gene regions might be subjected to filtering. SplAdder will automatically ignore all introns in the input data that could be assigned to more than one gene in the annotation. (This decision is strand-specific, that is, both genes and the intron need to be on the same strand.)
   (c) As SplAdder will sample all possible events from the graph, there can be a certain redundancy in the events (although all events are unique as a whole). For instance, if there are three possible donors paired with an acceptor in a gene on the positive strand, SplAdder will output all three possible pairs of alternative 3 prime splice site events. This "over-sampling" will influence the number of events detected as well as the number of events tested (and corrected for) in the differential analysis.
2. The confidence level of a SplAdder run determines how strongly input alignments are filtered before new nodes and edges are added to the splicing graphs. In general, there are four confidence levels, with confidence increasing from 0 to 3. The default level is 3 and applies the highest level of filtering. To adapt this choice, e.g., to confidence level 2, one can use:

```
spladder build ... --confidence 2 ...
```

The read filter criteria are dependent on the read length. A short overview of the criteria for each of the levels is provided in Table 1. The maximum number of mismatches is used to remove reads that have low quality alignments, the minimum number of alignments is the number of split/spliced alignments necessary to confirm a new intron edge for being taken into the graph, the minimum anchor length is the shortest overlap to an exon segment that a split/spliced alignment

Table 1
Summary of the four different confidence levels supported by SplAdder and the settings of the predefined quality criteria

Level	Criteria	Value
3	Maximum number of mismatches	0
3	Minimum number of alignments	2
3	Minimum anchor length	ceil(0.25 * readlength)
3	Maximum intron length	350,000
...	...	...
2	Maximum number of mismatches	max(1, floor(0.01 * readlength)
2	Minimum number of alignments	2
2	Minimum anchor length	ceil(0.20 * readlength)
2	Maximum intron length	350,000
...	...	...
1	Maximum number of mismatches	max(1, floor(0.02 * readlength)
1	Minimum number of alignments	2
1	Minimum anchor length	ceil(0.15 * readlength)
1	Maximum intron length	350,000
...	...	...
0	Maximum number of mismatches	max(2, floor(0.03 * readlength)
0	Minimum number of alignments	1
0	Minimum anchor length	ceil(0.10 * readlength)
0	Maximum intron length	350,000

needs to have to be counted toward confirming an intron, and the maximum intron length is the upper threshold for new introns to be counted.

References

1. Wang ET, Sandberg R, Luo S et al (2008) Alternative isoform regulation in human tissue transcriptomes. Nature 456:470–476
2. Pan Q, Shai O, Lee LJ et al (2008) Deep surveying of alternative splicing complexity in the human transcriptome by high-throughput sequencing. Nat Genet 40:1413–1415
3. Kahles A, Ong CS, Zhong Y, Rätsch G (2016) SplAdder: identification, quantification and testing of alternative splicing events from RNA-Seq data. Bioinformatics 32:1840–1847
4. Dobin A, Davis CA, Schlesinger F et al (2013) STAR: ultrafast universal RNA-seq aligner. Bioinformatics 29:15–21

5. Kim D, Paggi JM, Park C et al (2019) Graph-based genome alignment and genotyping with HISAT2 and HISAT-genotype. Nat Biotechnol 37:907–915
6. Frankish A, Diekhans M, Ferreira A-M et al (2019) GENCODE reference annotation for the human and mouse genomes. Nucleic Acids Res 47:D766–D773
7. Yates AD, Achuthan P, Akanni W et al (2020) Ensembl 2020. Nucleic Acids Res 48:D682–D688
8. Li H, Handsaker B, Wysoker A, Fennell T (2009) The sequence alignment/map format and SAMtools. Bioinformatics 25:2078–2079

Chapter 12

PipeIT2: Somatic Variant Calling Workflow for Ion Torrent Sequencing Data

Andrea Garofoli, Désirée Schnidrig, and Charlotte K. Y. Ng

Abstract

The detection of somatic genetic variants forms the cornerstone of precision oncology. Molecular diagnostics laboratories require economical and rapid protocols to generate accurate variant calls for clinical purposes. The Ion Torrent sequencing platform is frequently used but lacks robust analysis workflows tailored to the needs of custom sequencing panels. We present PipeIT2, a somatic variant calling workflow for both matched tumor-germline and tumor-only analyses. PipeIT2 is enclosed in a Singularity container to ensure reproducibility and ease of deployment.

Key words Ion Torrent, Somatic mutations, Variant calling, Singularity, Tumor-only variant calling

1 Introduction

One of the premises of precision oncology is the targeting of specific molecular characteristics of tumors, with somatic genetic alterations forming a major class of such molecular targets. The detection of somatic genetic alterations through the use of next-generation sequencing in diagnostic laboratories requires cost-effective, rapid, and reproducible assays. The typical diagnostic assay would target a selected range of genomic regions covering genes and mutations that are therapeutically actionable and/or have prognostic value. The Ion Torrent sequencing platforms are frequently used in the diagnostic setting due to their relatively low cost, rapid turnaround, and modest requirement in terms of genetic material.

Data analysis of Ion Torrent sequencing data, however, presents particular challenges. Variant calling is usually performed using the Ion Reporter platform, which is tightly integrated with the sequencers. In the molecular diagnostic laboratories, variant calls from the Ion Reporter platform typically require extensive manual review and curation, given its high false-positive rate.

Charlotte K. Y. Ng and Salvatore Piscuoglio (eds.), *Variant Calling: Methods and Protocols*,
Methods in Molecular Biology, vol. 2493, https://doi.org/10.1007/978-1-0716-2293-3_12,

Moreover, the default solution lacks optimized analysis workflows for custom targeted sequencing panels, which are increasingly common given the diversity of genomic cancer variants. Alternative solutions would involve creating custom analysis pipelines that sequentially execute a series of tools with many parameters, leading to poor reproducibility and portability.

Here we present PipeIT2, an automated somatic variant calling pipeline for Ion Torrent sequencing data enclosed in a Singularity container image [1]. PipeIT2 has been optimized to detect somatic mutations with high sensitivity and specificity from Ion Torrent sequencing data with or without the sequencing of a matched germline sample. The Singularity container ensures portability and reproducibility, locking down software versions and configurations and giving the same results regardless of the hardware configurations and operating systems [2]. PipeIT2 can be rapidly deployed and executed from end-to-end using a single command.

2 Software

The PipeIT2 Singularity image was built on a CentOS7 Docker image and can be obtained at https://github.com/ckynlab/PipeIT2. Singularity has been chosen over Docker to allow the execution of the analysis in high-performance computing environments [2]. Aside from the Singularity environment (*see* **Note 1**), no additional software installation is required. This chapter describes PipeIT2 v2.0.0.

PipeIT2 is composed of two analysis pipelines, allowing somatic variant calling with and without matched germline samples. In the tumor-germline workflow, the sequencing of the matched germline control is used to filter out germline variants as well as recurrent sequencing artefacts. In the more clinically realistic scenario, where matched germline is not sequenced due to costs and/or availability issues, the tumor-only pipeline leverages unmatched control samples and population genetic databases to define somatic variants. PipeIT2 offers standard parameters previously optimized from benchmark studies but also provides various options for customization.

2.1 Tumor-Germline Workflow

The tumor-germline workflow performs somatic variant calling using the sequencing data of a cancer sample together with its matched germline sample. The workflow consists of the following steps (Fig. 1):

1. Variant calling: Variant calling is performed using the Torrent Variant Caller (TVC) version 5.12 (https://github.com/iontorrent/TS), using a set of previously benchmarked low-stringency parameters [1]. The list of parameters is

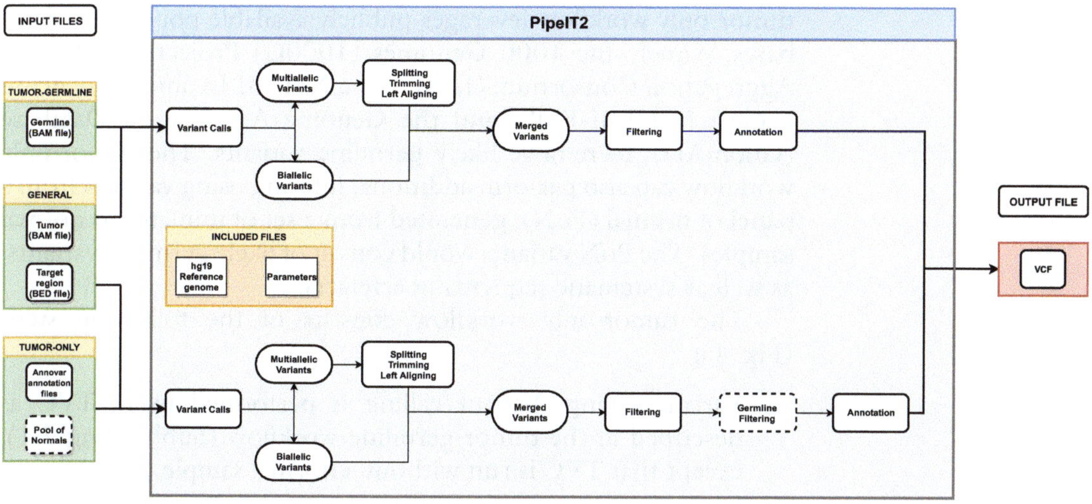

Fig. 1 PipeIT2 workflow

contained in a JSON file present within the container. However, PipeIT2 also allows user-specified TVC parameters provided as a JSON file.

2. Post-processing of multiallelic variants: Multiallelic variants are frequent in TVC output, particularly in homopolymer or repeat regions, which may also include variants with zero variant allele fraction (VAF). To facilitate downstream filtering and annotation, PipeIT2 splits multiallelic variants into monoallelic variants with `norm` in BCFtools (http://samtools.github.io/bcftools) and left-aligns the monoallelic variants with the `LeftAlignAndTrimVariants` function in the Genome Analysis ToolKit (GATK) [3].

3. Variant filtration: The GATK `VariantFiltration` function is used for identifying variants that are outside the target regions or do not meet minimum requirements based on read depth, variant quality score, the number of reads supporting the variants or the VAF ratio between tumor and normal. Given the biological and clinical significance of many hotspot variants, hotspot variants [4] are annotated and whitelisted even when they do not meet the thresholds. We recommend manually reviewing hotspot variants.

4. Annotation (optional): Filtered variants, as well as whitelisted hotspot variants, are annotated by the SnpEff `ann` function [5], using the canonical transcripts (defined as the longest protein coding transcript) from the genome version GRCh37.75.

2.2 Tumor-Only Workflow

The tumor-only workflow performs somatic variant calling using the sequencing data of a cancer sample without a matched germline sample. Instead of the matched germline sample, the PipeIT2 tumor-only workflow leverages publicly available population databases, namely the 1000 Genomes (1000G) Project, the Exome Aggregation Consortium (ExAC), the NHLBI Exome Sequencing Project (NHLBI-ESP), and the Genome Aggregation Database (GnomAD), to remove likely germline variants. The tumor-only workflow can also perform additional filtering using variants from a panel of normal (PoN), generated from a set of unmatched normal samples. The PoN variants would consist of likely germline variants, as well as systematic sequencing artefacts.

The tumor-only workflow consists of the following steps (Fig. 1):

1. Variant calling: Variant calling is performed using TVC as described in the tumor-germline workflow (Subheading 2.1), except that TVC is run without a normal sample.
2. Post-processing of multiallelic variants: Normalization of variants is as described in the tumor-germline workflow (Subheading 2.1).
3. Variant filtration: Similar to the tumor-germline workflow (Subheading 2.1), the GATK `VariantFiltration` function is used to identify variants outside the target regions and variants which do not meet minimum requirements based on read depth, variant quality, the number of reads supporting the variant, and variant allele fraction. Additionally, variants are filtered out if they are in homopolymeric regions. Variants are then annotated for their population frequencies in 1000G, ExAC, NHLBI-ESP, and GnomAD using ANNOVAR [6]. Variants with VAF between 40% and 60% which are present in any of the four databases are considered likely germline variants and filtered out. Common variants (variants with a MAF above the population threshold specified with *-k*) are also removed regardless of the VAF. Hotspot variants [4] are annotated and whitelisted as with the tumor-germline workflow (Subheading 2.1).
4. Panel of normal filtering (optional): Variants derived from a panel of normal samples (PoN) sequenced using the same sequencing assay can be useful for the filtering of likely germline variants and recurrent sequencing and alignment artefacts. Whitelisted hotspot variants are not filtered out by the PoN filtering.
5. Annotation (optional): Filtered variants, as well as whitelisted hotspot variants, are annotated by the SnpEff `ann` function [5], as described in the tumor-germline workflow (Subheading 2.1).

3 Methods

PipeIT2 includes both the tumor-germline and tumor-only variant calling workflows described in Subheading 2 and can be executed as follows:

```
singularity run PipeIT_<version>.img [options]
```

The complete list of parameters can be displayed using the following command:

```
singularity run PipeIT_<version>.img --help
```

The PipeIT2 version can be printed on screen using the following command:

```
singularity run PipeIT_<version>.img --version
```

3.1 Input Files

The basic command needed to perform the tumor-germline workflow is:

```
singularity run PipeIT_<version>.img -t path/to/tumor.bam -n
path/to/normal.bam -e path/to/region.bed
```

The mandatory input files for the matched tumor-germline pipeline are the tumor and the normal BAM files (the `-t` and `-n` parameters, respectively), and the BED file of the targeted regions (the `-e` parameter). Paths to input files must be relative paths (*see* **Note 2**) (Fig. 1).

The basic command needed to perform the tumor-only workflow is:

```
singularity run PipeIT_<version>.img -t path/to/tumor.bam -e
path/to/region.bed -c path/to/annovar/humandb/folder
```

The mandatory input files for the tumor-only pipeline are the tumor BAM file (the `-t` parameter) and the BED file of the targeted regions (the `-e` parameter, Fig. 1). The path to ANNOVAR population databases (the `-c` parameter) is also required for variant filtering. Paths to input files must be relative paths (*see* **Note 2**).

3.2 Parameters

The PipeIT2 pipeline includes the following parameters. Those specific to either the tumor-germline workflow or the tumor-only workflow are indicated.

Mandatory parameters:

- `-t, --tumor <path>`: Path to the BAM file of the tumor sample.
- `-n, --normal <path>` (tumor-germline workflow only): Path to the BAM file of the germline sample (*see* **Note 3**).
- `-c, --humandb <path>` (tumor-only workflow only): Path to the folder where the ANNOVAR population data is stored (*see* **Note 4**).
- `-e, --bed <path>`: Path to the BED file of the targeted genomic regions.

Optional parameters:

- `-o, --output <string>`: The prefix for the output and intermediate files. When -o is specified, the output file will be *PipeIT/results/<output>/<output>.PipeIT.vcf.* If this is not provided, PipeIT2 will use the name of the tumor BAM file, minus the bam file extension.
- `-u, --unmerged <path>`: The path to the unmerged BED file (*see* **Note 5**). If this is not provided, PipeIT2 will create it internally.
- `-j, --TVC_json <path>`: The path to the JSON file that includes the parameters for TVC. If not specified, PipeIT2 will use a set of parameters included within the container image.
- `-l, --black_list <path>`: The path to the blacklist bed file to be used by TVC (*see* **Note 6**).
- `-p, --threads <n>`: The number of parallel threads used by TVC. The default is 4.
- `-a, --run_annotation <true/false>`: Whether to annotate the variants with SnpEff. The default is true.
- `-i, --intermediate_files <true/false>`: Whether PipeIT2 should keep the intermediate files produced during the analysis. The intermediate files may be useful for debugging and for testing parameters. The default is false.

Optional parameters to customize filtering:

- `-q, --quality <n>`: The minimum quality score of the variant. The default is 0.
- `-s, --min_supporting_reads <n>`: The minimal number of reads supporting the variant, used for filtering. The default is 8.
- `-m, --min_tumor_depth <n>`: The minimum read depth for the tumor sample. The default is 10 for the tumor-germline workflow and 20 for the tumor-only workflow.

- `-r, --min_normal_depth <n>` (tumor-germline workflow only): The minimum read depth for the normal sample. The default is 10.
- `-f, --min_TN_VAF_ratio <n>` (tumor-germline workflow only): The minimum VAF ratio between tumor and normal samples. The default is 10.
- `-g, --min_allele_fraction <n>` (tumor-only workflow only): The minimum VAF in the tumor sample. The default is 0.1.
- `-d, --pon <path>` (tumor-only workflow only): This parameter can point either to a VCF file containing variants from a panel of normal samples (PoN) or to a .txt file containing the paths to BAM files of a panel of normal samples, one per line. If the .txt file is provided, PipeIT will generate the PoN VCF file used for the analysis (*see* **Note 7**).
- `-b, --homopolymer_run <n>` (tumor-only workflow only): The maximum homopolymer length accepted. Variants in homopolymer regions exceeding this length will be filtered out. The value 0 will disable this filter. The default is 4.
- `-k, --max_pop_af <n>` (tumor-only workflow only): The maximum minor allele frequency of the variant in population databases. The default is 0.005.
- `-y, --non_coding <true/false>`: Whether PipeIT2 should report non-coding variants in the final output. The default is false.

3.3 Output File

The final output is a VCF file of the variants (*PipeIT/results/<output>/<output>.PipeIT.vcf*). Tables 1 and 2 list the INFO fields of particular relevance for variant filtering for the tumor-germline and tumor-only pipelines, respectively. In both cases, the output VCF file contains the allele counts information (*see* **Note 8**) and, if relevant, SnpEff gene annotations. The output VCF of the tumor-only pipeline also contains information regarding homopolymer length of the flanking region and population frequencies.

4 Notes

1. A Singularity framework is mandatory in order to use PipeIT2. Please refer to the Singularity official documentation (https://sylabs.io/docs/) for more detailed information regarding the installation process.
2. Because of the way Singularity treats relative and absolute paths, paths to input files have to be relative paths. Relative paths will appear to PipeIT2 as files defined outside of the

Table 1
Table of selected INFO fields in the output VCF file of the tumor-germline workflow

Tumor-germline workflow	
Annotation	**Description**
FAO	Flow space alternate allele depth at position
AO	Alternate allele depth at position
FRO	Flow space reference allele depth at position
RO	Reference allele depth at position
AF	Allele frequency

Table 2
Table of selected INFO fields in the output VCF file of the tumor-only workflow

Tumor-only workflow	
Annotation	**Description**
FAO	Flow space alternate allele depth at position
AO	Alternate allele depth at position
FRO	Flow space reference allele depth at position
RO	Reference allele depth at position
AF	Allele frequency
FSAF	Flow alternate allele frequency on the forward strand
FSAR	Flow alternate allele frequency on the reverse strand
SAF	Alternate allele frequency on the forward strand
SAR	Alternate allele frequency on the reverse strand
HRUN	Length of the homopolymer region in the reference genome
ExAC_ALL	Alternate allele frequency in the Exome Aggregation Consortium dataset
esp6500siv2_all	Alternate allele frequency in the NHLBI-ESP project dataset
ANN_1000g2015aug	Alternate allele frequency in the 1000 Genomes Project dataset
gnomAD_genome_ALL	Alternate allele frequency in the gnomAD dataset

container, while fully qualified paths refer to files inside the container. Depending on the high-performance computing environment, it may be necessary to mount additional files and paths. By default, Singularity automatically mounts certain directories such as the user's home directory, */tmp*, */proc*, */sys*, and */dev* inside the container. If the input files are located outside of these mounted directories, the paths of the input files must be specified by using the *-B* flag. For example, if the BAM and BED files are stored in */myHPC/a/different/folder/*:

```
singularity run -B /myHPC/a/different/folder/ PipeIT_<ver-
sion>.img -t path/to/tumor.bam -n path/to/normal.bam -e path/
to/region.bed
```

Further information can be found in the Singularity official documentation (https://sylabs.io/docs/).

3. PipeIT2 determines whether to run the tumor-germline or the tumor-only workflow based on if `-n` is provided.
4. The ANNOVAR population databases required for the tumor-only workflow are large and therefore not included in the PipeIT2 Singularity image. We have provided a script in PipeIT2 to facilitate the automatic downloading of the databases. The following commands will execute the downloads:

```
singularity exec PipeIT_<version>.img annotate_variation.pl
-downdb -webfrom annovar -buildver hg19 esp6500siv2_all hu-
mandb/
singularity exec PipeIT_<version>.img annotate_variation.pl
-downdb -webfrom annovar -buildver hg19 1000g2015aug humandb/
singularity exec PipeIT_<version>.img annotate_variation.pl
-downdb -webfrom annovar -buildver hg19 exac03 humandb/
singularity exec PipeIT_<version>.img annotate_variation.pl
-downdb -webfrom annovar -buildver hg19 gnomad_genome humandb/
```

5. The unmerged bed can be manually created using the `tvcutils validate_bed --unmerged-detail-bed` option from the Torrent Suite (https://github.com/iontorrent/TS) by providing it with the BED file of the target regions.
6. The blacklist typically consists of recurrent artefacts identified through the sequencing of normal samples. Some commercially available gene panels come with a blacklist, typically included in the hotspot BED file, and these variants are tagged with "BSTRAND=F" (on the forward strand), "BSTRAND=R" (on the reverse strand), or "BSTRAND=B" (on both strands).
7. Variants derived from a panel of normal samples (PoN) sequenced using the same sequencing assay can be useful for the filtering of likely germline variants and recurrent sequencing and alignment artefacts. A PoN VCF file can be directly used by PipeIT2 to remove variants present in the file. If a list of BAM files is given, PipeIT2 will call variants as per the variant calling and the post-processing of multiallelic variants steps in the tumor only workflow (Subheading 2.2) and merge the

variants from these files into a single PoN VCF to be used for filtering.

8. TVC provides two types of allele counts: the conventional allele counts and the flow evaluator allele counts (Tables 1 and 2). In general, we compute VAF using the flow evaluator allele counts (FAO and FRO), except when they are not reported by TVC, in which case we compute VAF using the conventional allele counts (AO and RO). FAO and FRO are usually equal to AO and RO, respectively, but may also differ due to complex alleles and/or downsampling. FDP and DP are not used for the calculation of VAF.

References

1. Garofoli A, Paradiso V, Montazeri H et al (2019) PipeIT: a singularity container for molecular diagnostic somatic variant calling on the ion torrent next-generation sequencing platform. J Mol Diagn 21:884–894
2. Kurtzer GM, Sochat V, Bauer MW (2017) Singularity: scientific containers for mobility of compute. PLoS One 12:e0177459
3. McKenna A, Hanna M, Banks E et al (2010) The genome analysis Toolkit: a MapReduce framework for analyzing next-generation DNA sequencing data. Genome Res 20:1297–1303
4. Chang MT, Bhattarai TS, Schram AM et al (2018) Accelerating discovery of functional mutant alleles in cancer. Cancer Discov 8:174–183
5. Cingolani P, Platts A, Wang LL et al (2012) A program for annotating and predicting the effects of single nucleotide polymorphisms, SnpEff. Fly 6:80–92
6. Wang K, Li M, Hakonarson H (2010) ANNOVAR: functional annotation of genetic variants from high-throughput sequencing data. Nucleic Acids Res 38:e164

Chapter 13

Variant Calling from RNA-seq Data Using the GATK Joint Genotyping Workflow

Jean-Simon Brouard and Nathalie Bissonnette

Abstract

The Genome Analysis Toolkit (GATK) developed at the Broad Institute provides state-of-the-art pipelines for germline and somatic variant discovery and genotyping. Unfortunately, the fully validated GATK pipeline for calling variant on RNAseq data is a Per-sample workflow that does not include the recent improvements seen in modern workflows, especially the possibility to perform joint genotyping analysis. Here, we describe how modern GATK commands from distinct workflows can be combined to call variants on RNAseq samples. We provide a detailed tutorial that starts with raw RNAseq reads and ends with filtered variants, of which some were shown to be associated with bovine paratuberculosis.

Key words GATK, RNA-seq, SNP, Variant calling, Joint genotyping, GVCF

1 Introduction

Since the introduction of RNAseq, many researchers have seen the opportunity to use this data not only to quantify gene expression levels but also for discover genomic variations [1]. Whereas trusted bioinformatic protocols exist for detecting sequence variants on a variety of DNAseq samples (germline DNA, whole-exome sequencing, etc.) that come from distinct contexts [2], protocols designed to handle RNAseq data are scarce [1]. Among all resources available, many rely on the Genome Analysis Toolkit (GATK), an industry standard for variant discovery from next-generation sequencing (NGS) data, developed and maintained by the Broad Institute.

At present, the gold standard for variant calling on RNAseq data is the GATK Per-sample workflow although an updated documentation for calling variants in RNAseq data is in the roadmap of the GATK experts [3]. Currently, the researchers interested in performing variant discovery on RNAseq data have the option of using the fully validated Per-sample workflow [4] or using an

Charlotte K. Y. Ng and Salvatore Piscuoglio (eds.), *Variant Calling: Methods and Protocols*,
Methods in Molecular Biology, vol. 2493, https://doi.org/10.1007/978-1-0716-2293-3_13,

advanced work-in-progress workflow designed for cloud computing (e.g., Google Cloud Platform) [5]. The Per-sample approach has several drawbacks, notably that only variable positions are reported and that the HaplotypeCaller engine cannot leverage population-wide information when calling variants [4, 6].

An appealing alternative would be to use guidelines from the GATK best practices relative to RNAseq data and to take advantage of joint genotyping approach which is available for the Germline short variant discovery (SNPs + Indels) workflow [7]. The joint genotyping method has proven to be more sensitive, more flexible, and reduce computational challenges relative to the traditional calling approach [8]. In addition, the latter approach has the advantage to facilitate the incremental discovery of variants that origin from distinct cohorts of samples. Technically, this can be achieved by combining parts of the GATK RNAseq workflow and parts of the GATK joint genotyping workflow (Fig. 1).

In spite that the protocol described here largely uses workflows and concepts developed by the GATK team, it should be pointed out that calling variants on RNAseq data with the joint genotyping workflow has still not been validated by GATK experts. We have shown previously that this approach yields similar if not better

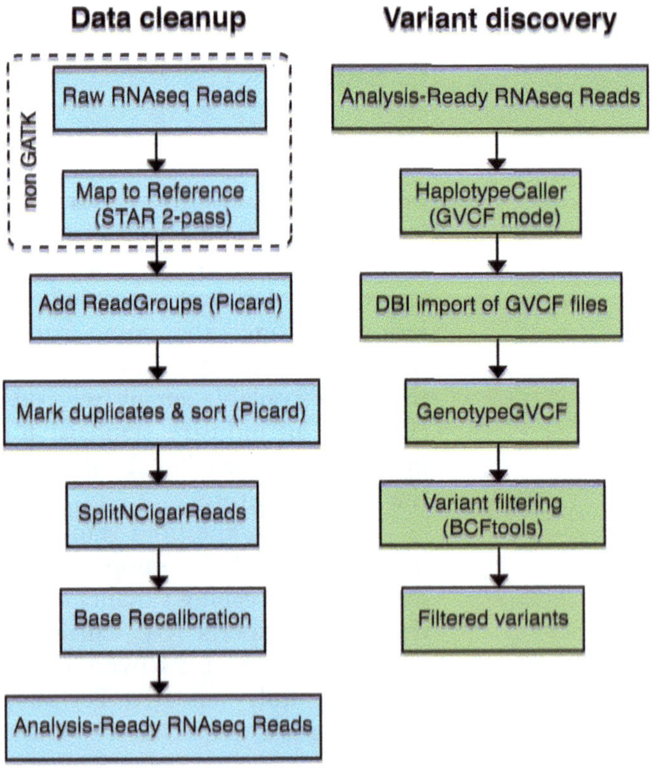

Fig. 1 Workflow for calling variants from RNA-seq data using the GATK joint genotyping approach

results when compared to a Per-sample method [6], but one would be advised to carefully and critically examine the results to detect potential artifacts. Here, we present a fully updated version of this approach in the form of an end-to-end analysis of RNAseq samples extracted from 50 dairy cows selected for studying the mechanism of resistance to bovine paratuberculosis.

As shown in Fig. 1, the workflow presented here involves several steps. In the data cleanup part, the raw RNAseq reads are prepared for RNAseq analysis according to the GATK best practices relative to RNAseq data. In the variant discovery part, all steps belong to the germline joint genotyping workflow up to the variant filtering step.

In the next section, we describe how the diverse programs required to perform the whole analysis can be installed.

2 Materials

2.1 Environment

The vast majority of commands used in this tutorial have been carefully tested and fully executed on a remote Linux server working with the Sun Grid Engine (SGE) workload manager. Obviously, Bash scripts presented here will need to be slightly adapted to be used with your own sequence data. In addition, other small changes will be required if another workload manager is in place on your computer cluster or if you intend to perform the analysis on a local machine.

2.2 Installing Bioinformatic Programs

2.2.1 The GATK Suite

It is suggested to install the GATK suite in a separate conda environment. Assuming that you are familiar with the conda package management system, you could install all GATK programs in an environment called "gatk4" with the following command:

```
conda create -n gatk4 -c bioconda gatk4
```

To generate optional plots in the Base Quality Score Recalibration subsection, additional libraries are required. Use the following commands to install them alongside GATK in the gatk4 environment:

```
conda activate gatk4
conda install -c conda-forge r-base
conda install -c r r-ggplot2
conda install -c r r-gplots
conda install -c bioconda r-gsalib
```

2.2.2 The NCBI SRA Toolkit Programs

You can easily download public sequences from the NCBI Sequence Read Archive (SRA) using the NCBI SRA toolkit. Detailed instructions about this tool can be found at https://trace.ncbi.nlm.nih.gov/Traces/sra/sra.cgi?view=toolkit_doc. However, before using it, do not forget to configure the SRA toolkit program (https://github.com/ncbi/sra-tools/wiki/03.-Quick-Toolkit-Configuration).

Once installed, export the SRA toolkit programs in you PATH:

```
export PATH=$PATH:$PWD/sratoolkit.2.10.9-ubuntu64/bin
```

As usual, you can make this change persistent, by adding the previous line to your .bashrc file.

2.2.3 The STAR Aligner

Although you can retrieve and install the STAR aligner with conda, it can be installed easily by just downloading the latest STAR source from releases:

```
wget https://github.com/alexdobin/STAR/archive/2.7.7a.zip
unzip 2.7.7a.zip
cd STAR-2.7.7a/
```

You can safely use the pre-compiled STAR executables located in the bin/ subdirectory. It is convenient to add the executables to your PATH:

```
export PATH=$PATH:$PWD/bin/Linux_x86_64
```

2.2.4 The Picard Tools

We will use the Picard tools (https://broadinstitute.github.io/picard/) to mark duplicated reads. One can download the pre-build Java program with

```
wget https://github.com/broadinstitute/picard/releases/down-
load/\
2.25.0/picard.jar
```

It is recommended to set up an environment variable to act as a shortcut. Again, to make it persistent in other sessions, simply, add a line to your .bashrc file:

```
export PICARD=$HOME/bioinfo_programs/picard.jar
```

Then, you would be able to call Picard tools with

```
java -jar $PICARD
```

2.2.5 Samtools, BCFtools, and HTSlib

The Samtools website (http://www.htslib.org/) contains plenty of information about the Samtools suite, a collection of widely used bioinformatic programs designed to read, write, edit, index, and view alignment files in the SAM, BAM, and CRAM formats [9]. Less known, but just as powerful, the BCFtools are the best option to manipulate sequence variants stored in the BCF2, VCF, and gVCF formats. The HTSlib is a C library designed to read and write high-throughput sequencing data that is used by both the Samtools and the BCFtools. It notably contains the tabix and bgzip indexing and compression utilities.

Since the Samtools, the BCFtools, and the HTSlib projects are now divided into three separated repositories, the most straightforward way to make use of these three distinct packages is to build them independently.

Use the commands below to install the Samtools (and similarly for the BCFtools and HTSlib):

```
wget  https://github.com/samtools/samtools/releases/download/
\
1.11/samtools-1.11.tar.bz2
bzip2  -d  samtools-1.11.tar.bz2
tar  -xvf  samtools-1.11.tar
cd   samtools-1.11
./configure  --prefix=$HOME/bioinfo_programs/samtools-1.11
```

And you may wish to add the bin directory to your $PATH:

```
export  PATH=$PATH:$HOME/bioinfo_programs/samtools-1.11/bin
```

2.3 Downloading Scripts Used in This Tutorial

A good starting point would be to download the following GitHub repository:

```
git  clone  https://github.com/soda460/RNAseq_GATK_JGW.git
```

It contains all scripts described in the next sections as well as several text files that allow one to reproduce the analyses presented here.

2.4 Downloading Sequences from NCBI SRA

In this tutorial, we choose to work with full-length RNAseq datasets derived from macrophage transcriptomes of cows in response to the infection by *Mycobacterium avium ssp. paratuberculosis* (MAP), the etiological agent of paratuberculosis, or Johne's disease (JD) [10]. A representative subset of 24 samples was selected for which the SRA identifiers can be found in the SRA.list file.

```
SRS2153774
SRS2153779
```

```
SRS2153781
...
SRS2153841
SRS2153845
```

To download these sequences (221 Gb), navigate to the /data directory and download the samples with the prefetch command from the SRAtoolkit:

```
prefetch --option-file SRA.list
```

To extract FASTQ files from .sra archives, the fasterq-dump command is required. For example, the following command will produce SRR5487396_R1.fastq.gz and SRR5487396_R2.fastq.gz:

```
fasterq-dump --split-files SRR5487396/SRR5487396.sra
```

In practice, you will want to extract all downloaded Sequence Read Archives, which are nested in distinct folders. Navigate to the /data folder and use the following qsub command to launch the faster-qdump commands sequentially:

```
qsub -V -S /bin/bash -cwd -j y -pe smp 12 faster-
qdump_sequential.sh
```

where faster-qdump_sequential.sh is a BASH script containing instructions for the SGE workload manager as well as the code to iterate on the downloaded archives and to produce the forward and reverse fastq files:

```
#! /bin/bash
#$ -N 'fasterq-dump'
#$ -o ./faster-qdump_log.txt
for i in 'ls -d SRR*'; do
cd $i; fasterq-dump --split-files $i.sra; cd ..
done
```

However, executing the latter script would take a lot of times. To take advantage of the parallel computing capacity of compute servers, one may consider launching SGE task array jobs (tasks in parallel). This can be done by using the SGE task array capabilities. Briefly, a single script is run multiple times with different values taken by the single environment variable $SGE_TASK_ID. Using a UNIX trick on a list of samples, each instance of the script can be run on a distinct dataset. Since many bioinformatic tasks used in this tutorial are computationally intensive, most of the scripts presented thereafter will be based on this model. For brevity, we did not include most of the required SGE options (lines beginning with #$ within BASH scripts). Therefore, it is important to include them in the qsub command. In this tutorial, all parallel tasks (with 24 samples) should be launched with command such as

```
qsub -V -S /bin/bash -cwd -j y -t 1-24 -pe smp 4
parallel_task.sh
```

whereas standard task should be launched with commands like

```
qsub -V -S /bin/bash -cwd -j y -pe smp 12
unparallel_task.sh
```

Before launching the parallel version of fasterq.dump.sh, create a simple list of the 24 SRR* folders present in the /data folder:

```
ls -d1 SRR5487??? > SRR.list
```

The SRR.list file should contain the 24 SRR identifiers (without the .sra extension) on separate lines. You are now ready to execute the faster-qdump_parallel.sh listing:

```
#!/bin/bash
#$ -N fasterq-dump
#$ -o ./fasterq-dump.$TASK_ID.log
input=$(head -n $SGE_TASK_ID SRR.list | tail -n 1)
cd $input
fasterq-dump --split-files $input.sra
cd ..
```

3 Methods

3.1 Mapping Reads to Reference (STAR 2-pass)

3.1.1 Generating Genome Index Files

The current recommendation of the GATK Per-sample RNAseq workflow is to perform 2-pass mapping with the STAR aligner [11]. STAR genomes are made available for a limited number of genomes by the Gingeras lab. Here, we will use the *Bos taurus* UMD3.1.87 annotation file which is compatible with the UMD 3.1.1/bosTau8 version of the cow genome. The GTF file describes all exons, whereas the .dna.toplevel.fa FASTA file contains the corresponding sequences.

```
wget http://labshare.cshl.edu/shares/gingeraslab/www-data/
dobin/\
STAR/STARgenomes/Old/ENSEMBL/bos_taurus/Bos_taurus.UMD3.1.87.
gtf;
```

```
wget  http://labshare.cshl.edu/shares/gingeraslab/www-data/
dobin/\
STAR/STARgenomes/Old/ENSEMBL/bos_taurus/\
Bos_taurus.UMD3.1.dna.toplevel.fa;
```

To prepare genome index files for STAR, use the genomeGenerate built-in STAR command. Note that you have to create a directory where STAR could build the index:

```
mkdir Bos_taurus.UMD3.1.87_index
```

Then, build the index with the star_genome_index_builder.sh:

```
#! /bin/bash
#$ -N 'STAR_genome_builder'
#$ -o ./star_genome_builder_log.txt
STAR --runThreadN 12 \
--runMode genomeGenerate \
--genomeDir Bos_taurus.UMD3.1.87_index \
--genomeFastaFiles ./Bos_taurus.UMD3.1.dna.toplevel.fa \
--sjdbGTFfile ./Bos_taurus.UMD3.1.87.gtf \
--sjdbOverhang 99
```

3.1.2 Aligning RNAseq Reads (1st Mapping Pass)

At this step, we align our raw FASTQ files with the STAR aligner:

```
#!/bin/bash
#$ -N 'STAR_aligner'
#$ -o STAR_aligner_1st_pass.$TASK_ID.log
input=$(head -n $SGE_TASK_ID SRR.list | tail -n 1)
mkdir -p ../analysis/STAR/output/$input
STAR --genomeDir ./Bos_taurus.UMD3.1.87_index \
--runThreadN 12 \
--readFilesIn  ./$input/$input"_1.fastq"  ./$input/$in-
put"_2.fastq" \
--outFileNamePrefix ../analysis/STAR/output/$input \
--outSAMtype BAM SortedByCoordinate \
--outSAMunmapped Within \
--outSAMattributes Standard
```

3.1.3 Aligning RNAseq Reads (2nd Mapping Pass)

The 2nd mapping pass is identical to the first alignment except that all splice junctions discovered in the first pass are given as input to the program, allowing to detect more reads mapping to novel junctions [11]. Simply specify a list of all splice junction files after the –sjdbFileChrStartEnd argument in the star_aligner_2nd_pass.sh listing:

```
#!/bin/bash
#$ -N 'STAR_aligner_2nd_pass'
#$ -o STAR_aligner_2nd_pass.$TASK_ID.log
input=$(head -n $SGE_TASK_ID SRR.list | tail -n 1)
mkdir -p ../analysis/STAR/output_2nd_pass/$input
STAR --genomeDir ./Bos_taurus.UMD3.1.87_index \
--runThreadN 4 \
--readFilesIn ./$input/$input"_1.fastq" ./$input/$in-
put"_2.fastq" \
--outFileNamePrefix ../analysis/STAR/output_2nd_pass/$input
\
--outSAMtype BAM SortedByCoordinate \
--outSAMunmapped Within \
--outSAMattributes Standard \
--outTmpDir ../analysis/STAR/output_2nd_pass/_STARtmp_$SGE_-
TASK_ID \
--sjdbFileChrStartEnd \
../analysis/STAR/output/SRR5487372SJ.out.tab \
../analysis/STAR/output/SRR5487384SJ.out.tab \
...
../analysis/STAR/output/SRR5487430SJ.out.tab \
../analysis/STAR/output/SRR5487442SJ.out.tab
```

3.1.4 Sorting Alignment Files

Once the RNA sequences are aligned, you will want your BAM files to be indexed and sorted by coordinates. As mentioned earlier, STAR allows the output to be sorted by coordinates. If you have carefully followed this tutorial, there is no need to sort again your BAM files and you can safely ignore the next step and jump to the next subheading. However, if for any reasons, your alignment files are not sorted adequately, the sorting by coordinates with the Samtools is as easy as

```
samtools sort file.bam -o file_sorted.bam
```

In a folder containing unsorted alignments, first prepare a list of bam files to be sorted with

```
ls -1 *.bam > bam.list
```

Then, you could sort your alignment files with the samtools_-sort_bam.sh script:

```
#!/bin/bash
#$ -V
#$ -N samtools_sort
#$ -o samtools_sort.$TASK_ID.log
input=$(head -n $SGE_TASK_ID bam.list | tail -n 1 |
```

```
xargs \
basename -s '.bam')
samtools sort $input.bam -o $input"_sorted.bam"
```

Note that this listing could be used as a template to perform a variety of Samtools subcommands that involve renaming output files.

3.1.5 Indexing Alignment Files

BAM files, the binary analog of Sequence Alignment Mapping (SAM) files, are very efficient bioinformatic files designed to store high-throughput alignment data. Typically, they contain the result of the alignment of millions of sequencing reads against a reference genome and greatly benefit from being indexed by genomic positions for faster random access to the aligned reads at a specific locus [12]. To index all BAM files present in the STAR output_2nd_pass folder, first prepare a list of files to be indexed with

```
ls -1 *.bam > bam.list
```

Then, you can index all alignments in parallel with the following samtools_batch_index.sh script:

```
#!/bin/bash
#$ -N samtools_index
#$ -o samtools_index.$TASK_ID.log
input=$(head -n $SGE_TASK_ID bam.list | tail -n 1)
samtools index $input
```

3.2 Adding Read Groups

Contrarily to FASTQ files, SAM files have the capacity to handle a large amount of metadata. A good practice would be to always include information about the reference genome and the samples in alignment files. Information about the processing steps could also be stored. Many programs, such as GATK, automatically add this kind of metadata information when manipulating SAM/BAM files.

Most GATK commands involving BAM files require that at least some read group fields have been defined to run without errors (*see* **Note 1**). Here, we will set the flowcells, sequencing lanes and sample barcodes in the PU (Platform Unit) tag. We will also set the PL (platform/technology used) and LB (DNA preparation library identifier) tags. Note that the ID (read group identifier) tag is overridden by the PU tag for base recalibration when the latter is defined.

Before running the main script, we will extract the first and fifth columns from our metadata file and place them in separate files, namely RGLB.txt and RGPU.txt, which will be used in the next listing to populate the LB and PU read group fields:

```
cut -f 1 -d ',' ../../metadata/metadata.csv | tail -n
```

```
+2 > RGLB.txt
cut -f 5 -d ',' ../../metadata/metadata.csv | tail -n
+2 > RGPU.txt
```

where /metadata/metadata.csv looks like

```
sample,cowID,SRAID,Run,RGPU
A_CTL24,cowA,SRS2153774,SRR5487372,C5NL3ACXX.1.CAGATC
A_MAP24,cowA,SRS2153779,SRR5487376,C5NL3ACXX.1.TGACCA
B_CTL24,cowB,SRS2153781,SRR5487378,C5NL3ACXX.3.TGACCA
B_MAP24,cowB,SRS2153786,SRR5487382,C5NL3ACXX.3.GTGAAA
C_CTL24,cowC,SRS2153787,SRR5487384,C5K8FACXX.3.AGTCAA
C_MAP24,cowC,SRS2153791,SRR5487388,C5K8FACXX.3.GTCCGC
D_CTL24,cowD,SRS2153793,SRR5487390,C5K8FACXX.1.TGACCA
...
K_CTL24,cowK,SRS2153835,SRR5487432,C547FACXX.5.AGTCAA
K_MAP24,cowK,SRS2153839,SRR5487436,C547FACXX.5.GTCCGC
L_CTL24,cowL,SRS2153841,SRR5487438,C547FACXX.7.AGTCAA
L_MAP24,cowL,SRS2153845,SRR5487442,C547FACXX.7.GTCCGC
```

The RGLB.txt file contains the library name for each 24 samples (e.g., C_MAP24, which designates a RNAseq library done on the cowC 24h post-infection with the MAP pathogen). The RGPU.txt stores the Platform Unit tag, which is made of three types of information: the flowcell, the sequencing lane, and the barcode identifiers, with the following format: FLOWCELL_BARCODE. LANE.SAMPLE_BARCODE. Use the following listing to add read groups to our alignment files:

```
#!/bin/bash
#$ -N AddOrReplaceReadGroups
#$ -o logfile.$TASK_ID.log
eval "$(conda shell.bash hook)"
conda activate gatk4
SAMPLES="$HOME/jsb/RNAseq_GATK_JGW/metadata/samples.list"
BAMPATH="$HOME/jsb/RNAseq_GATK_JGW/analysis/STAR/output_2nd_-
pass"
OUTPUT="$HOME/jsb/RNAseq_GATK_JGW/analysis/addReadGroups"
readarray -t RGLB < ./RGLB.txt
readarray -t RGPU < ./RGPU.txt
input=$(head -n $SGE_TASK_ID $SAMPLES | tail -n 1)
java -jar $PICARD AddOrReplaceReadGroups \
I=$BAMPATH/$input"Aligned.sortedByCoord.out.bam" \
O=$OUTPUT/$input".bam" \
RGLB=${RGLB[$SGE_TASK_ID -1]} \
RGPL=ILLUMINA \
RGPU=${RGPU[$SGE_TASK_ID -1]} \
RGID=${RGPU[$SGE_TASK_ID -1]} \
```

```
RGSM=$input
conda deactivate
```

Note that the value given by $SGE_TASK_ID -1 was used to fetch the correct values in RGLB and RGPU arrays because Bash arrays are zero-indexed.

Upon completion, you may want to validate that the read groups have been correctly set (*see* **Note 2**).

3.3 Marking Duplicate Reads

Duplicate reads can arise from PCR duplication artifacts that take place during the library construction or from reading errors that occur during the sequencing process (optical duplicates). Regardless of their origin, these reads need at least to be identified in alignment files. The MarkDuplicate program from the Picard tools has many other options to deal with these issues. For example, the program offers the possibility to completely remove the duplicate reads and to make the distinction between the optical and PCR duplicates. In the MarkDuplicates.sh listing, we simply identify the duplicate reads in our BAM files before proceeding to the next step:

```
#!/bin/bash
#$ -N MarkDuplicates
#$ -o logfile.$TASK_ID.log
SAMPLES="$HOME/jsb/RNAseq_GATK_JGW/metadata/samples.list"
BAMPATH="$HOME/jsb/RNAseq_GATK_JGW/analysis/addReadGroups"
OUTPUT="$HOME/jsb/RNAseq_GATK_JGW/analysis/MarkDuplicates"
input=$(head -n $SGE_TASK_ID $SAMPLES | tail -n 1)
java -jar $PICARD MarkDuplicates \
I=$BAMPATH/$input".bam" \
O=$OUTPUT/$input"_marked_duplicates.bam" \
M=$OUTPUT/$input"_marked_dup_metrics.txt"
```

3.4 Splitting RNAseq Reads

With the alignment of RNAseq reads, the SplitNCigarReads step (formerly called SplitNTrim step in previous versions of GATK workflows) is highly specific to RNAseq analysis. During this process, reads that span splice junctions (for example, reads that align over distinct exons) are split into intronic and exonic segments and map to the right location in new alignment files.

3.4.1 Preparing Reference Genome Files

Until now, we have worked with STAR reference genome which only contains sequences that are transcribed in RNA. In contrast, the GATK SplitNCigarReads command requires a complete reference sequence as input. At this point, one should be advised to download the same version of the reference genome that was used to prepare the STAR genome. In addition, entries in the reference genome need to be identical to the corresponding ones in the STAR genome (*see* **Note 3**). Here, we use the Bovine Genome

Database (https://bovinegenome.elsiklab.missouri.edu/) to get the UMD3.1 assembly.

```
wget http://128.206.12.216/drupal/sites/bovinegenome.org/\
files/data/umd3.1/UMD3.1_chromosomes.fa.gz
gunzip UMD3.1_chromosomes.fa.gz
```

Note that we have renamed the edited assembly file to refGenome.fasta (*see* **Note 3**). The bovine genome database also contains several annotation files for the UMD3.1 assembly. We will use the Ensembl annotation file of protein-coding and non-protein-coding genes when visualizing alignments with genome browsers at the end of this tutorial.

```
wget http://128.206.12.216/drupal/sites/bovinegenome.org/\
files/data/umd3.1/Ensembl79_UMD3.1_genes.gff3.gz
gunzip Ensembl79_UMD3.1_genes.gff3.gz
```

Along with the reference genome, many GATK tools will need a dictionary file (.dict) as well as an index file (.fai) (https://gatk.broadinstitute.org/hc/en-us/articles/360035531652-FASTA-Reference-genome-format). It is mandatory that these files share the same basename as the reference genome. You can prepare the index with

```
samtools faidx refGenome.fasta
```

On the other hand, use the GATK CreateSequenceDictionary tool to create the dictionary file:

```
gatk CreateSequenceDictionary -R refGenome.fasta
```

3.4.2 Running SplitNCigarReads

Having prepared the reference genome, we are ready to launch the SplitNCigarRead step:

```
#!/bin/bash
#$ -N SplitNCigarReads
#$ -o SplitNCigarReads.$TASK_ID.log
eval "$(conda shell.bash hook)"
conda activate gatk4
SAMPLES="$HOME/jsb/RNAseq_GATK_JGW//metadata/samples.list"
BAMPATH="$HOME/jsb/RNAseq_GATK_JGW/analysis/MarkDuplicates"
OUTPUT="$HOME/jsb/RNAseq_GATK_JGW/analysis/SplitNCigarReads"
input=$(head -n $SGE_TASK_ID $SAMPLES | tail -n 1)
gatk SplitNCigarReads \
-R refGenome.fasta \
-I $BAMPATH/$input"_marked_duplicates.batm" \
-O $OUTPUT/$input"_SplitNCigarReads.bam" \
--tmp-dir $OUTPUT/gatk_tmp
```

```
conda deactivate
```

3.5 Performing Base Quality Score Recalibration

Base Quality Score Recalibration (BQSR) is an optional but highly recommended step that is part of DNAseq and RNAseq GATK best practices workflows. This complex procedure, which involves binning and machine learning, consists of the re-evaluation of all base quality scores assigned by sequencing machines which are prone to systematic technical errors. Several features are taken into account, notably the reported quality score, the position within the read, and the sequence context. This procedure leads to more accurate base calls, a situation that improves the accuracy of variant calling overall. More details about how the model is build and how new quality scores are computed can be found at https://gatk.broadinstitute.org/hc/en-us/articles/360035890531-Base-Quality-Score-Recalibration-BQSR-.

3.5.1 Using BaseRecalibrator

In the first key step, a model of covariation is built with two components: the alignment given as input (BAM files) and a set of known variants (VCF files). The BaseRecalibrator will then produce a recalibration file which is used in the second key step of the procedure. To perform this task, one would need a source of reliable variants in the investigated genome. With human data and other common model organisms, it is easy to find such sources of variants, thanks to the existence of variation databases. Here, we will use a single VCF file containing the results of the BovineSNP50 genotyping BeadChip to train the model, but you can feed the model with several VCF files. As a matter of fact, the best would be to use as many true variants as possible since every other mismatch will be considered as an error.

As the first step, index the SNP50 VCF file with

```
gatk IndexFeatureFile -I snp50.vcf.gz
```

Then, run the BaseRecalibrator command with

```
#!/bin/bash
#$ -N BaseRecalibrator
#$ -o BaseQualityRecalibration_$TASK_ID.log
eval "$(conda shell.bash hook)"
conda activate gatk4
SAMPLES="$HOME/jsb/RNAseq_GATK_JGW/metadata/samples.list"
BAMPATH="$HOME/jsb/RNAseq_GATK_JGW/analysis/SplitNCigarReads"
OUTPUT="$HOME/jsb/RNAseq_GATK_JGW/analysis/BaseQualityRecali-
bration"
input=$(head -n $SGE_TASK_ID $SAMPLES | tail -n 1)
gatk BaseRecalibrator \
-R ../SplitNCigarReads/refGenome.fasta \
-I $BAMPATH/$input"_SplitNCigarReads.bam" \
--known-sites ./snp50.vcf.gz \
-O $OUTPUT/$input"_recal_data.table"
```

```
conda deactivate
```

3.5.2 Using ApplyBQSR to Adjust the Scores

In the second step of the BaseQuality recalibration process, the GATK ApplyBQSR command assigns new scores to each base based on the model produced in the first step and produces new BAM files that reflect these changes. Here, the ApplyBQSR_second_step.sh listing:

```
#!/bin/bash
#$ -N ApplyBQSR
#$ -o ApplyBQSR_$TASK_ID.log
eval "$(conda shell.bash hook)"
conda activate gatk4
SAMPLES="$HOME/jsb/RNAseq_GATK_JGW/metadata/samples.list"
BAMPATH="$HOME/jsb/RNAseq_GATK_JGW/analysis/SplitNCigarReads"
OUTPUT="$HOME/jsb/RNAseq_GATK_JGW/analysis/BaseQualityRecali-
bration"
echo $SAMPLES
input=$(head -n $SGE_TASK_ID $SAMPLES | tail -n 1)
gatk ApplyBQSR \
-R ../SplitNCigarReads/refGenome.fasta \
-I $BAMPATH/$input"_SplitNCigarReads.bam" \
--bqsr-recal-file $OUTPUT/$input"_recal_data.table" \
-O $OUTPUT/$input"_recal.bam"
conda deactivate
```

3.5.3 Comparing Pre- and Post-recalibration Metrics

To visualize the effect of the procedure, one can build a second model and generate plots with data before and after recalibration. The GATK command AnalyzeCovariates was created for that purpose as in the following example:

```
gatk AnalyzeCovariates \
-before recal1.table \
-after recal2.table \
-plots AnalyzeCovariates.pdf
```

The base recalibration being the final step in the data cleanup part of the workflow (Fig. 1), we are now ready for discovering variants from our analysis ready RNAseq reads with the joint genotyping approach.

3.6 Joint Genotyping Variant Calling

3.6.1 Calling Variants Per-sample (GVCF Mode)

In this step, the GATK HaplotypeCaller engine identifies candidate variation sites and records them in Genomic VCF (GVCF) files. This is why this step has been called "GVCF workflow." GVCF files act as intermediate between analysis ready reads (BAM files) and the final joint analysis of multiple samples [8]. Such intermediate files facilitate the incremental discovery of variants and should be kept since they can be reused in novel rounds of joint genotyping analysis. Our listing to produce GVCF files for each sample reads as follows:

```
#!/bin/bash
#$ -N GVCF
#$ -o singleSampleCallingGVCF_$TASK_ID.log
eval "$(conda shell.bash hook)"
conda activate gatk4
SAMPLES="$HOME/jsb/RNAseq_GATK_JGW/metadata/samples.list"
BAMPATH="$HOME/jsb/RNAseq_GATK_JGW/analysis/BaseQualityReca-
libration"
OUTPUT="$HOME/jsb/RNAseq_GATK_JGW/analysis/GVCF"
echo $SAMPLES
input=$(head -n $SGE_TASK_ID $SAMPLES | tail -n 1)
gatk --java-options "-Xmx4g" HaplotypeCaller \
  --dont-use-soft-clipped-bases true \
  -R ../SplitNCigarReads/refGenome.fasta \
  -I $BAMPATH/$input"_recal.bam" \
  -O $OUTPUT/$input".g.vcf.gz" \
  -ERC GVCF
conda deactivate
```

3.6.2 Merging of Files

Merging of intermediate files has been revised recently by GATK experts by the addition of the novel GenomicsDBImport command. It resembles the former CombineGVCFs command but promises better performances for parsing of VCF files and computing values stored in the VCF INFO fields (https://gatk.broadinstitute.org/hc/en-us/articles/360036883491-GenomicsDBImport). Interestingly, under the hood, this command makes use of a novel data management system called TileDB which is particularly useful for representing huge sparse 2D arrays such as the genomic data found in VCF files [13].

The command requires map files that list file paths of all files to be included in the resulting database. The -L option specifies the range of genomic intervals on which the command operates and may be used several times. Here, to save time, we will call variants solely on bovine chromosomes 4 and 11:

```
#!/bin/bash
#$ -N GenomicsDBImport
#$ -o GenomicsDBImport.log
eval "$(conda shell.bash hook)"
conda activate gatk4
SAMPLES="$HOME/jsb/RNAseq_GATK_JGW/metadata"
OUTPUT="$HOME/jsb/RNAseq_GATK_JGW/analysis/GenomicsDBImport"
```

```
gatk --java-options "-Xmx4g -Xms4g" \
GenomicsDBImport \
--genomicsdb-workspace-path $OUTPUT/"my_database" \
-L 4 \
-L 11 \
--sample-name-map $SAMPLES/"cohort_1.sample_map" \
--tmp-dir $OUTPUT/tmp \
--reader-threads 5
conda deactivate
```

3.6.3 Joint Genotyping

The final step in this variant calling workflow is performed with the GATK GenotypeGVCFs command (https://gatk.broadinstitute.org/hc/en-us/articles/360037057852-GenotypeGVCFs). During this operation, all samples for which variants have been pre-called with HaplotypeCaller are jointly genotyped. This step could be rerun when new samples are available, allowing the so-called incremental discovery of variants.

The command takes as argument a GenomicsDB workspace created by GenomicsDBImport, which can be seen as a sophisticated database that facilitates the parsing of the genomic data present in GVCF files. Alternatively, a single multi-sample file produced by CombineGVCFs can also be given as input.

```
#!/bin/bash
#$ -N GenotypeGVCF
#$ -o GenotypeGVCF.log
eval "$(conda shell.bash hook)"
conda activate gatk4
gatk --java-options "-Xmx4g" GenotypeGVCFs \
-R ../SplitNCigarReads/refGenome.fasta \
-V gendb://my_database \
-O output.vcf.gz
conda deactivate
```

3.7 Querying VCF Files

As mentioned earlier, BCFtools is optimized by design, to query and manipulate compressed VCF files. Therefore, it is worth the pain to familiarize with these tools and to avoid working with plain VCF files with UNIX tricks (*see* **Note 4**). That said, the BCFtools commands are highly versatile and can be used in several circumstances ranging from performing simple queries to applying complex filters. Here, we will start by getting a list of the 10 samples present in our VCF file:

```
bcftools query -l output.vcf.gz
```

To get useful metrics, like the sequencing depth or the ratio of transitions and transversions, use the built-in stats subcommand:

```
bcftools stats output.vcf.gz > stats.txt
```

To quickly see the numbers of variants that fall in each category, use the following command:

```
bcftools stats output.vcf.gz | grep -E "^SN"
output:
SN 0 number of samples: 10
SN 0 number of records: 147755
SN 0 number of no-ALTs: 0
SN 0 number of SNPs: 128421
SN 0 number of MNPs: 0
SN 0 number of indels: 19534
SN 0 number of others: 0
SN 0 number of multiallelic sites: 1520
SN 0 number of multiallelic SNP sites: 89
```

Again, one should keep in mind that we have imported only chromosomes 4 and 11 in the genomic database, a situation that reduces the number of variants reported.

The BCFtools query subcommand allows to fetch VCF fields and format them to the desired output. For example, here we retrieve chromosome numbers, positions, overall quality scores, overall read depths, and genotypes from variants present in the three selected samples in a 1000 bp interval:

```
bcftools query -r 11:21401262-21402262 -s SRR5487372,
SRR5487376,\
SRR5487378 -f 'chr%CHROM pos:%POS Q:%QUAL DP=%INFO/DP GT=
[ %GT]\n' \
output.vcf.gz
output:
chr11 pos:21401262 Q:304.11 DP=25 GT= 0/1 0/0 1/1
chr11 pos:21401911 Q:152.97 DP=13 GT= 0/0 0/0 ./.
```

3.8 Filtering Variants

Contrary to DNAseq workflows where sophisticated methods for filtering variants are available (e.g., VQSR and CNNScoreVariants), the sole option when working with RNAseq data is to use hard filters [4].

3.8.1 Hard-Filtering with Annotation Fields

We base our filtering criteria on the GATK generic hard-filtering recommendations (https://gatk.broadinstitute.org/hc/en-us/articles/360035890471-Hard-filtering-germline-short-variants) and on a more practical guide to hard-filter a cohort with the GATK

VariantFiltration command (https://gatk.broadinstitute.org/hc/en-us/articles/360035531112-How-to-Filter-variants-either-with-VQSR-or-by-hard-filtering). Since variant filtering is an active area of research, the reader is invited to consult these resources to be aware of all the novelties. The first recommendation is to hard-filter SNPs and indels separately. Therefore, we start by splitting our initial dataset:

```
gatk SelectVariants \
-V output.vcf.gz \
-select-type SNP \
-O snps.vcf.gz


gatk SelectVariants \
-V output.vcf.gz \
-select-type INDEL \
-O indels.vcf.gz
```

We will then use the GATK VariantFiltration command which is designed to hard-filter variants on the basis of annotations present in INFO or FORMAT VCF fields (https://gatk.broadinstitute.org/hc/en-us/articles/360037269391-VariantFiltration). But, before going further, we will have a look at the actual distribution of six kinds of annotation present in our VCF files (Fig. 2). An explanation of how these plots were prepared and a brief description of the different annotations are given in **Note 5**. Overall, the plots displayed in Fig. 2 indicate that we should retain most of our variants when hard-filtering our data with the generic thresholds suggested for six highly informative annotations: QualByDepth, Fisher strand, StrandOddsRatio, RMSMappingQuality, MappingQualityRankSumTest, and ReadPosRankSumTest (Table 1).

In the following listings (for SNPs and indels), we will also remove SNP clusters, which are clusters of at least 3 SNPs that are within a window of 35 bases between them because the alignment of sequences in the vicinity of indels can be problematic and may lead to false positives. Note that the following commands can be run interactively in the gatk4 environment:

```
gatk VariantFiltration \
-V indels.vcf.gz \
--cluster-window-size 35 \
--cluster-size 3 \
-filter "QD < 2.0" -filter-name "QD2" \
-filter "FS > 30.0" -filter-name "FS30" \
-filter "ReadPosRankSum < -20.0" -filter-name "Read-
PosRankSum-20" \
```

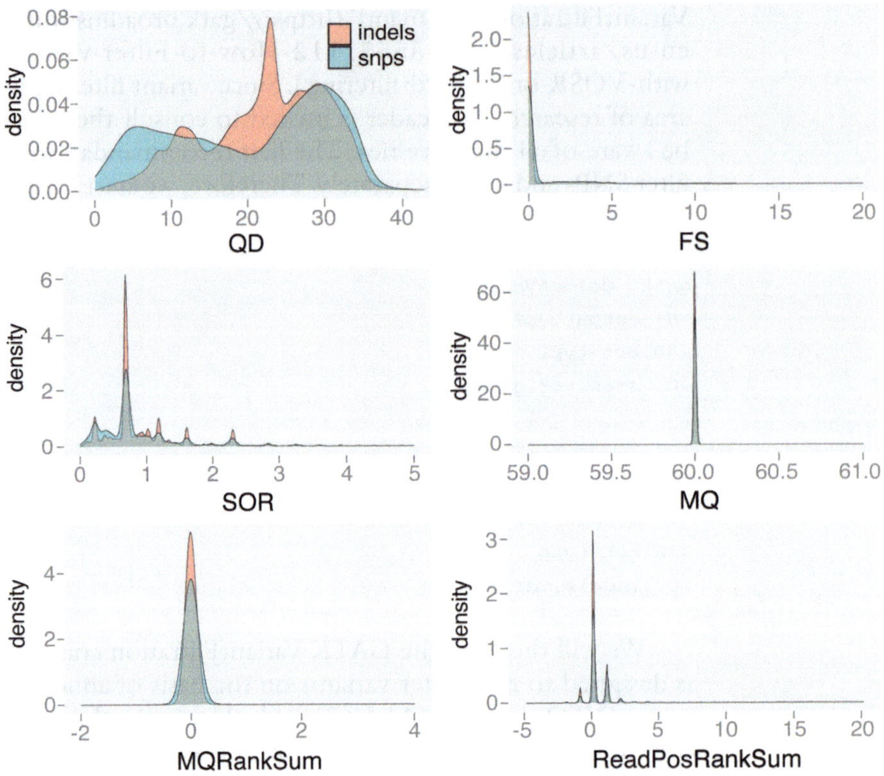

Fig. 2 Distribution of values from six highly informative annotations used for hard-filtering variants

Table 1
GATK recommendations for hard-filtering germline short variants according to six important annotations

Annotation	Symbol	Suggested threshold
QualByDepth	QD	<2.0
Fisher strand	FS	>60.0
StrandOddsRatio	SOR	>3.0
RMSMappingQuality	MQ	<40.0
MappingQualityRankSumTest	MQRankSum	< – 12.5
ReadPosRankSumTest	ReadPosRankSum	< – 8.0

```
-O  filtered.indels.vcf.gz
```

```
gatk  VariantFiltration  \
-V  snps.vcf.gz  \
--cluster-window-size  35  \
```

```
--cluster-size 3 \
-filter "QD < 2.0" -filter-name "QD2" \
-filter "FS > 30.0" -filter-name "FS30" \
-filter "SOR > 3.0" -filter-name "SOR3" \
-filter "MQ < 40.0" -filter-name "MQ40" \
-filter "MQRankSum < -12.5" -filter-name "MQRankSum-12.5"
\
-filter "ReadPosRankSum < -8.0" -filter-name "Read-
PosRankSum-8" \
-O filtered.snps.vcf.gz
```

Notice that the previous commands do not actually remove the variants that do not meet the specified criteria. Instead, depending on whether the variant passes or fails the filter tests, the strings 'PASS', "SnpCluster," "QD2," "FS30," etc. are written in the FILTER field of the output VCF file. A description of the applied filters is also written in the header:

```
bcftools view -h filtered.snps.vcf.gz | grep -i
"##filter"
output:
##FILTER=<ID=PASS,Description="All filters passed">
##FILTER=<ID=FS30,Description="FS > 30.0">
##FILTER=<ID=LowQual,Description="Low quality">
##FILTER=<ID=MQ40,Description="MQ < 40.0">
##FILTER=<ID=MQRankSum-12.5,Description="MQRankSum < -12.5">
##FILTER=<ID=QD2,Description="QD < 2.0">
##FILTER=<ID=ReadPosRankSum-8,Description="ReadPosRankSum <
-8.0">
##FILTER=<ID=SOR3,Description="SOR > 3.0">
##FILTER=<ID=SnpCluster,Description="SNPs found in clus-
ters">
```

In order to actually get rid of these variants, we can use the BCFtools view command with the -apply-filter option:

```
bcftools view --apply-filters .,PASS filtered.snps.vcf.gz
| bgzip -c \
> filtered.snps_2.vcf.gz
```

Note how the BCFtools is fully integrated with other UNIX commands. Here, the filtered VCF file is given to the HTSlib bgzip program that compresses it before the UNIX file redirection operator writes the result to a new file.

To learn how the BCFtools filter command can be used to determine how many variants are removed by a specific filter, see our **Note 6**.

3.8.2 Selecting Variants for Association Studies

As mentioned earlier, RNA samples used in this tutorial have been used for performing a genome-wide association study [10]. This kind of analysis benefits from high-quality genotype calls [14]. To select variants (and genotypes) for such purpose, we will first mask calls (not variants) with poor quality (GQ < 20) or very low read depth (less than 3 reads), a strategy that has proven to be useful with variants derived from Genotyping-By-Sequencing [15]. This can be done with the BCFtools filter command with the -S option. The effect of the following command will be to replace low-confidence genotypes by missing data (./.) in the GT field of the VCF file, while retaining all other information:

```
bcftools filter -e 'FMT/DP<3 | FMT/GQ<20' -S . \
filtered.snps_2.vcf.gz | bgzip -c > filtered.snps_3.vcf.gz
```

We can use a query subcommand to validate that everything has been set correctly:

```
bcftools query -s SRR5487372,SRR5487376,SRR5487378 \
-i 'FMT/DP>3 | FMT/GQ>20' \
-f'%POS[ %SAMPLE:\tGT=%GT DP=%DP GQ=%GQ]\n' filtered.
snps_3.vcf.gz | head

output:
77529 SRR5487372: GT=0/1 DP=4 GQ=37
105638 SRR5487378: GT=./. DP=5 GQ=15
132369 SRR5487376: GT=./. DP=5 GQ=15
133743 SRR5487378: GT=./. DP=4 GQ=12
133958 SRR5487378: GT=./. DP=4 GQ=12
133982 SRR5487378: GT=0/1 DP=3 GQ=25
134806 SRR5487376: GT=./. DP=6 GQ=18
134844 SRR5487376: GT=0|1 DP=9 GQ=63 SRR5487378: GT=./.
DP=6 GQ=18
134845 SRR5487376: GT=0|1 DP=9 GQ=63 SRR5487378: GT=./.
DP=6 GQ=18
138430 SRR5487372: GT=./. DP=5 GQ=15 SRR5487378: GT=./.
DP=5 GQ=15
```

The masking of poor-quality genotypes has the consequence that some sites no longer have called genotypes (-u print sites without a called genotype):

```
bcftools view -H -u filtered.snps_3.vcf.gz | wc -l
56260
```

In a single step, we will remove variants based on their minor allele frequency (MAF) and on the proportion of missing data. Setting MAF cutoffs is a hard task that we will not discuss here, but there are increasing evidences that low-frequency and rare variants contribute to complex traits and should be kept in GWAS studies [16]. In contrast, here we will artificially set MAF to a high value of 0.1 in order to minimize singletons, that is, variants present in a single sample. With 10 biallelic samples (20 alleles) and a MAF set to 0.1, variants will need to harbor at least 2 minor alleles to be retained. Therefore, all singletons present in our dataset should have been called homozygous for the alternate allele. In the same step, we will also remove variants for which missing genotypes exceed 20%.

```
bcftools filter -e 'F_MISSING > 0.2 || MAF <= 0.05' \
filtered.snps_3.vcf.gz | bgzip -c > filtered.snps_4.vcf.gz
```

At this point, our latest VCF file still contains few sites with three or more alleles:

```
bcftools view -H -m3 filtered.snps_4.vcf.gz | wc -l
output:
21
```

These sites can be used as a starting point to inspect potentially misaligned regions in genome browsers. To kept only biallelic sites in the final dataset, use the following BCFtools view subcommand:

```
bcftools view -m2 -M2 filtered.snps_4.vcf.gz | \
bgzip -c > filtered.snps_5.vcf.gz
```

We can validate that there are no variants with only one alternate allele (-c stand for minimum allele count, default nref):

```
bcftools view -H -c1 -C1 filtered.snps_5.vcf.gz | wc -l
output:
0
```

Our filtering steps have retained 5712 high-quality SNPs that lie on chromosomes 4 and 11 out of the 128,421 variants originally reported at the end of the joint genotyping step.

3.9 Examining and Visualizing Alignment Files

Visualizing alignments allows one to see the distribution of reads on the complete reference genome. This step is helpful to detect potential misaligned regions or other artifacts. Tablet (https://ics.hutton.ac.uk/tablet/download-tablet/) and the Integrated Genome viewer (https://software.broadinstitute.org/software/igv/) are good choices for such tasks.

3.9.1 Visualizing Reads from Sample SRR5487372 at a Specific Locus

IGV browsers can read annotation files and display their content in genome tracks. If you are interested to add a genome track representing annotated features, the Ensembl UMD 3.1 GFF3 file mentioned in Subheading 3.4 is your best option (*see* **Note 7**).

Here, we will explore a region of the *SOS* gene known to contain the rs43668743 SNP that has been shown to be highly associated with susceptibility to bovine paratuberculosis [10]. In order to work with a smaller BAM file, we could extract the desired region to be investigated with the Samtools view command:

```
samtools view -h -b SRR5487372Aligned.sortedByCoord.out.
bam 11 \
-o SRR5487372_11_STAR.bam
```

Index this small BAM file as previously and load it in the Integrated Genome viewer.

Additionally, you will need to provide a compatible reference genome (refGenome.fasta) as well as its index (refGenome.fasta.fai). You may also want the Ensembl79_UMD3.1_genes_e.gff3 file to be loaded in an annotation track as well as the filtered.snps_3.vcf.gz VCF file.

If you jump to coordinates 11:21,401,262 and zoom in, you should be able to see why the GATK engine has called a heterozygous call at this locus. When zooming out, you should be able to see an organization of reads, exons, splice junctions, and variants similar to the one displayed in Fig. 3. This representation is typical of variant calling on RNAseq data with a high density of reads in

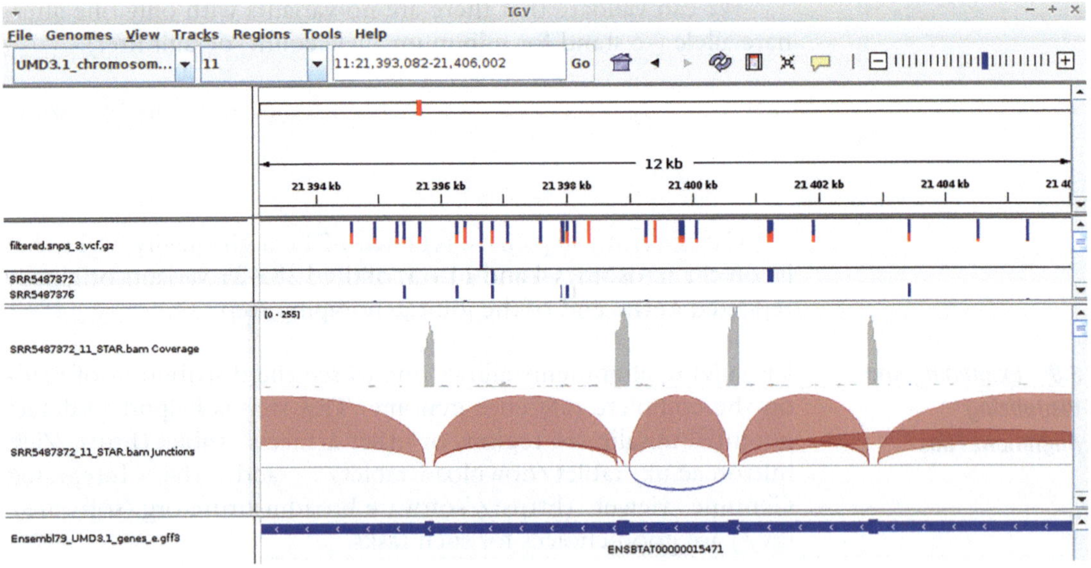

Fig. 3 Screenshot from the Integrated Genome viewer. The region displays 4 exons of the *SOS1* gene from sample SRR5487372

exonic regions and much less reads sparse in intronic sequences where most variants are actually found.

4 Notes

1. This step ensures that relevant information about the sequencing process follows each read in downstream processing. In addition, when this step is done appropriately, it allows to mitigate the consequences of artifacts associated with sequencing features in the duplicate marking and base recalibration steps. For example, when samples are multiplexed, it is important to known which reads origin from which library, which were sequenced on which flowcell, etc. To learn more about the read groups as understand by the GATK team and to learn how to derive this information from read names, one should consult this page: https://gatk.broadinstitute.org/hc/en-us/articles/360035890671-Read-groups.
2. When present, read groups are written in the header of BAM files. Thus, this simple UNIX trick allows one to quickly inspect the read groups associated with a BAM file:

```
samtools view -H sample.bam | grep '@RG'
```

   To see read groups on all our BAM files, use something like

```
for i in 'ls *.bam | xargs basename -s '.bam'';\
do samtools view -H $i.bam | grep '@RG'; done
```

```
@RG ID:C5NL3ACXX.1.CAGATC LB:A_CTL24 PL:ILLUMINA SM:SRR5487372
PU:C5NL3ACXX.1.CAGATC
@RG ID:C5NL3ACXX.1.TGACCA LB:A_MAP24 PL:ILLUMINA SM:SRR5487376
PU:C5NL3ACXX.1.TGACCA
@RG ID:C5NL3ACXX.3.TGACCA LB:B_CTL24 PL:ILLUMINA SM:SRR5487378
PU:C5NL3ACXX.3.TGACCA
@RG ID:C5NL3ACXX.3.GTGAAA LB:B_MAP24 PL:ILLUMINA SM:SRR5487382
PU:C5NL3ACXX.3.GTGAAA
```

3. If you inspect the FASTA headers of the UMD3.1_chromosomes.fa, you will notice that each entry contains many fields (e.g., gnl , UMD3.1 Accession numbers):

```
grep ">" UMD3.1_chromosomes.fa
output:
>gnl|UMD3.1|GK000010.2 Chromosome 10 AC_000167.1
>gnl|UMD3.1|GK000011.2 Chromosome 11 AC_000168.1
```

```
>gnl|UMD3.1|GK000012.2 Chromosome 12 AC_000169.1
...
>gnl|UMD3.1|GK000009.2 Chromosome 9 AC_000166.1
>gnl|UMD3.1|AY526085.1 Chromosome MT NC_006853.1
>gnl|UMD3.1|GK000030.2 Chromosome X AC_000187.1
>gnl|UMD3.1|GJ057137.1 GPS_000341577.1 NW_003097882.1
>gnl|UMD3.1|GJ057138.1 GPS_000341578.1 NW_003097883.1
...
```

In contrast, the chromosome entries in the STAR genome are quite different:

```
cat Bos_taurus.UMD3.1.87_index/chrName.txt | head -n 32
output:
1
10
11
...
9
MT
X
GJ058422.1
```

Therefore, the chromosome identifiers found in the *Bos taurus* UMD3.1 genome need to be changed to match their counterparts in the STAR genome. This could be achieved with UNIX SED:

```
sed -r s'/^>.+Chromosome\s+(\S+)\s+.+/>\1/' UMD3.1_chromo-
somes.fa \
> temp1.fa
grep ">" temp1.fa | head -n 40

# Handle unassigned scaffolds (accessions that begin
with "GJ").
sed -r s'/^>gnl\|UMD3\.1\|(\S+)+\s+.+/>\1/' temp1.fa >
temp2.fa
grep ">" temp2.fa | tail -n +30 | head -n 10
mv temp2.fa refGenome.fasta
rm temp1.fa
```

4. Since BCFtools works well with compressed files, it would be a waste of time to compress or extract VCF files not to mention the risk of corrupting or erasing a VCF file. In addition, because one can easily erase a file by mistake when attempting to redirect the result of a command, it is a good practice to keep a copy of the primary VCF output file in a safe place.

5. The following commands were used to prepare two tab-separated files containing annotation values from six important annotation fields: QD, FS, SOR, MQ, MQRankSum, and ReadPosRankSum:

```
bcftools query snps.vcf.gz -f '%FS\t%SOR\t%MQRankSum\t \
%ReadPosRankSum\t%QD\t%MQ\n' > snps.metrics.txt
bcftools query indels.vcf.gz -f '%FS\t%SOR\t%MQRankSum\t \
%ReadPosRankSum\t%QD\t%MQ\n' > indels.metrics.txt
echo -e "FS\tSOR\tMQRankSum\tReadPosRankSum\tQD\tMQ" >
header
cat header indels.metrics.txt > indels.metrics.tsv
cat header snps.metrics.txt > snps.metrics.tsv
```

We then process these tables with R (see the data_wrangling.R in our GitHub repository) to produce the density plots presented in Fig. 2. The QualByDepth (QD) annotation is an indicator of the variant quality independent of the coverage depth at the site. It provides a better estimate than using directly the primary QUAL or QD fields. Both the FisherStrand (FS) and the StrandOddsRatio (SOR) annotations estimate, with distinct statistical tests, the probability that a strand bias exists at a site. For example, when there is no strand bias, the frequencies of alternate alleles that reside on the forward and reverse strands should be similar and the FS should be close to 0. The RMSMappingQuality (MQ) is a superior indicator (the square root of the average of the squares) than the average mapping quality because it includes the variation in the dataset. Finally, the MappingQualityRankSumTest (MQRankSum) allows to compare the mapping qualities of the reads that support the reference allele relative to those that support the alternate alleles. For an in-depth description of these annotations, consult the GATK site (https://gatk.broadinstitute.org/hc/en-us/articles/360035890471-Hard-filtering-germline-short-variants).

6. For example, we could be interested to know how many variants will be excluded by hard-filtering with FS > 60.0 and "QD < 2.0."

   We begin by counting how many variants will be excluded with FS > 60.0:

```
bcftools filter -i 'FS > 60.0' filtered_2.vcf.gz | \
bcftools view -H | wc -l
2610
```

and how many variants will be excluded with QD < 2.0:

```
bcftools filter -i 'QD < 2.0' filtered_2.vcf.gz | \
```

```
bcftools view -H | wc -l
15089
```

To correctly translate the previous recommendation and combine the two criterions, one would be advised to use the logical "||" operator (OR) (rather than "&&" (AND)) because one will want to throw out any variants with "FS > 60" or "QD < 2.0." Therefore, to know how many variants will be excluded, use

```
bcftools filter -i 'FS > 60 || QD < 2.0' filtered_2.
vcf.gz \
| bcftools view -H | wc -l
17273
```

Note that the Boolean expressions "&&" vs "&" and "||" vs "|" may have different meanings when used with the BCFtools (https://samtools.github.io/bcftools/howtos/filtering.html).

7. Chromosome names in this file differ from those present in the STAR genome and consequently from those present in our alignment files. With the command below, one can see the identifiers present in the GFF3 file as well as the number of features associated with them :

```
cat Ensembl79_UMD3.1_genes.gff3 | cut -f 1 | uniq
-cd
```

output:

```
21287 GK000001.2
18819 GK000030.2
26235 GK000002.2
29793 GK000003.2
...
3 GJ060027.1
3 GJ058256.1
```

We therefore need to replace the pattern used for the chromosome IDs (GK + 0000 + chromosome number) by solely the chromosome number. This could be done easily with SED:

```
sed -r s'/^GK[0]+([0-9]+).2/\1/'g Ensembl79_UMD3.1_genes.
gff3 \
> Ensembl79_UMD3.1_genes_e.gff3
```

As always, the edited file should be inspected to ensure that the desired changes have been done properly.

References

1. Piskol R, Ramaswami G, Li JB (2013) Reliable identification of genomic variants from RNA-seq data. Am J Hum Genet 93:641–651. https://doi.org/10.1016/j.ajhg.2013.08.008
2. Koboldt DC (2020) Best practices for variant calling in clinical sequencing. Genome Med 12:91
3. GATK (2021) Are there best practices for calling variants in RNAseq data? https://gatk.broadinstitute.org/hc/en-us/articles/360035889711-Are-there-Best-Practices-for-calling-variants-in-RNAseq-data-
4. GATK (2021) RNAseq short variant discovery (snps + indels). https://gatk.broadinstitute.org/hc/en-us/articles/360035531192-RNAseq-short-variant-discovery-SNPs-Indels-
5. GATK (2021) gatk4-rnaseq-germline-snps-indels. https://github.com/gatk-workflows/gatk4-rnaseq-germline-snps-indels
6. Brouard JS, Schenkel F, Marete A, Bissonnette N (2019) The GATK joint genotyping workflow is appropriate for calling variants in RNA-seq experiments. J Anim Sci Biotechnol 10:44. https://doi.org/10.1186/s40104-019-0359-0
7. GATK (2021) Germline short variant discovery (snps + indels). https://gatk.broadinstitute.org/hc/en-us/articles/360035535932-Germline-short-variant-discovery-SNPs-Indels-
8. GATK (2021) The logic of joint calling for germline short variants. https://gatk.broadinstitute.org/hc/en-us/articles/360035890431-The-logic-of-joint-calling-for-germline-short-variants
9. Danecek P, Bonfield JK, Liddle J, Marshall J, Ohan V, Pollard MO, Whitwham A, Keane T, McCarthy SA, Davies RM, Li H (2021) Twelve years of SAMtools and BCFtools. GigaScience 10. https://doi.org/10.1093/gigascience/giab008
10. Ariel O, Brouard JS, Marete A, Miglior F, Ibeagha-Awemu E, Bissonnette N (2021) Genome-wide association analysis identified both RNA-seq and DNA variants associated to paratuberculosis in Canadian Holstein cattle 'in vitro' experimentally infected macrophages. BMC Genomics 22:162
11. Dobin A, Davis CA, Schlesinger F, Drenkow J, Zaleski C, Jha S, Batut P, Chaisson M, Gingeras TR (2013) Star: ultrafast universal RNA-seq aligner. Bioinformatics (Oxford, England) 29:15–21. https://doi.org/10.1093/bioinformatics/bts635
12. Buffalo V (2015) Bioinformatics data skills. O'Reilly, Sebastopol
13. Papadopoulos S, Datta K, Madden S, Mattson T (2016) The TileDB array data storage manager 10:349–360. https://doi.org/10.14778/3025111.3025117
14. Nielsen R, Paul JS, Albrechtsen A, Song YS (2011) Genotype and SNP calling from next-generation sequencing data. Nat Rev Genet 12:443–51
15. Brouard JS, Boyle B, Ibeagha-Awemu EM, Bissonnette N (2017) Low-depth genotyping-by-sequencing (GBS) in a bovine population: strategies to maximize the selection of high quality genotypes and the accuracy of imputation. BMC Genet 18(1):1–14. https://doi.org/10.1186/s12863-017-0501-y
16. Tam V, Patel N, Turcotte M, Bossé Y, Paré G, Meyre D (2019) Benefits and limitations of genome-wide association studies. Nat Rev Genet 20(8):467–484. https://doi.org/10.1038/s41576-019-0127-1

Chapter 14

UMI-Varcal: A Low-Frequency Variant Caller for UMI-Tagged Paired-End Sequencing Data

Vincent Sater, Pierre-Julien Viailly, Thierry Lecroq, Élise Prieur-Gaston, Élodie Bohers, Mathieu Viennot, Philippe Ruminy, Hélène Dauchel, Pierre Vera, and Fabrice Jardin

Abstract

The rapid transition from traditional sequencing methods to Next-Generation Sequencing (NGS) has allowed for a faster and more accurate detection of somatic variants (Single-Nucleotide Variant (SNV) and Copy Number Variation (CNV)) in tumor cells. NGS technologies require a succession of steps during which false variants can be silently added at low frequencies. Filtering these artifacts can be a rather difficult task especially when the experiments are designed to look for very low frequency variants. Recently, adding unique molecular barcodes called UMI (Unique Molecular Identifier) to the DNA fragments appears to be a very effective strategy to specifically filter out false variants from the variant calling results (Kukita et al. DNA Res 22(4):269–277, 2015; Newman et al. Nat Biotechnol 34(5):547–555, 2016; Schmitt et al. Proc Natl Acad Sci U S A 109(36):14508–14513). Here, we describe UMI-VarCal (Sater et al. Bioinformatics 36:2718–2724, 2020), which can use the UMI information from UMI-tagged reads to offer a faster and more accurate variant calling analysis.

Key words NGS, UMI, Sequencing, Bioinformatics, Variant calling

1 Introduction

Generally, raw-read-based variant callers do a good job finding variants with relatively high frequencies. However, their accuracy starts decreasing when they are used in experiments designed to detect very low-frequency variants such as in cell-free DNA samples. To work around this problem and to obtain better results, some variant callers require a matched normal sample to act as a control, allowing to filter artifacts. However, matched normal samples are not always available which is why using UMI tags is a more reliable method. Analyzing the UMI tags in the reads would allow to obtain far more accurate results than those obtained with raw--read-based variant callers without the need for a matched sample.

Charlotte K. Y. Ng and Salvatore Piscuoglio (eds.), *Variant Calling: Methods and Protocols*,
Methods in Molecular Biology, vol. 2493, https://doi.org/10.1007/978-1-0716-2293-3_14,

Unfortunately, implementing a good UMI analysis in UMI-based variant caller has led to developing tools that are relatively slower than their raw-read-based counterparts.

UMI-VarCal is both faster and more accurate than actual UMI-based and raw-read-based variant callers. This is achieved by:

1. Not relying on SAMtools [5] to do the pileup step. Instead, UMI-VarCal uses a new homemade and very efficient algorithm that is precisely developed to analyze the UMI information carried out by the reads.
2. Using a Poisson statistical test that is specifically used to study the probabilities of rare events.
3. Implementing a UMI analysis that will treat the UMI tag of a read and compare it to all the other reads with same UMI tag, allowing to create UMI-tagged set of reads.
4. Applying post-analysis filters to remove strand-biased variants and variants that occur in long homopolymer regions.

2 Software

UMI-VarCal is a tool for analyzing BAM/SAM files produced by an Illumina MiSeq sequencer using UMI (Unique Molecular Identifiers). The software starts by building a pileup based on the regions specified in the BED file. This software integrates an extraction tool that can be used to extract UMI tags from the start of the reads. The software also has a calling tool for calling variants and that outputs the results under VCF format. Although SAM files generated by an Illumina MiSeq sequencer were used to develop this tool, it should work on BAM/SAM files generated by any other sequencer as long as they respect the Sequence Alignment/Map Format Specification. UMI-VarCal is freely available under MIT open source license. The source code is located at https://gitlab.com/vincent-sater/umi-varcal with test data.

2.1 Dependencies

UMI-VarCal is available for Python2 and Python3 users. However, the Python2 version is no longer maintained and was not updated since January 22, 2020. We tried to make UMI-VarCal as much independent and easy to use as possible.

Therefore, to run UMI-VarCal, only a few tools or packages are required:

1. a UNIX-based operating system.
2. bwa (http://bio-bwa.sourceforge.net/) is needed for the extraction tool.
3. Python3.3 or later.
4. Python package: msgpack (version 0.5.6).

5. Python package: pyfasta (up to date).
6. Python package: pysam (up to date).
7. Python package: scipy (up to date).
8. Python package: statsmodels (up to date).

2.2 Download and Install UMI-VarCal

Source code of UMI-VarCal is available via GitLab repository: https://gitlab.com/vincent-sater/umi-varcal. The latest source code can be cloned and ready for use with the following commands:

1.

```
git clone --recursive https://gitlab.com/vincent-sater/umi-
varcal
```

2. Go to the `functions/` folder in the downloaded directory.
3. Delete the 3 files: `functions.so`, `functions.c` and `functions.cpython-35 m-x86_64-linux-gnu.so`.
4.

```
python3 setup.py build_ext --inplace.
```

2.3 Usage

Two tools are available in this software:

1. `extract`
2. `call`

To get help on a specific tool, type:

```
python3 umi-varcal.py <tool>
```

or

```
python3 umi-varcal.py <tool> --help
```

or

```
python3 umi-varcal.py <tool> -h
```

To use a specific tool, type:

```
python3 umi-varcal.py <tool> [tool arguments]
```

3 Methods

3.1 The Extraction Tool

The `extraction` tool allows to extract the UMI sequence from the SAM/BAM sequence and add it to the end of the read ID, preceded by an "_" (Fig. 1). The UMI is always extracted from the start of the sequence. Extracting the UMI from within the sequence is not handled by this tool.

3.1.1 Required Arguments

`-f, --fasta=FILE` (case sensitive): The reference genome in FASTA format with its index present in the same directory. It is recommended that you enter the FASTA file full path. For example: `/home/my_dir/my_subdir/hg19.fasta`.

`-i, --input=FILE` (case sensitive): The input file SAM/BAM generated by the sequencer. It is recommended that you enter the input file full path. For example: `/home/my_dir/my_subdir/ind1.sam`.

`-l, --umi_length=INT`: The length of the UMI sequence to extract. The value must be a valid integer. If not specified, the default value (`umi_length=12`) will be used.

3.1.2 Optional Arguments

`-t, --bwa_threads=INT`: The number of cores to be used for the bwa alignment. Using more cores will significantly reduce the alignment time. The value of this parameter defaults to 1.

3.1.3 Example

```
python3 umi-varcal.py extract -i /home/.../test.sam -f /home/
.../hg19.fa -l 12 -t 5
```

3.1.4 Output

The tool will automatically create a directory in the current directory called `EXTRACTED/`. It will contain a folder named after the sample input. In this directory, you can find the extracted R1 FASTQ, the extracted R2 FASTQ, the extracted BAM (with its index), and the extracted SAM file.

3.2 The Variant Calling Tool

The call tool starts by building a pileup based on the regions specified in the BED file (*see* **Note 1**). Given that the UMI tags had already been extracted correctly, this tool will call variants and output the results under VCF format. Although BAM/SAM files generated by an Illumina MiSeq sequencer were used to develop this tool, it should work on BAM/SAM files generated by any other sequencer as long as they respect the Sequence Alignment/Map Format Specification.

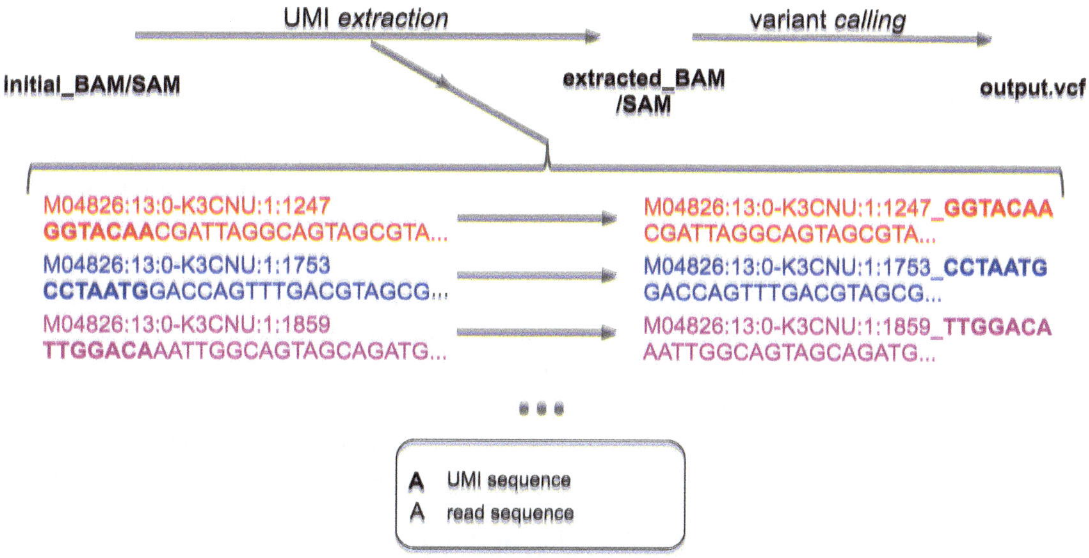

Fig. 1 UMI-VarCal's extraction tool can be used on raw SAM/BAM files in order to remove UMI tags from the start of the read sequence and add it to the read ID. It converts the reads to the correct format expected by the calling tool that will produce the results in a VCF file

3.2.1 Required Arguments

`-f, --fasta=FILE` (case sensitive): The reference genome in FASTA format with its index present in the same directory. It is recommended that you enter the FASTA file full path. For example: `/home/my_dir/my_subdir/hg19.fasta`.

`-b, --bed=FILE` (case sensitive): The panel design file in a BED format. It is recommended that you enter the BED file full path. For example: `/home/my_dir/my_subdir/design.bed`.

`-i, --input=FILE` (case sensitive): The input file SAM/BAM generated by the sequencer and from which the UMI tags have already been extracted from the sequences and added at the end of the read ID, preceded by an "_". This format is the only format accepted by the variant caller. It is recommended that you enter the input file full path. For example: `/home/my_dir/my_subdir/ind1.sam`.

3.2.2 Optional Arguments

`-p, --pileup=FILE` (case sensitive): The PILEUP file for the SAM file previously dumped during an older analysis with the same BED and FASTA files (*see* **Note 4**). This is not the pileup format generated by SAMtools. If specified, the tool will attempt to retrieve the counts of each position from the PILEUP file instead of rebuilding the pileup from scratch, a process that is relatively fast and can save time. It is recommended that you enter the PILEUP file full path. For example: `/home/my_dir/my_subdir/ind1.pileup`.

`-o, --output=DIR` (case sensitive): The output directory in which your files will be generated. It is recommended that you enter the output dir full path. For example: `/home/my_dir/my_subdir/out_dir`.

`-c, --cores=INT`: The number of cores to be used for the analysis. Using more cores will significantly reduce analysis time but RAM usage will be higher (up to 3× higher). The value of this parameter defaults to 1.

`--min_base_quality=INT`: The minimum value for which a base with a certain quality is considered as "good." Only bases with a quality ≥ `min_base_quality` will be considered. The value of this parameter defaults to 10.

`--min_read_quality=INT`: The minimum value for which a read with a certain mean quality is considered as "good." Only reads with a quality ≥ `min_read_quality` will be considered. The value of this parameter defaults to 20.

`--min_mapping_quality=INT`: The minimum value for which a read with a certain mapping quality is considered as "good." Only bases with a mapping quality ≥ `min_mapping_quality` will be considered. The value of this parameter defaults to 20.

`--min_variant_umi=INT`: The minimum count for a variant with a certain unique UMI count to be called. Only variants with a unique UMI count ≥ `min_variant_umi` will be called. The value of this parameter defaults to 5.

`--alpha=INT/FLOAT`: This is the type I error probability known as alpha level. Positions with a p-value/q-value ≥ `alpha` will be considered as noisy and variants in these positions will not be called. The value of this parameter defaults to 0.05 (5%).

`--strand_bias_method=STR`: This parameter lets the user specify the desired method used to calculate variant strand bias (SB, *see* **Note 2**). The value can only be "`default`" or "`torrent_suite`." Choosing any of the methods can have minimal effect on the final variants called and the SB value differs: the torrent suite method generates SB values [0.5;1] but the default method generates SB values [0;+Inf]. Obviously, this parameter defaults to the value "`default`."

`--max_strand_bias=INT/FLOAT`: The minimum strand bias value for a variant with a certain negative and positive coverage to be called (*see* **Note 2**). Only variants with a strand bias ≤ `max_strand_bias` will be called. The value of this parameter defaults to 1.0 for the default method and to 0.743 for the "`torrent_suite`" method.

`--max_hp_length=INT`: The maximum homopolymer length for a variant at a certain position in an homopolymer region to be called (*see* **Note 3**). Only variants with a homopolymer length

≤ `max_hp_length` will be called. The value of this parameter defaults to 7.

`--gvcf=BOOL`: This is an experimental feature that is turned off by default and that, if set to `True`, will generate a gVCF file along with the normal VCF file.

3.2.3 Example

```
python3 umi-varcal.py call -i /home/.../test.sam -b /home/.../
design.bed -f /home/.../hg19.fa -p /home/.../test.pileup -o /
home/.../out -c 3 --min_base_quality 15 --min_read_quality
25 --min_mapping_quality 40 --min_variant_umi 7 --strand_-
bias_method default --max_strand_bias 0.8 --alpha 0.01 --
max_hp_length 5 --gvcf True
```

3.2.4 Output

1. A VCF file (`<INPUT_NAME>.vcf`) containing all the variants that were successfully called (Fig. 2). The VCF file contains eight columns:
   (a) The first column has the name of the chromosome of the variant.
   (b) The second column gives the exact position on the chromosome of the variant.
   (c) The third column is the ID of the variant that follows the HGVS nomenclature.
   (d) The fourth and fifth columns contain the reference and the alternative alleles, respectively, at this position.
   (e) The sixth column has the q-value of the Poisson test performed at this position. The Poisson test is performed to check whether the variation at a specific position is simply due to chance or if it is caused by the presence of a true variant. A value <0.05 means that the variant at this position passed the Poisson test.
   (f) The seventh column represents the quality filter check for each variant.
   (g) The eighth column represents the INFO field and contains specific details for each variant: AF = Allelic Frequency, AO = Allele observations, DP = Depth, TYPE = type of the variation <SNV/INS/DEL > and CONF = confidence <LOW / MEDIUM / HIGH / STRONG / CERTAIN>.
2. A gVCF file (`<INPUT_NAME>.gvcf`) containing all the variants that were successfully called and information for the blocks in which no variant was detected (Fig. 3).

#CHROM	POS	ID	REF	ALT	Q-VALUE	FILTER	INFO
chr1	2488101	chr1:2488101G>A	G	A	0	PASS	AF=0.919387755102;AO=901;DP=980;HP=2;TYPE=SNV;CONF=CERTAIN
chr1	2489200	chr1:2489200C>A	C	A	0	PASS	AF=0.782778864971;AO=800;DP=1022;HP=1;TYPE=SNV;CONF=CERTAIN
chr1	2491260	chr1:2491260A>G	A	G	0	PASS	AF=0.696428571429;AO=702;DP=1008;HP=1;TYPE=SNV;CONF=CERTAIN
chr1	2493201	chr1:2493201T>A	T	A	0	PASS	AF=0.577884615385;AO=601;DP=1040;HP=2;TYPE=SNV;CONF=CERTAIN
chr1	2494300	chr1:2494300G>A	G	A	0	PASS	AF=0.514583333333;AO=494;DP=960;HP=1;TYPE=SNV;CONF=CERTAIN
chr1	23885600	chr1:23885600C>A	C	A	0	PASS	AF=0.409183673469;AO=401;DP=980;HP=3;TYPE=SNV;CONF=STRONG
chr1	23885800	chr1:23885800A>T	A	T	0	PASS	AF=0.30612244898;AO=300;DP=980;HP=2;TYPE=SNV;CONF=CERTAIN
chr1	27022900	chr1:27022900C>A	C	A	0	PASS	AF=0.195694716243;AO=200;DP=1022;HP=2;TYPE=SNV;CONF=CERTAIN
chr1	27023200	chr1:27023200C>A	C	A	0	PASS	AF=0.0941064638783;AO=99;DP=1052;HP=1;TYPE=SNV;CONF=CERTAIN
chr1	27093001	chr1:27093001G>A	G	A	0	PASS	AF=0.0534979423868;AO=52;DP=972;HP=1;TYPE=SNV;CONF=CERTAIN
chr1	27100350	chr1:27100350C>A	C	A	0	PASS	AF=0.0103519668737;AO=10;DP=966;HP=4;TYPE=SNV;CONF=STRONG
chr1	120458000	chr1:120458000C>CTA	C	CTA	0	PASS	AF=0.093984962406;AO=100;DP=1064;HP=1;TYPE=INS;CONF=CERTAIN
chr1	120466600	chr1:120466600TGTC>T	TGTC	T	0	PASS	AF=0.108225108225;AO=100;DP=924;HP=1;TYPE=DEL;CONF=CERTAIN

Fig. 2 The output VCF file format

#CHROM	POS	ID	REF	ALT	Q-VALUE	FILTER	INFO
chr1	2488028	.	A	<NON_REF>	.	.	END=2488100;MEAN_DP=443.0;MEAN_QSCORE=16
chr1	2488101	chr1:2488101G>A	G	A,<NON_REF>	0	PASS	AF=0.9193877551020408;AO=901;DP=980;HP=2;TYPE=SNV;CONF=CERTAIN
chr1	2488102	.	G	<NON_REF>	.	.	END=2488202;MEAN_DP=970.0;MEAN_QSCORE=36
chr1	2489134	.	C	<NON_REF>	.	.	END=2489199;MEAN_DP=867.0;MEAN_QSCORE=31
chr1	2489200	chr1:2489200C>A	C	A,<NON_REF>	0	PASS	AF=0.7827788649706457;AO=800;DP=1022;HP=1;TYPE=SNV;CONF=CERTAIN
chr1	2489201	.	G	<NON_REF>	.	.	END=2489299;MEAN_DP=754.0;MEAN_QSCORE=27
chr1	2489769	.	C	<NON_REF>	.	.	END=2489922;MEAN_DP=915.0;MEAN_QSCORE=32
chr1	2491197	.	T	<NON_REF>	.	.	END=2491259;MEAN_DP=304.0;MEAN_QSCORE=11
chr1	2491260	chr1:2491260A>G	A	G,<NON_REF>	0	PASS	AF=0.6964285714285714;AO=702;DP=1008;HP=1;TYPE=SNV;CONF=CERTAIN
chr1	2491261	.	G	<NON_REF>	.	.	END=2491487;MEAN_DP=780.0;MEAN_QSCORE=29
chr1	2492024	.	G	<NON_REF>	.	.	END=2492195;MEAN_DP=563.0;MEAN_QSCORE=20
chr1	2493106	.	C	<NON_REF>	.	.	END=2493200;MEAN_DP=1015.0;MEAN_QSCORE=36
chr1	2493201	chr1:2493201T>A	T	A,<NON_REF>	0	PASS	AF=0.5778846153846153;AO=601;DP=1040;HP=2;TYPE=SNV;CONF=CERTAIN
chr1	2493202	.	T	<NON_REF>	.	.	END=2493327;MEAN_DP=776.0;MEAN_QSCORE=27
chr1	2494277	.	C	<NON_REF>	.	.	END=2494299;MEAN_DP=83.0;MEAN_QSCORE=3
chr1	2494300	chr1:2494300G>A	G	A,<NON_REF>	0	PASS	AF=0.5145833333333333;AO=494;DP=960;HP=1;TYPE=SNV;CONF=CERTAIN
chr1	2494301	.	C	<NON_REF>	.	.	END=2494400;MEAN_DP=966.0;MEAN_QSCORE=35
chr1	2494550	.	T	<NON_REF>	.	.	END=2494737;MEAN_DP=864.0;MEAN_QSCORE=31
chr1	23885399	.	T	<NON_REF>	.	.	END=23885521;MEAN_DP=1064.0;MEAN_QSCORE=36

Fig. 3 The output gVCF file format

chr	pos	amplicon	reference	idx	A	C	G	T	N	=	+	-	depth	n_UMI_wt	n_UMI_mt	n_unique_UMI_wt	n_unique_UMI_mt	n_singleton_UMI_wt
chr1	2488101	-	G	chr1-2488101	901	0	79	0	0	0	0	0	980	79	901	32	451	1
chr1	2489200	-	C	chr1-2489200	800	222	0	0	0	0	0	0	1022	222	800	61	400	0
chr1	2491260	-	A	chr1-2491260	303	2	702	1	0	0	0	0	1008	303	702	58	352	1
chr1	2493201	-	T	chr1-2493201	601	2	0	437	0	0	0	0	1040	437	601	65	301	0
chr1	2494300	-	G	chr1-2494300	494	1	464	1	0	0	0	0	960	464	494	68	250	0
chr1	23885600	-	C	chr1-23885600	401	578	1	0	0	0	0	0	980	578	401	70	201	0
chr1	23885800	-	A	chr1-23885800	677	1	2	300	0	0	0	0	980	677	300	70	150	0
chr1	27022900	-	C	chr1-27022900	200	821	0	1	0	0	0	0	1022	821	200	73	100	0
chr1	27023200	-	C	chr1-27023200	99	952	0	1	0	0	0	0	1052	952	99	66	51	0
chr1	27093001	-	G	chr1-27093001	52	1	919	0	0	0	0	0	972	919	52	81	27	0
chr1	27100350	-	C	chr1-27100350	10	956	0	0	0	0	0	0	966	956	10	69	5	0
chr1	120458000	-	C	chr1-120458000	962	1	1	0	0	0	100	0	1064	962	100	76	50	0
chr1	120466600	-	TGTC	chr1-120466600	1	0	823	0	0	0	0	100	924	823	100	77	50	0

Fig. 4 The output VARIANTS file format

3. A VARIANTS file (`<INPUT_NAME>.variants`) containing all the variants that were successfully called (Fig. 4). This file

contains the same variants of the VCF file, in addition to detailed metrics about the variants.

4. A PILEUP file (`<INPUT_NAME>.pileup`) corresponding to the entire pileup dumped. This file is produced in the `pileups/` directory. If the pileups directory is not available, it will be created automatically. This file can be used to skip the pileup building and load it instead if the analysis was already done.

4 Notes

1. Performance: UMI-VarCal was developed with performance in mind, and we tried to assure optimal execution time and memory consumption for most scenarios. In our tests, UMI-VarCal was capable of beating even two of the best raw-read-based low-frequency variant callers (SiNVICT [6] and OutLyzer [7]) and one UMI-based variant caller (DeepSNVMiner [8]) in terms of execution time (Fig. 5). The test was performed on a machine running under Ubuntu distribution with a CPU running at 2.20 GHz and 16 GB of total memory. The average depth of the dataset is around 1000× and the UMI tags had a length of 7. Using longer UMI tags, sequencing at higher depths can all significantly affect execution time and memory consumption. Running UMI-VarCal on multiple cores can significantly reduce execution time but will considerably increase the memory consumption.
2. Strand bias filter: A high strand bias (SB) proved to reflect a possible high false-positive rate, especially in Illumina short-read sequencing data [9]. The strand bias filter calculates the strand bias score for each variant and aims to remove all strand biased variants by applying a threshold that can be customized by the user. Guo et al. (2012) [10] demonstrated that the traditional SB calculation is a very effective method to detect false positives by using a threshold of 1.0 which is the parameter's default value in UMI-VarCal.
3. Homopolymer region length: Generally, most sequencers find it harder to call correctly the bases that are found in long homopolymer-containing regions. It was proved by Ivády et al. [11] that the base calling precision drops considerably when the size of the homopolymer region increases (higher than four identical bases). SomaticSniper [12] is a variant calling software that uses a homopolymer length filter to filter out variants that are found in long homopolymer regions because it considers that they have a high likelihood of being artifacts caused by sequencing errors. The default value of this parameter in UMI-VarCal is set to 7 but it can easily be changed by the user.

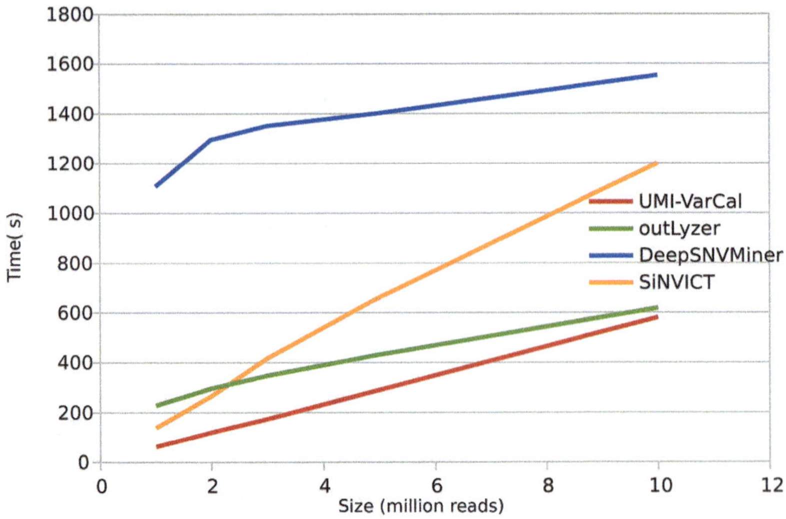

Fig. 5 Performance comparison between UMI-VarCal, OutLyzer, DeepSNVMiner, and SiNVICT on datasets with increasing sizes

4. The PILEUP file: When running UMI-VarCal for the first time on a file, a PILEUP file is automatically generated in the `pileups/` directory. If the user wishes to redo an analysis on the same file but with different parameters (that does not concern the pileup), this file can be given to UMI-VarCal at execution. This will skip the pileup generation step and use directly the pileup generated for the first time. It will allow the user to save much significant time as the pileup generation is undoubtedly the heaviest step in the tool.

References

1. Kukita Y, Matoba R, Uchida J, Hamakawa T, Doki Y, Imamura F, Kato K (2015) High-fidelity target sequencing of individual molecules identified using barcode sequences: de novo detection and absolute quantitation of mutations in plasma cell-free DNA from cancer patients. DNA Res 22(4):269–277
2. Newman AM, Lovejoy AF, Klass DM, Kurtz DM, Chabon JJ, Scherer F, Stehr H, Liu CL, Bratman SV, Say C, Zhou L, Carter JN, West RB, Sledge GW, Shrager JB, Loo BW, Neal JW, Wakelee HA, Diehn M, Alizadeh AA (2016) Integrated digital error suppression for improved detection of circulating tumor DNA. Nat Biotechnol 34(5):547–555
3. Schmitt MW, Kennedy SR, Salk JJ, Fox EJ, Hiatt JB, Loeb LA (2012) Detection of ultra-rare mutations by next-generation sequencing. Proc Natl Acad Sci U S A 109(36): 14508–14513
4. Sater V, Viailly P-J, Lecroq T, Prieur-Gaston E, Bohers E, Viennot M, Ruminy P, Dauchel H, Vera P, Jardin F (2020) UMI-VarCal: a new UMI-based variant caller that efficiently improves low-frequency variant detection in paired-end sequencing NGS libraries. Bioinformatics 36:2718–2724. https://doi.org/10.1093/bioinformatics/btaa053
5. Li H, Handsaker B, Wysoker A, Fennell T, Ruan J, Homer N, Marth G, Abecasis G, Durbin R (2009) The Sequence Alignment/Map format and SAMtools. Bioinformatics 25: 2078–2079. https://www.ncbi.nlm.nih.gov/pmc/articles/PMC2723002/. https://doi.org/10.1093/bioinformatics/btp352
6. Kockan C, Hach F, Sarrafi I, Bell RH, McConeghy B, Beja K, Haegert A, Wyatt AW, Volik SV, Chi KN, Collins CC, Sahinalp SC (2017) SiNVICT: ultra-sensitive detection of

single nucleotide variants and indels in circulating tumour DNA. Bioinformatics 33(1):26–34

7. Muller E, Goardon N, Brault B, Rousselin A, Paimparay G, Legros A, Fouillet R, Bruet O, Tranchant A, Domin F, San C, Quesnelle C, Frebourg T, Ricou A, Krieger S, Vaur D, Castera L (2016) OutLyzer: software for extracting low-allele-frequency tumor mutations from sequencing background noise in clinical practice. Oncotarget 7(48):79485–79493
8. Andrews TD, Jeelall Y, Talaulikar D, Goodnow CC, Field MA (2016) DeepSNVMiner: a sequence analysis tool to detect emergent, rare mutations in subsets of cell populations. PeerJ 4:e2074
9. Guo Y, Li J, Li C-I, Long J, Samuels DC, Shyr Y (2012a) The effect of strand bias in Illumina short-read sequencing data. BMC Genomics 13:666
10. Guo Y, Cai Q, Samuels DC, Ye F, Long J, Li C-I, Winther JF, Tawn EJ, Stovall M, Lähteenmäki P, Malia N, Levy S, Shaffer C, Shyr Y, Shu X-O, Boice JD (2012b) The use of next generation sequencing technology to study the effect of radiation therapy on mitochondrial DNA mutation. Mutat Res 744(2): 154–160
11. Ivády G, Madar L, Dzsudzsák E, Koczok K, Kappelmayer J, Krulisova V, Macek M, Horváth A, Balogh I (2018) Analytical parameters and validation of homopolymer detection in a pyrosequencing-based next generation sequencing system. BMC Genomics 19:158
12. Larson DE, Harris CC, Chen K, Koboldt DC, Abbott TE, Dooling DJ, Ley TJ, Mardis ER, Wilson RK, Ding L (2012) SomaticSniper: identification of somatic point mutations in whole genome sequencing data. Bioinformatics 28(3):311–317

Chapter 15

Alignment-Free Genotyping of Known Variations with MALVA

Giulia Bernardini, Luca Denti, and Marco Previtali

Abstract

The discovery and characterization of sequence variations in human populations are crucial in genetic studies. Standard methods for addressing this problem are computationally expensive and highly time consuming, thus impractical for clinical applications, where time is often an issue. When the task is to genotype variations that have been previously annotated, alignment-free methods come to the aid. Here, we describe **MALVA**, an alignment-free approach for genotyping a set of known variations. **MALVA** is the first mapping-free tool which is able to genotype multi-allelic SNPs and indels, even in high-density genomic regions, and to effectively handle a huge number of variations.

Key words Variant genotyping, Mapping-free, *k*-mers, Next-generation sequencing, Bioinformatics

1 Introduction

Current approaches to call variations from NGS samples can be summarized in three categories. Alignment-based approaches, such as **GATK** [1] and **BCFtools** [2], that first align the reads against a reference genome and infer genotypes from the alignments; assembly-based approaches, such as **DiscoSNP++** [3], that assemble the reads in an assembly graph and then detect variations by analyzing such graph; and variation graph-based approaches such as **GraphTyper** [4] and **vg** [5]. However, all these approaches are highly time consuming, and they can be hardly applied in medical contexts, where it is necessary to be accurate and extremely fast.

When the organism under investigation is well characterized, it is possible to reduce the task of variant calling to the task of variant genotyping, that is, to call the genotype only in those positions where variants have been previously annotated. Variant genotyping is particularly relevant to medical applications, which often require to know the genotype at certain loci that are known to be of medical importance.

Charlotte K. Y. Ng and Salvatore Piscuoglio (eds.), *Variant Calling: Methods and Protocols*,
Methods in Molecular Biology, vol. 2493, https://doi.org/10.1007/978-1-0716-2293-3_15,

The task of genotyping known variants has been recently addressed by alignment-free methods that genotype variations directly from raw sequencing reads, i.e., without mapping them to the reference genome [6–8]. Such approaches are extremely fast and accurate, but they have large memory requirements, which can easily exceed hundreds of GB of RAM. An even more significant limitation of such tools is that they focus on biallelic SNPs, providing very limited support for multi-allelic SNPs and short insertions and deletions of nucleotides (indels).

MALVA is an alignment-free tool that addresses all the limitations of previous alignment-free approaches: it is a lightweight and fast method, which is effectively able to handle close and multi-allelic SNPs as well as indels. At a high-level, the peculiarities of **MALVA** are the following.

1. It represents each allele of a variation using a set of k-mers (length-k substrings), allowing to easily model multi-allelic variations, i.e., those variations for which multiple alleles are known (different individuals may express a different allele for the same variation).
2. It allows for modeling and genotyping indels, again in virtue of using a set of k-mers instead of a single k-mer for representing variations.
3. By using lightweight and efficient data structures, it is extremely fast and requires little memory.
4. It uses k-mer counting to extract information from the input raw NGS sample, thus making it extremely fast if compared to alignment-based approaches.
5. It uses a multinomial distribution (instead of the binomial distribution used by previous methods) to call the genotypes of known variations, allowing for genotyping multi-allelic variations.

2 Materials

MALVA is distributed under the GPL-3.0 open-source license, and the source code is available at https://github.com/algolab/malva. **MALVA** is also available on the Bioconda repository, and therefore, it can be easily installed via the `conda` package management system.

2.1 Software

MALVA was developed in C++, and it has been developed and tested on 64bit Linux system. Indeed, **MALVA** cannot be run on MacOS or Windows system due to dependencies incompatibility.

MALVA depends on the following tools and packages:

1. C++11-compliant compiler. We recommend using GCC 7.4 or newer.
2. **sdsl-lite** library (v2.1.1) [9], a C++11 library implementing succinct data structures. **MALVA** uses this library to implement Bloom filters, which constitute its core data structure.
3. **htslib** (v1.10.2 or newer), a C library for accessing common file formats used for high-throughput sequencing data. This library is required to read and parse the input VCF file containing the set of known variations.
4. **KMC** (v3.1.1 or newer), an efficient k-mer counter. This is used to perform the k-mer counting step of the **MALVA** pipeline.

2.2 Download and Install MALVA

The source code of **MALVA** is freely available at https://github.com/algolab/malva. The latest version of **MALVA** can be retrieved using the following command:

```
git clone --recursive https://github.com/AlgoLab/malva.git
```

For convenience, all the required dependencies are included in the repository. If **sdsl-lite**, **htslib**, and **KMC3** are not already installed, they can be installed by running the following commands:

```
cd sdsl-lite
./install.sh ..
cd ../KMC
make
cd ../htslib
make
cd ..
```

Finally, **MALVA** can be compiled from the root of the local copy of the repository simply by running:

```
make
```

To simplify its installation, **MALVA** is also available on Bioconda. Therefore, once the `conda` package management system is installed and the Bioconda channel is enabled, **MALVA** and all its dependencies can be installed with a single command:

```
conda install malva
```

3 Methods

The alignment-free approach implemented by **MALVA** can be summarized in the following three steps (also illustrated in Fig. 1). Further details can be found in the supplemental material of [10].

1. Each input variation is characterized by two sets of strings of fixed lengths $k < r$, respectively: the shorter k-mers are used to represent the alleles of the variation, whereas the longer r-mers give information about the genomic region surrounding the variation's locus. Intuitively, the longer r-mers are used to tell apart the variants that are actually expressed at some locus from other genome repetitions. Both sets are stored in a Bloom filter.
2. All r-mers are extracted from the input raw read sample along with their frequencies (i.e., the number of times they occur in the sample). These frequencies are then used to assign a weight to each of the k-mers computed in **step 1**.
3. The input variants are genotyped using a probabilistic framework based on the Bayes' theorem that combines allele (a priori) frequencies and k-mers weights.

3.1 Inputs, Outputs, and Parameters

To see required and optional parameters for **MALVA**, run:

```
MALVA -h
```

Indeed, to ease the execution of **MALVA**, we provide a bash script (named *MALVA*) that runs the entire **MALVA** pipeline. The required input files to run **MALVA** successfully are:

1. The reference genome in FASTA format.
2. The set of know variations of the population under investigation in VCF format (*see* **Notes 1** and **2**).
3. A NGS read sample in FASTA/Q format.

Therefore, the command

```
MALVA /path/to/genome /path/to/vcf /path/to/sample
```

runs the **MALVA** pipeline and prints to standard output a new VCF file containing the input variations genotyped with respect to the input sample. For the sake of convenience, we suggest to redirect **MALVA** output to file using output redirection:

```
MALVA /path/to/genome /path/to/vcf /path/to/sample > /path/to/
output/vcf
```

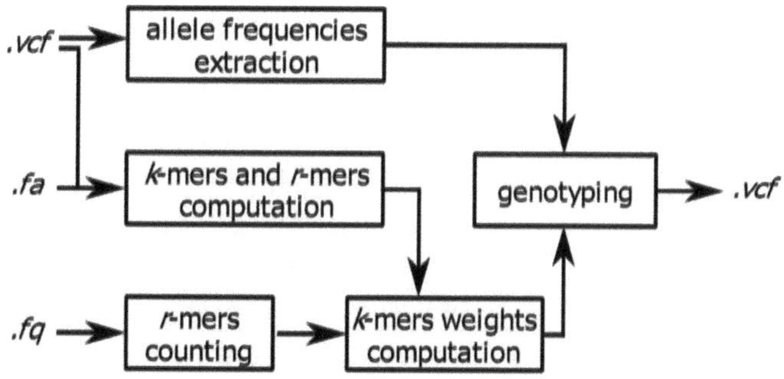

Fig. 1 The approach of MALVA. Input files are the reference genome (FASTA format) along with known variations (VCF format) and a raw reads sample (FASTQ format). First, known alleles are characterized using k-mers and r-mers. Then, r-mer counting from input sample is used to assign a weight to each r-mer (and k-mer). Finally, the allele frequencies and the k-mer weights are used to genotype the input variation, producing a new VCF file

Optional parameters can be tweaked to adapt the execution of **MALVA** to user needs:

- `-k <int>` and `-r <int>`: **MALVA** characterizes each input variation by two sets of strings of different fixed lengths `k` < `r` (*see* **Note 3**). The default values are 35 and 43, respectively. In principle, larger values of `k` and `r` improve the precision of the results while making the method computationally heavier, although using longer `k`-mers will increase the probability of covering erroneous bases in them. The default values are optimized for NGS data.
- `-e <float>` is the expected error rate of the input sample. The default value is 0.001.
- `-s <filepath>` is a file containing the samples of the input VCF to consider. Use this if you want to genotype the set of variations with respect to a subset of the population used to build the VCF. By default, **MALVA** considers all the samples contained in the input VCF (*see* **Note 4**).
- `-f <string>` is the name of the INFO field of the VCF containing the allele frequency. The default value is `AF` but users should set this value in accordance to the name of the field containing the allele frequency in their VCF (*see* **Note 5**).
- `-c <int>` is the maximum coverage allowed per allele. Each variation with alleles that are covered too much (more than `c`) is genotyped as 0/0, as extremely high coverages are a hint of genome repetitions. In these cases, **MALVA** is conservative and prefers not to genotype the variation. The default value is 200. Larger values of `c` lead, in general, to a larger recall, while lowering the method's precision.

- `-b <int>` is the size in GigaByte of the Bloom filter used to store the k-mers (*see* **Note 6**). The default value is 4. Larger values of b lower the Bloom filter's false-positive rate but make the method more memory-intensive.
- `-m <int>` is a parameter of KMC. It indicates that KMC will attempt to use a maximum amount of m GigaBytes during k-mer counting, though the bound could be exceeded in the case of very large datasets. The default value is 4.
- `-u` tells **MALVA** to use uniform a priori probabilities. By default, **MALVA** derives such probabilities from the allele frequencies provided by the input VCF (*see* **Note 5**).
- `-1` tells **MALVA** to run in haploid mode. The default **MALVA** mode is diploid.

3.2 Interpreting the Output File

The output of **MALVA** is a VCF file containing the input variations that have been genotyped and their genotype, computed with respect to the input sample of raw reads. For each of the considered variations (*see* **Note 7**), **MALVA** reports the most likely (unphased) genotype along with its likelihood. For example, consider the following variation (a line from the output of **MALVA**):

```
20 66370 rs6054257 G A,T 100 PASS . GT:GQ 0/2:100
```

There are five relevant fields in this record. It describes a multi-allelic SNP occurring on chromosome 20 (first field) at position 66370 (second field) with reference allele G (fourth field) and two possible alternative alleles, A and T (fifth field). **MALVA** genotyped such SNP as 0/2 with likelihood 100% (last field). Since this is a multi-allelic variation, **MALVA** computed six different genotypes (i.e., 0/0, 0/1, 0/2, 1/1, 1/2, and 2/2) and then produced as output the one with the highest likelihood.

3.3 Running MALVA on Humans (Diploid Organism)

Let us assume we have a human sample we want to genotype. Since **MALVA** relies on an input set of variations to be efficient and accurate, it is necessary to retrieve such a set before running **MALVA**. If the sample under investigation is from a human individual, the VCF provided by the 1000 Genomes Project is a good candidate, especially when the individual under analysis is from one of the populations targeted by the project (*see* **Note 4**). Unfortunately, the 1000 Genomes Project provides a single VCF per chromosome. Therefore, once downloaded, it is necessary to concatenate them into a single VCF, e.g., using `bcftools concat` [11]. Finally, **MALVA** needs an input reference genome: assuming we have downloaded the VCFs produced by mapping reads against GRCh37, that will do (*see* **Note 2**).

Once we have all the required input files, we can genotype the individual under investigation with the following command:

```
MALVA /path/to/GRCh37.fa /path/to/1kgp.vcf /path/to/human-
sample.fq > genotypes.vcf
```

Although MALVA is already fast (*see* **Note 8**), additionally, to speed up its computation while (possibly) lowering the overall genotype accuracy, we can limit **MALVA** to only consider the annotated genotypes of the individuals that belong to the same population as the sample under investigation (*see* **Note 4**). To do so, we need to select such individuals and produce a text file containing their IDs (as appearing in the VCF file), one name per line. In the case of the 1000 Genomes Project, it is possible to easily produce this file using the sample panel provided along with the VCFs. For instance, once downloaded the panel file (`integrated_call_samples_v3.20130502.ALL.panel`) associated with the phase 3 callset provided by the 1000 Genomes Project, it is possible to extract the list of European samples using command-line utilities usually available on any unix-based system:

```
grep "EUR" integrated_call_samples_v3.20130502.ALL.panel | cut
-f 1 > eur.panel
```

We can then run **MALVA**:

```
MALVA -s eur.panel /path/to/GRCh37.fa /path/to/1kgp.vcf /path/
to/human-sample.fq > genotypes.vcf
```

3.4 Running MALVA on Viruses (Haploid Organism)

As stated above, one of the most important prerequisites of **MALVA** is the creation or retrieval of the input set of variations to be genotyped. However, when the sample under investigation is not from a well-characterized species, a VCF of known variations may not be always readily available. This is the case, for example, of viruses. Even in this case, though, it is still possible to take advantage of the efficiency of **MALVA**. Indeed, if multiple complete assemblies of the species of interest are available, it is possible to create a VCF file of known variations by multi-aligning the assemblies and then extracting the variations from the multiple alignments. This is a common and simple pipeline, but the user must implement it by himself. Alternatively, it is possible to exploit more sophisticated pipelines, such as those proposed and implemented in **MALVIRUS** [12], that relieve the user from this burden and provide a simple to use interface. Differently from humans, that are diploids, viruses are haploids, meaning they have a single haplotype per chromosome. Therefore, in this new scenario, the genotyping module of **MALVA** must be tweaked accordingly: the genotypes to compute are not 0/0, 0/1, and so on, but they are simply 0, 1, 2... Switching the genotyping module to the haploid one is as easy as adding the `-1` argument to **MALVA** command:

```
MALVA -1 /path/to/reference.fa /path/to/variations.vcf /path/
to/virus-sample.fq > genotypes.vcf
```

4 Notes

1. The input VCF must contain the set of variations observed in a population along with the observed genotypes (*see* for example the VCF files provided by the 1000 Genomes Project[1]). Indeed, **MALVA** leverages on the known genotypes to build the *k*-mers associated with each allele. For this reason, **MALVA** cannot work with VCF containing variations only (such those provided by dbSNP).
2. The reference genome and the input VCF must use the same naming conventions (i.e., if the names of the chromosome start with `chr` in the FASTA file, they must start in the same way in the VCF as well). Moreover, the reference genome must be the same as the one used to produce the VCF (i.e., if the input calls were called against GRCh38, then GRCh38 must be used as reference genome).
3. As previously mentioned, **MALVA** identifies each variant allele with a set of *k*-mers, for a fixed value of *k* chosen by the user. Roughly speaking, each such *k*-mer is the *k*-long substring centered in the variant that would occur in a genome in which that allele is expressed (possibly in combination with other close variants). **MALVA** then calls the genotypes by carefully analyzing the frequency in the input sample of each *k*-mer describing variants alleles. Choosing a small *k* makes the method less computationally heavy, but the smaller the value of *k*, the more probable that the same *k*-mer occurs at different locations of the genome, potentially leading **MALVA** to call the wrong genotype due to these repetitions. To avoid most of such wrong calls, when a *k*-mer associated to a variant allele is repeated in other places of the reference genome, **MALVA** makes use of a larger context around the allele to distinguish it from the other occurrences. Contexts are identified by longer *r*-mers, whose length $r > k$ can also be tuned by the user. Since, in general, there are fewer contexts to be checked then variants to genotype, in this way **MALVA** achieves a convenient trade-off between memory usage and precision. For technical

[1] ftp.1000genomes.ebi.ac.uk/vol1/ftp/release/20130502/

reasons, it is advisable that both *r* and *k* are chosen to be odd numbers.

4. Since **MALVA** genotypes a new sample using a VCF built from a population of individuals of the same species, it is possible to speed the computation up by subsampling the population or by directly considering a subpopulation. For instance, if the sample we are interested in is from a European individual, we may consider genotyping it with respect to the other European individuals contained in the input VCF only. For this reason, **MALVA** provides the `-s` parameter. However, focusing on a single (sub)population may lower the overall accuracy of **MALVA**, especially in the event of contaminations among different (sub)populations.
5. To better genotype input variations, **MALVA** combines the information contained in the VCF (i.e., the frequency of each allele of a variation), and the information extracted by the read sample (i.e., the coverage of each allele). Typically, the allele frequencies are contained in the `INFO` column of the VCF file, encoded using the `AF` key. If a VCF encodes this information with a different key, it is possible to set the new key using the `-f` parameter. Otherwise, if a VCF does not contain this information, it is possible to use uniform frequencies using the `-u` parameter.
6. **MALVA** stores sets of *k*-mers in two Bloom filters, which are very efficient probabilistic data structures that allow membership queries. The size of a Bloom filter is fixed, and it is decided before its creation: a bigger Bloom filter should reduce the probability of false positives in membership queries. **MALVA** uses by default two Bloom filters of 4 GB each, thus requiring at least 8 GB RAM (plus overhead). However, users must set the corresponding parameter to suit their needs: if 4 GB is too high a requirement, we have experimentally assessed that, when the task is to genotype known human variations, Bloom filters of 2 GB each are large enough to produce reasonably good results.
7. **MALVA** genotypes all the variations contained in the input VCF except those variations with symbolic alleles (e.g., non-nucleotides alleles such as `<INS>` and `<DEL>`).
8. **MALVA**, thanks to its mapping-free nature, is very fast: it genotyped the full callset provided by the 1000 Genomes Project (2504 individuals and 84,739,838 variants) from a 30× human sample (696168435150-bp-long reads) in less than 10 h and using 39 GB RAM. For comparison, **GATK** and **BCFtools**, two widely used alignment-based approaches, completed their analysis in more than 2 and 3 days requiring 9 GB and 33 GB RAM, respectively.

References

1. McKenna A, Hanna M, Banks E et al (2010) The genome analysis toolkit: a MapReduce framework for analyzing next-generation DNA sequencing data. Genome Res 20(9): 1297–1303
2. Li H (2011) A statistical framework for SNP calling, mutation discovery, association mapping and population genetical parameter estimation from sequencing data. Bioinformatics 27(21):2987–2993
3. Peterlongo P, Riou C, Drezen E, Lemaitre C (2017) DiscoSnp++: de novo detection of small variants from raw unassembled read set (s). BioRxiv
4. Eggertsson HP, Jonsson H, Kristmundsdottir S et al (2017) Graphtyper enables population-scale genotyping using pangenome graphs. Nat Genet 49:1654–1660
5. Garrison E, Sirén J, Novak AM et al (2018) Variation graph toolkit improves read mapping by representing genetic variation in the reference. Nat Biotechnol 36(9):875–879
6. Pajuste FD, Kaplinski L, Möls M et al (2017) FastGT: an alignment-free method for calling common SNVs directly from raw sequencing reads. Sci Rep 7(1):1–10
7. Shajii A, Yorukoglu D, William YY, Berger B (2016) Fast genotyping of known SNPs through approximate k-mer matching. Bioinformatics 32(17):i538–i544
8. Sun C, Medvedev P (2019) Toward fast and accurate SNP genotyping from whole genome sequencing data for bedside diagnostics. Bioinformatics 35(3):415–420
9. Gog S, Beller T, Moffat A, Petri M (2014) From theory to practice: plug and play with succinct data structures. In: International symposium on experimental algorithms. Springer, Cham, pp 326–337
10. Denti L, Previtali M, Bernardini G et al (2019) MALVA: genotyping by mapping-free ALlele detection of known VAriants. IScience 18:20–27
11. Danecek P, Bonfield JK, Liddle J et al (2021) Twelve years of SAMtools and BCFtools. Gigascience 10(2):giab008
12. Ciccolella S, Denti L, Bonizzoni P et al (2020) MALVIRUS: an integrated web application for viral variant calling. BioRxiv

Chapter 16

Kmer2SNP: Reference-Free Heterozygous SNP Calling Using k-mer Frequency Distributions

Yanbo Li, Hardip Patel, and Yu Lin

Abstract

DNA sequencing technologies enable the generation of genetic profiles from many individuals at a rapid rate. Identifying single-nucleotide polymorphism (SNP) between biological samples is fundamental in genetics with various applications, such as disease diagnosis and associations and ancestry and relationship inference. Most methods use a species-specific reference genome for aligning raw sequenced reads for accurate SNP calling. However, high-quality reference genomes may not be available for all species. Therefore, we developed a reference-free algorithm, Kmer2SNP, to identify heterozygous SNPs from raw sequenced reads to facilitate genetic studies in species without the reference genome. Kmer2SNP first calculates the k-mer frequency distribution from reads to determine k-mers containing heterozygous SNPs. Next, these k-mers are rapidly matched with each other to identify pairs of exact heterozygous k-mers that belong to one of the two possible haplotypes in a diploid organism. Finally, using overlapping neighboring k-mers, weights are assigned for SNP assignments; higher weights increase SNP discovery confidence.

Key words Genomics, SNP calling, Reference-free, K-mer, Alignment-free, Genetic variants

1 Introduction

Next-generation sequencing (NGS) has revolutionized healthcare, biomedical research, and environmental monitoring using genetics. Systematic evaluation of individual-specific DNA sequence allows detecting inter-individual genetic variations for disease diagnosis, population genetic studies, and environmental monitoring. Genetic variations can manifest as single-nucleotide polymorphism (SNP) on one or both chromosomes of homologous pair inherited from each parent. These SNPs can be homozygous where both alleles are identical but different to the reference genome or other individuals. In contrast, heterozygous SNPs contain two non-identical alleles at a locus. Typically, short DNA sequence reads (100–250 nucleotide long) obtained using NGS workflows are efficiently aligned to a high-quality reference genome to detect SNPs. A combination of alignment software (e.g., bwa-mem [1]

Charlotte K. Y. Ng and Salvatore Piscuoglio (eds.), *Variant Calling: Methods and Protocols*,
Methods in Molecular Biology, vol. 2493, https://doi.org/10.1007/978-1-0716-2293-3_16,

and bowtie2 [2]) and variation detection tools (SNPdetector [3], novoSNP [4], PolyScan [5], MAQ [6], SOAPsnp [7], and GATK [8]) is used to derive SNP profiles for individuals. However, many species do not have high-quality reference genomes available, limiting the application of reference-based approaches for detecting genetic variations in those organisms. We developed a reference-free algorithm to efficiently detect heterozygous SNPs using raw sequenced reads, thus avoiding the need for alignments to the reference genome and enabling genetic studies in species without the availability of high-quality reference genomes. In this chapter, we describe Kmer2SNP [9], our software to call heterozygous SNPs. Kmer2SNP has two main advantages:

1. Kmer2SNP infers candidate heterozygous k-mers using their frequency distribution in raw sequence reads efficiently.
2. Kmer2SNP assigns edge weights from the heterozygous k-mer graph and finds the k-mer pair with maximum weight to infer heterozygous k-mer pairs accurately.

2 Materials

Kmer2SNP is freely available at GitHub repository (https://github.com/yanboANU/Kmer2SNP) under the permissive MIT License (*see* **Note 6**).

2.1 Software

Kmer2SNP is developed in Python with the following packages and software as requirements to run in a Linux or Unix environment:

- Kmer2SNP was developed in *Python 3* in Linux environment. We recommend using *Python 3.6* or newer versions for compatibility. *Networkx* (v2.4 or newer versions) and the latest version of *Biopython* libraries are required for network and DNA sequence data manipulations, respectively.
- K-mers in read data are enumerated using the *DSK* [10] software. DSK can be replaced by Jellyfish [11] for counting k-mers, however, it has not been implemented as part of the Kmer2SNP yet. The input for DSK is FastQ files containing raw sequenced reads. DSK will output numeric histogram of k-mer frequencies for a fixed size of the *k*.
- *FindGSE* [12] is used to infer candidate heterozygous k-mers by using k-mer frequencies fitted with a skew normal distribution model. Note that *FindGSE* is an *R* package and thus *R* (v3.5 or newer) is required. FindGSE uses the output of DSK as input and generates the upper and lower bounds of k-mer frequency for heterozygous k-mers as output (*see* **Notes 3** and **4**).

Once the dependencies are installed and available in the PATH, the latest source code of Kmer2SNP can be cloned with the following git command:

```
git clone https://github.com/yanboANU/Kmer2SNP.git
```

3 Methods

3.1 Kmer2SNP: Single Sample Mode

The users can familiarize themselves with the available parameter settings by running the following command:

```
python3 kmer2snp.py single --help
```

Typical command for running Kmer2SNP in single sample mode is as follows:

```
python3 kmer2snp.py single --k 31 --c 30 --fastaq example.fastq
```

3.1.1 Inputs and Parameters

- `--k`: the k-mer size. Raw sequenced reads are divided into overlapping k-mers of length k for further analysis. For a given read of length l, there are $l - k + 1$ possible k-mers. Although DSK can handle larger k-mers ($k > 32$), typically, the k-length should be an odd number between 17 and 31 for best results and efficiency. Ideally, smaller values of k (17 to 25) are preferred for high heterozygosity data and smaller genomes, while larger k values (27 to 31) are recommended for low heterozygosity data and larger genomes for memory footprint considerations and optimal results. For the human genome, k can be set to 31, and for fungi genomes, k can be set to 21. Note: All k-mers are converted to their canonical forms for computations; for a k-mer and its reverse complement, the k-mer with smaller lexicographic order is chosen to represent both k-mers. For example, for the k-mer *TGCA* and its reverse complement *ACGT* is chosen.
- `--c`: the homozygous genome coverage. The homozygous coverage is the estimated coverage (sequencing depth) of homozygous k-mers that forms the major peak in k-mer frequency histogram. Heterozygous k-mer coverage peak is expected to be half of the homozygous coverage corresponding to the coverage of each haplotype in diploid genomes (*see* **Note 4**). Kmer2SNP performs best with at least 30x homozygous coverage. For an estimated genome size G with a mean sequenced read length l and T total reads, the homozygous coverage can be approximated to be $(l \times T)G$ (*see* **Note 3**).
- `--fastaq`: path to raw sequence read file(s) from next-generation sequencing (NGS) experiments in the fasta or fastq format. Multiple files can be passed as input by separating the

file paths with a comma (",") without spaces. These files are typically large in size and therefore stored as compressed files (.gz extension). Kmer2SNP can use these compressed files for computation.

- `--output-dir`: the path to output directory (default value: current working directory). All computed temporary and final result files will be written to this directory.
- `--b`: the flag (0 or 1) for isolated or non-isolated k-mers (default value = 1). Kmer2SNP can identify at most 2 heterozygous SNPs in a pair of k-mers. If b is set to 0, Kmer2SNP calls only isolated SNPs (one SNP in a pair of k-mers). If b is set to 1, Kmer2SNP will call two heterozygous SNPs if they exist in a given k-mer. Occasionally, genomes may contain two heterozygous SNPs next to each other in a given k-mer. Kmer2SNP is able to handle such cases and report appropriately the existence of both SNPs in a k-mer.
- `--t1`: the k-mer frequency file (default value: calculated using DSK from supplied fasta/fastq files). If the user has used alternate program (Jellyfish or similar) to calculate k-mer frequencies, then Kmer2SNP can use that file for detecting heterozygous SNPs. Each line of this file must consist of the canonical k-mer sequence and the frequency of that k-mer separated by a space.
- `--c1`: the estimated minimum coverage of heterozygous k-mer (optional). The user can choose to provide the minimum cutoff (> 0) for heterozygous k-mer frequency to avoid noisy k-mers introduced due to sequencing errors or to increase specificity in the output. By default, the value of c1 is computed by using FindGSE using k-mer frequency histogram implemented in Kmer2SNP.
- `--c2`: the estimated maximum coverage of heterozygous k-mer (optional). Similar to c1 option, the user can choose to provide the maximum cutoff (must be < homozygous coverage) for heterozygous k-mer frequency to increase sensitivity in the output. By default, the value of c2 is computed by using FindGSE.
- `--r`: the estimated heterozygous rate (optional). By default, this rate is automatically computed by FindGSE.

3.1.2 Output Files and Their Interpretations

The list of output files generated by Kmer2SNP is shown in Table 1. The main output of the Kmer2SNP is the `kmer_pair.snp` text file (exemplified in Table 2) containing pairs of heterozygous k-mers with corresponding edge weights in a space-separated format (*see* **Note 2**). These files can be further processed to prioritise relevant heterozygous SNP pairs for downstream analysis.

Table 1
List of output files generated by Kmer2SNP

Output Files	Contents	File type	Description
`chr_k.histo`	k-mer histogram	text	generated by DSK
`chr_k.txt`	frequency of each k-mer	text	generated by DSK
`hete.para`	estimated heterozygous rate and coverage range of heterozygous k-mers	text	generated by FindGSE
`kmer_pair.snp`	isolated SNP k-mer pair	text	Kmer2SNP output
`kmer_pair.snp.extend`	extended isolated SNP k-mer pair	text	Kmer2SNP output
`kmer_pair.non.sep`	non-isolated SNP k-mer pair	text	Kmer2SNP output with `--b 1` option
`kmer_pair.non.extend`	extended non-isolated SNP k-mer pair	text	Kmer2SNP output with `--b 1` option
`kmer_pair.population`	SNP k-mer pair collated for all samples	text	only for Kmer2SNP population mode

Table 2
An example of first ten lines of Kmer2SNP's output kmer_pair.snp ($k = 17$)

```
AAAAAAAAAAACAATTT AAAAAAAATAACAATTT 8
AAAAAAAAAAAGCACGG AAAAAAAACAAGCACGG 9
AAAAAAAAAAAGCCACC AAAAAAAATAAGCCACC 11
AAAAAAAAAAAGTAGAA AAAAAAAACAAGTAGAA 10
AAAAAAAAAATTTCTGA AAAAAAAAGATTTCTGA 7
AAAAAAAAAAGTTCTTA AAAAAAAAGAGTTCTTA 16
AAAAAAAAACTAAGAGA AAAAAAAAGCTAAGAGA 15
AAAAAAAAAGAAATTAC AAAAAAAACGAAATTAC 8
AAAAAAAAAGAACAATG AAAAAAAATGAACAATG 9
AAAAAAAAAGTAAAACT AAAAAAAAGGTAAAACT 9
```

Table 3
An example of first ten lines of Kmer2SNP's output kmer_pair.snp.extend ($k = 17$)

```
AAAAAAAAAAACAATTTTCATTATT AAAAAAAATAACAATTTTCATTATT 8
AAAAAAAAAAAAGCACGGAATACTGA AAAAAAAAACAAGCACGGAATACTGA 9
AAAAAAAAAAAAAAGCCACCACATCTAA AAAAAAAAAAATAAGCCACCACATCTAA 11
TGAAAAAAAAAAAGTAGAACAAGAAGT TGAAAAAAAACAAGTAGAACAAGAAGT 10
GAAAAAAAAAATTTCTGAAATCAA GAAAAAAAAGATTTCTGAAATCAA 7
TTTAATTCAAAAAAAAAAGTTCTTATTCATAAT TTTAATTCAAAAAAAAGAGTTCTTATTCATAAT 16
ATAAAGCAAAAAAAAACTAAGAGAGGATTATT ATAAAGCAAAAAAAAGCTAAGAGAGGATTATT 15
AAAAAAAAAGAAATTACTTTGTCCT AAAAAAAACGAAATTACTTTGTCCT 8
GAAAAAAAAAGAACAATGTTGAAACA GAAAAAAAATGAACAATGTTGAAACA 9
AAAAAAAAAAGTAAAACTGAAAAGTT AAAAAAAAAGGTAAAACTGAAAAGTT 9
```

Heterozygous k-mers contain SNPs in the middle of the k-mer. Therefore, there can be additional overlapping k-mers containing the same SNP. For a heterozygous k-mer pair, Kmer2SNP walks out in either direction to extend the sequence containing the same SNP. Extended pairs of sequences containing heterozygous SNPs are reported in the `kmer_pair.snp.extend` text file (exemplified in Table 3).

If the users have selected the `--b 1` option, then Kmer2SNP generates two additional output text files: `kmer_pair.non.sep` and `kmer_pair.non.extend`. These files contain k-mer pairs with two SNPs in them. They are formatted similar to the `kmer_-pair.snp` and `kmer_pair.snp.extend` files.

3.2 Kmer2SNP: Population Mode

Ideally, the users will have multiple samples for comparisons. Therefore, we have implemented Kmer2SNP such that multiple samples can be studied collectively in `population` mode (*see* **Notes 1** and **2**). To familiarize with the available parameter settings, the users can run the following command:

```
python3 kmer2snp.py population --help
```

Typical command for population mode is as follows:

```
python3 kmer2snp.py population --k 31 --cfile sam-
ple.coverage.txt --faqfile <file with per sample
fastq/fasta file>
```

3.2.1 Inputs and Parameters

- `--k`: the k-mer size. Similar to the single sample mode, this parameter defines the length of k-mer used for detecting heterozygous k-mer pairs. For the large human genome samples, `k` can be set to `31`, and for small fungi genomes, the `k` value can be set to `21`.
- `--cfile`: the homozygous coverage file. This file contains homozygous coverage information for each sample (*see* **Notes 3** and **4**). Each row of this file must contain information about one sample; the first column must list the sample identifier (no spaces allowed), and the second column lists the numeric homozygous coverage value.
- `--faqfile`: a text file containing list of NGS sequence reads file of all samples. Each row of the file must contain information about one sample; the first column must list the sample identifier (no spaces allowed), and the second column must contain path to at least one sequence read file in fastq/fasta format. If the sample has multiple files, file paths can be separated by a comma (","). Sample identifiers must match with sample identifiers supplied to the `--cfile`.
- `--b`: the flag (0 or 1) for isolated or non-isolated k-mers (default value = 0). This flag is similar in behavior to the single sample mode. However, the default value is set to 0 instead of 1 in the population mode to reduce run time.
- `--output-dir`: the path to output directory (default value: current working directory). All computed temporary and final result files will be written to this directory.

Table 4
An example of the first ten lines of Kmer2SNP's output kmer_pair.population (*k* = 31)

kmer	sample1	sample2
ACGCCGCGCCTGGGG[C/T]ATGACTGCCCTCAGC	0/1:44,26:30	0/1:46,48:30
ATAATGCTAACACTC[C/G]ATGCTGGTAACCAGC	0/1:39,26:30	0/1:42,44:30
ATACAGGTCAGACGA[C/G]AGCCACTGGGGCGTC	0/1:28,38:30	0/1:44,50:30
ATACCACTGTTGGAA[C/T]ACTGTCAGGCTTGCC	0/1:27,28:24	0/1:26,49:24
ATACTTTTTTTTTTT[G/T]TTTTTTGAGACGATA	0/1:29,41:30	0/1:45,53:30
ATATCAAATCATTCA[A/C]ATTAAATCAGGATCA	0/1:21,28:30	0/1:45,49:30
ATATGTGCCGAGCTC[C/T]GGAGCTTCCTGCGGG	0/1:29,39:30	0/1:58,54:30
ATAGGGTGTAAAATT[C/T]CCACCTACTCTGATA	0/1:36,32:30	0/1:38,42:30
ATCCTGAACCAACCA[A/T]GGTGGCCCAGCTGGG	0/1:35,35:30	0/1:37,33:30

3.2.2 Output Files and Their Interpretations

For each sample, Kmer2SNP reports a `kmer_pair.snp` file and a `kmer_pair.snp.extend` file (Table 1). The format and content of these two files are the same as the single sample mode. The `--b` option can be toggled to generate non-isolated SNPs (`kmer_pair.non.sep` and `kmer_pair.non.extend`). In addition to the single sample mode, population mode generates a file (`kmer_pair.population`) with heterozygous k-mer pairs collated for all samples for efficient downstream comparisons between samples (Table 4). This is a text file with the first column listing heterozygous k-mers and subsequent columns with information from each sample included in the analysis (*see* **Notes 2** and 7).

- The first column contains heterozygous k-mer pair information in ACATG[A/C]CCATG format for `k = 11`, where the middle position with two possible heterozygous alleles is listed in square parenthesis separated by the forward slash ("/").
- Subsequent columns represent sample-level information about the heterozygous k-mer pair (*see* **Note 1**). These columns are formatted similar to the popular variant call format (VCF) representation containing (a) inferred genotype values, (b) counts of individual k-mer of the pair (allele depth), and (c) the edge weights of heterozygosity calculated from the k-mer graph; the three values are separated by ":" symbol similar to the VCF file.

  (a) Genotype field can contain 0/0, 0/1, or 1/1 as one of the three possible values:

  0/0 indicates that the sample is homozygous containing only ACATG[A]CCATG allele.

  1/1 indicates that the sample is homozygous containing only ACATG[C]CCATG allele.

0/1 indicates that the sample is heterozygous containing both ACATG[A/C]CCATG alleles.

(b) k-mer count (allele depth) field contains two comma-separated integer values to enumerate the counts of the first (ACATG[A]CCATG) and the second (ACATG[C]CCATG) k-mer, respectively, as observed in the sample data.

(c) Edge weight field contains numeric values. If the value is > 0, it indicates the edge weight computed from the k-mer graph for a heterozygous k-mer pair. If the value is -1, then it indicates that no edge weight was calculated for that sample for the k-mer pair. Further details about edge weight are available in [9]. Note: A higher edge weight reflects greater confidence in the heterozygous call.

3.3 Resource Usage

Kmer2SNP is faster and more efficient in memory than other state-of-the-art methods for a relatively small data [1]. For example, Kmer2SNP used 1.35 h of CPU time and 2.51 GB memory to process 62.73 million reads (148 bp long) aligned to the human chromosome 22 (300x coverage) from NA12878 sample (*see* **Note 5**). Please see paper [9] for more information on run time and memory usage. We welcome feedback and input from the user community for their requirements and suggestions for improving the efficiency of our software in the future (*see* **Note 6**).

4 Notes

1. One of the limitations of the Kmer2SNP is that it identifies only heterozygous SNPs by design. However, we have attempted to include the homozygous status of a sample where possible in the population mode of the Kmer2SNP.
2. Kmer2SNP is useful for inferring heterozygous SNPs for species without the reference genome. Therefore, the software cannot assign these SNPs to genomic positions. This limitation is widely acknowledged, and international efforts are underway to overcome these limitations in the long run.
3. FindGSE used in the workflow to infer limits for heterozygous k-mer frequency relies on accurate estimation of the homozygous coverage provided by the user. The users are encouraged to provide accurate estimation of homozygous coverage or attempt running the Kmer2SNP with multiple values to improve accuracy.
4. Kmer2SNP is applicable for diploid organisms only. K-mer frequency spectra are less well understood for polyploid

genomes and therefore deserves further investigation for extending to polyploid genomes.

5. Kmer2SNP does not support multiple threaded execution except for DSK part of the workflow. Future improvements can include multi-threaded support to reduce run times of Kmer2SNP.
6. The users can request help, post their statistics, and seek feature improvements through the GitHub repository.
7. Our software is complementary to other similar software such as DiscoSNP++ [13, 14] and EBWT2SNP [15]. The users may benefit from evaluating results from Kmer2SNP against other software for comparisons for the best outcome.

References

1. Li H, Durbin R (2009) Fast and accurate short read alignment with burrows–wheeler transform. Bioinformatics 25(14):1754–1760
2. Langmead B, Salzberg SL (2012) Fast gapped-read alignment with Bowtie 2. Nat Methods 9(4):357–359
3. Zhang J, Wheeler DA, Yakub I, Wei S, Sood R, Rowe W, Liu PP, Gibbs RA, Buetow KH (2005) Snpdetector: a software tool for sensitive and accurate SNP detection. PLoS Comput Biol 1(5):e53
4. Weckx S, Del-Favero J, Rademakers R, Claes L, Cruts M, De Jonghe P, Van Broeckhoven C, De Rijk P (2005) novoSNP, a novel computational tool for sequence variation discovery. Genome Res 15(3):436–442
5. Chen K, McLellan MD, Ding L, Wendl MC, Kasai Y, Wilson RK, Mardis ER (2007) PolyScan: an automatic indel and SNP detection approach to the analysis of human resequencing data. Genome Res 17(5):659–666
6. Li H, Ruan J, Durbin R (2008) Mapping short DNA sequencing reads and calling variants using mapping quality scores. Genome Res 18(11):1851–1858
7. Li R, Li Y, Fang X, Yang H, Wang J, Kristiansen K, Wang J (2009) SNP detection for massively parallel whole-genome resequencing. Genome Res 19(6):1124–1132
8. DePristo MA, Banks E, Poplin R, Garimella KV, Maguire JR, Hartl C, Philippakis AA, Del Angel G, Rivas MA, Hanna M et al (2011) A framework for variation discovery and genotyping using next-generation DNA sequencing data. Nat Genet 43(5):491
9. Li Y, Patel H, Lin Y (2020) Kmer2snp: reference-free SNP calling from raw reads based on matching. In: 2020 IEEE international conference on bioinformatics and biomedicine (BIBM). IEEE, Piscataway, pp 208–212
10. Rizk G, Lavenier D, Chikhi R (2013) DSK: k-mer counting with very low memory usage. Bioinformatics 29(5):652–653
11. Marçais G, Kingsford C (2011) A fast, lock-free approach for efficient parallel counting of occurrences of k-mers. Bioinformatics 27(6): 764–770
12. Sun H, Ding J, Piednoël M, Schneeberger K (2017) FindGSE: estimating genome size variation within human and Arabidopsis using k-mer frequencies. Bioinformatics 34(4): 550–557
13. Uricaru R, Rizk G, Lacroix V, Quillery E, Plantard O, Chikhi R, Lemaitre C, Peterlongo P (2014) Reference-free detection of isolated SNPs. Nucleic Acids Res 43(2):e11–e11
14. Peterlongo P, Riou C, Drezen E, Lemaitre C (2017) DiscoSnp++: de novo detection of small variants from raw unassembled read set (s). BioRxiv, p 209965
15. Prezza N, Pisanti N, Sciortino M, Rosone G (2019) SNPs detection by eBWT positional clustering. Algorithms Mol Biol 14(1):3

Chapter 17

Somatic Single-Nucleotide Variant Calling from Single-Cell DNA Sequencing Data Using SCAN-SNV

Sajedeh Bahonar and Hesam Montazeri

Abstract

SCAN-SNV is a recent computational tool for somatic single-nucleotide variant (SNV) identification from the single-cell DNA sequencing data. The workflow of the SCAN-SNV package is as follows. First, candidate somatic SNVs and credible heterozygous single-nucleotide polymorphisms (hSNP) are obtained by analyzing single-cell and matched bulk sequencing data, respectively. Subsequently, SCAN-SNV estimates genome-wide allele-specific amplification balance (AB) at any position of DNA sequencing data using a probabilistic spatial statistical model. Finally, candidate somatic SNVs that are likely artifacts according to the AB predictions are further removed to obtain putative mutations. This chapter provides a step-by-step practical guide of the package by explaining how to install and use the variance caller in a real-world example.

Key words SCAN-SNV, Somatic variant calling, Single-cell DNA sequencing, Allele-specific amplification balance, Gaussian process, Heterozygous SNPs, Conda environment

1 Introduction

Single-cell DNA sequencing (scDNA-seq) technology provides an unprecedented opportunity to study genetic heterogeneity and somatic evolution at the resolution of individual cells [1]. The emerging field of single-cell genomics enables re-investigating long-standing problems in genomics—including delineating evolutionary dynamics of cancer and tumor heterogeneity [2–5] and biomarker discovery [6, 7] at the cellular level. A central computational task in single-cell genomics, which is a necessary step for virtually all applications of scDNA-seq data, is the identification of somatic variants. However, experimental errors and biases that arise during sequencing pose formidable computational challenges for distinguishing artifacts from actual somatic variants. These errors mainly occur during whole-genome amplification (WGA) extracted DNA from a single cell.

Charlotte K. Y. Ng and Salvatore Piscuoglio (eds.), *Variant Calling: Methods and Protocols*,
Methods in Molecular Biology, vol. 2493, https://doi.org/10.1007/978-1-0716-2293-3_17,

There are two single-cell amplification techniques: PCR-based and multiple displacement amplification (MDA) methods [1]. PCR-based methods are suitable for copy number variant (CNV) calling [8], while MDA-based methods are predominantly used for somatic SNV calling. This approach uses $\Phi -29$ polymerase with lower error rates and more extensive genome coverage than PCR-based methods [9]. Single-cell WGA using MDA-based methods leads to imbalanced allelic amplification due to the underrepresentation of one of the homologous alleles in the amplification process [10, 11] and makes the variant calling a challenging computational task [12, 13].

Several computation tools have been developed to identify somatic single-nucleotide variants through single-cell DNA sequencing data (Table 1). SCAN-SNV [13] and SCcaller [14] build allelic amplification balance models by accounting for adjacent heterozygous single-nucleotide polymorphisms obtained by analyzing matched bulk sequencing data. Monovar [15] performs variant calling and eliminates likely artifacts by sharing information among cells without enforcing a dependency structure across neighboring genomic locations. SCIΦ [16] also integrates information across cells but through inference of cell lineage tree. This book chapter provides a practical guide on performing somatic variant calling using SCAN-SNV; this tool has outperformed Monovar and SCcaller by reducing more than threefold in false discovery rate and obtaining a similar sensitivity performance [13]. SCAN-SNV was developed as a conda package and was distributed for Linux-flavored operating systems. The software is publicly available in the Github repository at https://github.com/parklab/scan-snv.

SCAN-SNV first builds a genome-wide amplification balance (AB) model to estimate allelic amplification balance at any genomic location through the Gaussian process machinery. The AB estimations are based on credible heterozygous SNPs (hSNPs) in the adjacent genomic regions, obtained by applying the GATK HaplotypeCaller on a matched normal bulk sequencing data. SCAN-SNV then incorporates a set of primary and secondary filters to distinguish artifacts from true somatic SNVs (sSNVs). The primary filters are based on the AB predictions and aim to remove artifacts that arise during pre-amplification and early amplification steps by examining the local AB predictions. Secondary filters eliminate candidate mutations, harboring an excessive number of reads with indels and clipping reads compared to the matched normal bulk sample [13].

Table 1
Summary of four recent single-cell somatic SNV callers

Variant caller	Programming language	Input data type	Main computational idea	References
SCAN-SNV	Python	scDNAseq BAM files	Estimating genome-wide AB models through Gaussian processes	[13]
SCcaller	Python	scDNAseq BAM files	Estimating genome-wide AB models through a Nadaraya–Watson kernel model	[14]
Monovar	Python	scDNAseq BAM files	Enhanced artifact removal through information sharing across cells	[15]
SCIΦ	C++	scDNAseq mpileup files	Information sharing across cells via a phylogenetic model	[16]

2 Software

This protocol explains the steps necessary for performing single-cell sSNV calling. A high-performance computer system with a Linux operating system is required to proceed with this protocol.

Installation of SCAN-SNV requires three steps: (1) installing Miniconda as SCAN-SNV is a conda distributed package, (2) activating the SCAN-SNV conda environment, and (3) registering the GATK installation. The steps necessary for the installation are detailed as follows:

1. Follow the below instructions to download and install Miniconda and to initialize the conda environment (skip this step if conda is already available on your system):

```
$ mkdir -p ~/miniconda3
$ wget https://repo.anaconda.com/miniconda/Miniconda3-lat-
est-Linux-x86_64.sh -O ~/miniconda3/miniconda.sh
$ bash ~/miniconda3/miniconda.sh -b -u -p ~/miniconda3
$ rm -rf ~/miniconda3/miniconda.sh
$ ~/miniconda3/bin/conda init bash
```

You may need to restart your shell for modifications to be affected.

2. Deactivate the conda base environment and then activate the `scansnvtest` environment.

```
$ conda deactivate
$ conda create -v -n scansnvtest -c bioconda -c conda-forge/
label/cf201901 -c jluquette -c soil -c dranew SCAN-SNV
$ conda activate scansnvtest
```

3. The last step is to register the GATK installation. To this aim, download the suitable version of GenomeAnalysisTK (we tried GenomeAnalysisTK-3.8-1-0-gf15c1c3ef.tar) from the GATK/ folder of the below cloud storage.

```
#download GATK: https://console.cloud.google.com/storage/
browser/gatk-software/package-archive/
```

Complete the GATK registration using the following commands.

```
$ tar -xvf package-archive_gatk_GenomeAnalysisTK-3.8-1-0-
gf15c1c3ef.tar
$ gatk-register GenomeAnalysisTK-3.8-1-0 gf15c1c3ef/GenomeA-
nalysisTK.jar
```

Make sure the installation is complete by checking the GATK version.

```
$ gatk -version
```

3 Methods

In this section, we demonstrate how to perform the single-cell variant calling using SCAN-SNV. We explain step-by-step instructions on a human dermal fibroblast sample. The single-cell and matched bulk sequencing data are available from the NCBI's Sequence Read Archive (SRA) database under submission numbers SRR2976573 and SRR2976567, respectively.

3.1 Downloading Required BAM Files

The SCAN-SNV requires BAM files as input data. First, download the SRA files from the SRA database using the SRA toolkit (*see* **Note 4**). The reader is referred to Subheading 3.6.1 if this package needs to be installed.

```
$ prefetch SRR2976573 --max-size 60G
$ prefetch SRR2976567 --max-size 60G
```

Then use the SRA toolkit to convert SRA files to the sequence alignment map (SAM) file format:

```
$ sam-dump SRR2976573.sra > SRR2976573.sam
$ sam-dump SRR2976567.sra > SRR2976567.sam
```

Lastly, use the samtools package (*see* Subheading 3.6.1) to convert SAM files to binary alignment map (BAM) files. BAM files are binary and compressed forms of SAM files.

```
$ samtools view -b -S SRR2976573.sam > SRR2976573.bam
$ samtools view -b -S SRR2976567.sam > SRR2976567.bam
```

3.2 Indexing BAM Files

Create BAM index files (.bai extension) of the BAM files using the samtools command:

```
$ samtools index SRR2976573.bam
$ samtools index SRR2976567.bam
```

3.3 Extraction of the SM Tag of a BAM File

Before running SCAN-SNV, we should extract the SM tag from the corresponding BAM file using the following command:

```
$ samtools view -H xxx.bam | grep '^@RG' | sed "s/.*SM:\([^\t]*
\).*/\1/g" | uniq
```

3.4 Important Arguments of SCAN-SNV

In this section, we introduce the main arguments of the `scansnv` command. `Scansnv -h` or `scansnv --help` provides a detailed explanation of all SCAN-SNV arguments.

3.4.1 Required Arguments

First, we explain the required arguments for running SCAN-SNV. There are no default values for any of the required arguments. The below arguments specify path to the reference genome FASTA file (*see* **Note 2**), an indexed dbSNP VCF file (*see* **Note 3**), and the SHAPEIT's haplotype reference panel (*see* Subheadings 3.6.2 and 3.6.3):

```
--ref PATH
--dbsnp PATH
--shapeit-panel DIR
```

Similarly, we set the output directory:

```
--output-dir DIR
```

The following argument specifies the path to BAM files (*see* **Note 3**):

```
--bam SAMPLE_ID BAM_FILE
```

where `SAMPLE_ID` is the SM tag of the BAM file. Repeated usage of this argument is permitted in order to input multiple BAM files.

The following arguments can be used to indicate the type of each input BAM file:

```
--bulk-sample SAMPLE_ID
--sc-sample SAMPLE_ID
```

3.4.2 Optional Arguments

The SCAN-SNV has some optional practical arguments. The `joblimit` argument specifies the number of simultaneous jobs to execute at any given time and `memlimit` determines the total available memory in MB.

```
--joblimit INT (default: 1)
--memlimit INT (default: None)
```

The following argument splits the AB model sampling stage into different jobs and may significantly shorten total runtime in a multiprocessor system.

```
--abmodel-chunks INT (default: 20)
```

The argument `abmodel-samples-per-chunk` specifies the number of samples for computing the AB model log-likelihood function for each chunk. The product of `abmodel-chunks` and `abmodel-samples-per-chunk` determines the total number of samples for each chromosome.

```
--abmodel-samples-per-chunk INT (default :1000)
```

Argument `regions` can be used to perform SCAN-SNV on specific regions of the genome. For example, the following command restricts the variant calling to a particular region of the first chromosome:

```
--regions 1:10000-200000
```

Another helpful feature of SCAN-SNV is its ability to restart a previously running analysis using the `--resume` argument. The default value of this argument is `False`.

3.4.3 Spiked-in Arguments

hSNP Spike-in mutations are used in the SCAN-SNV package to estimate the sensitivity of variant calling and the effects of various filters. A determined number of mutations are randomly placed in reads of BAM files as actual variants in the spike-in process. There are two arguments for the simulation of spike-ins. The argument hsnp-spikein-replicates specifies the number of separate spike-in replicates, each with hsnp-spikein-replicate-size spiked-in hSNPs.

```
--hsnp-spikein-replicates INT (default: 100)
--hsnp-spikein-replicate-size INT (default: 40)
```

3.5 Performing Variant Calling

We perform variant calling using SCAN-SNV on the input BAM files using the below command (*see* **Note 1**, Table 2):

```
$ scansnv \
--ref /path/to/human_g1k_v37_decoy.fasta \
--dbsnp /path/to/dbsnp_138.b37.vcf \
--shapeit-panel /path/to/1000GP_Phase3 \
--output-dir sample_one \
--bam hunamp SRR2976567.bam \
--bam hdf-con-2 SRR2976573.bam \
--sc-sample hdf-con-2 \
--bulk-sample hunamp \
--abmodel-chunks 1 \
--abmodel-samples-per-chunk 10000 \
--abmodel-hsnp-chunk-size 50 \
--hsnp-spikein-replicates 5 \
--joblimit 1 --resume
```

where /path/to/ specifies an actual path of the subsequent file in your environment. The execution of this command creates various files in the specified output directory. The putative variants are reported in the file somatic_genotypes.rda as an R data file. The file is located at /output-path/scanscnv/<sc-SAMPLE-Tag>/ and can be loaded as an R data frame named *somatic* using the R software. The *somatic* data frame contains all candidate sSNVs, of which variants passed SCAN-SNV's thresholds are indicated in the *pass* column (54th column). The following R script demonstrates how to load and see the data.

```
> load("/path/to/<output_folder>/scanscnv/hdf-con-2/ soma-
tic_genotypes.rda")
> dim(somatic)
# 35406 54
> pass <- which(somatic[,'pass']==TRUE)
> head(somatic[pass,c("chr","pos","refnt","altnt","hdf.
con.2","pass")]
# chr pos refnt altnt hdf.con.2 pass
120 1 11222556 G A 0/1 TRUE
162 1 116152795 A G 0/1 TRUE
```

Table 2
Description of the selected columns of the SCAN-SNV output R data frame for putative somatic variants

Column name	Description
chr	The chromosome on which the variant is located
pos	The genomic position of the variant
refnt	The base (or bases in case of insertion or deletion) at the reference genome
altnt	Alternate base(s)
dbsnp	Indicates whether the variant is in the set of credible SNPs
hdf.con.2 (the tag of single-cell bam file)	Indicates the heterozygous or homozygous status of the variant allele at the corresponding position of the single-cell sequence
muttype	The mutation type of the variant
dp	Sequencing depth
ab	Amplification balance estimate at the genomic position of the variant
pass	Indicates whether the variant passes all the SCAN-SNV filters

```
186 1 118796575 C T 0/1 TRUE
270 1 15050036 G A 0/1 TRUE
530 1 164119177 G T 0/1 TRUE
683 1 175812710 A T 0/1 TRUE
```

3.6 Supplementary Materials

3.6.1 Installing SRA Toolkit and Samtools

We download the SRA toolkit [17] from the following FTP address:

```
https://ftp-trace.ncbi.nlm.nih.gov/sra/sdk/2.10.9/sratoolk-
it.2.10.9-ubuntu64.tar.gz
```

Extract the compressed file and add SRA toolkit path to the Linux $Path environment variable using `export` command or by editing .bashrc file in your home directory.

```
$ tar -zxvf sratoolkit.2.10.9-ubuntu64.tar.gz
$ export PATH="$PATH:/path/to/sratoolkit/sratoolkit.2.10.9-
ubuntu64/bin"
```

Check whether the installation is successful by:

```
$ sam-dump -V
# 2.10.9
```

The installation command for the samtools is [18]:

```
$ sudo apt install samtools
```

3.6.2 Download Reference Genome and SHAPEIT's Haplotype Reference Panel

We used the human reference genome from the NCBI (build 37) for the example presented in this protocol. The `wget` command in Linux can be used for downloading. Direct downloading using the following links is also possible [19].

```
$ wget ftp://gsapubftp-anonymous@ftp.broadinstitute.org/bundle/b37/human_g1k_v37_decoy.fasta.gz
$ wget ftp://gsapubftp-anonymous@ftp.broadinstitute.org/bundle/b37/human_g1k_v37_decoy.fasta.fai.gz
$ wget ftp://gsapubftp-anonymous@ftp.broadinstitute.org/bundle/b37/human_g1k_v37_decoy.dict.gz
```

One can find 1000 genomes project phase 3 SHAPEIT panel at the following links:

```
$ wget https://mathgen.stats.ox.ac.uk/impute/1000GP_Phase3.tgz
$ wget https://mathgen.stats.ox.ac.uk/impute/1000GP_Phase3_chrX.tgz
```

We then unzip all the downloaded files and place them in the appropriate folders:

```
$ gunzip *.gz
$ tar xzvf 1000GP_Phase3.tgz
$ tar xzvf 1000GP_Phase3_chrX.tgz
# move chrX SHAPEIT file into main path of SHAPEIT
$ mv genetic_map_chrX_* 1000GP_Phase3_chrX* 1000GP_Phase3
```

3.6.3 Downloading and Indexing dbSNP VCF Files

dbSNP consisting of common variants is required for the SCAN-SNV package. The dbSNP build 147, as suggested in the original publication, is used in this protocol. If you want to use another dbSNP, *igvtools* can be used to index the dbSNP VCF file [19]:

```
$ conda install -c bioconda igvtools
$ igvtools index /path/to/dbsnp.vcf
```

4 Notes

1. Before running the variant caller, run `conda activate scansnvtest` command. Check whether the `scansnvtest` environment is activated using the command: `scansnv --h`. In case of successful activation, the command displays the arguments and the help of the package.
2. Note the reference genome, index (.fai) and dictionary (.dict) files must be in the same path.
3. It is required that dbSNP VCFs and input BAM files are indexed as explained in Subheading 3.6.
4. If you work on a remote SSH session, consider executing lengthy tasks, such as downloading SRA files, as nohupping background jobs using the `nohup` command to avoid their termination upon logging off. For example, consider the following command for downloading an SRA file:

```
$ nohup prefetch SRR2976573 &
```

References

1. Lähnemann D, Köster J, Szczurek E et al (2020) Eleven grand challenges in single-cell data science. Genome Biol 21:31
2. Rodriguez-Meira A, Buck G, Clark SA et al (2019) Unravelling Intratumoral heterogeneity through high-sensitivity single-cell mutational analysis and parallel RNA sequencing. Mol Cell 73:1292–1305.e8. https://doi.org/10.1016/j.molcel.2019.01.009
3. Lawson DA, Kessenbrock K, Davis RT et al (2018) Tumour heterogeneity and metastasis at single-cell resolution. Nat Cell Biol 20:1349–1360
4. Roerink SF, Sasaki N, Lee-Six H et al (2018) Intra-tumour diversification in colorectal cancer at the single-cell level. Nature 556:437–462. https://doi.org/10.1038/s41586-018-0024-3
5. Bozic I, Wu CJ (2020) Delineating the evolutionary dynamics of cancer from theory to reality. Nat Cancer 1:580–588. https://doi.org/10.1038/s43018-020-0079-6
6. Li X, Liu L, Goodall GJ et al (2020) A novel single-cell based method for breast cancer prognosis. PLoS Comput Biol 16:e1008133. https://doi.org/10.1371/journal.pcbi.1008133
7. Gawel DR, Serra-Musach J, Lilja S et al (2019) A validated single-cell-based strategy to identify diagnostic and therapeutic targets in complex diseases. Genome Med 11:47. https://doi.org/10.1186/s13073-019-0657-3
8. Zhang L, Cui X, Schmitt K et al (1992) Whole genome amplification from a single cell: implications for genetic analysis. Proc Natl Acad Sci U S A 89:5847–5851. https://doi.org/10.1073/pnas.89.13.5847
9. Zhou X, Xu Y, Zhu L et al (2020) Comparison of multiple displacement amplification (MDA) and multiple annealing and looping-based amplification cycles (MALBAC) in limited DNA sequencing based on tube and droplet. Micromachines 11:645. https://doi.org/10.3390/mi11070645
10. Bäumer C, Fisch E, Wedler H et al (2018) Exploring DNA quality of single cells for genome analysis with simultaneous whole-genome amplification. Sci Rep 8:7476. https://doi.org/10.1038/s41598-018-25895-7
11. Picher ÁJ, Budeus B, Wafzig O et al (2016) TruePrime is a novel method for whole-genome amplification from single cells based on TthPrimPol. Nat Commun 7:1–16. https://doi.org/10.1038/ncomms13296

12. Zhang CZ, Adalsteinsson VA, Francis J et al (2015) Calibrating genomic and allelic coverage bias in single-cell sequencing. Nat Commun 6:4. https://doi.org/10.1038/ncomms7822
13. Luquette LJ, Bohrson CL, Sherman MA, Park PJ (2019) Identification of somatic mutations in single cell DNA-seq using a spatial model of allelic imbalance. Nat Commun 10:1–14. https://doi.org/10.1038/s41467-019-11857-8
14. Dong X, Zhang L, Milholland B et al (2017) Accurate identification of single-nucleotide variants in whole-genome-amplified single cells. Nat Methods 14:491–493. https://doi.org/10.1038/nmeth.4227
15. Zafar H, Wang Y, Nakhleh L et al (2016) Monovar: single-nucleotide variant detection in single cells. Nat Methods 13:505–507. https://doi.org/10.1038/nmeth.3835
16. Singer J, Kuipers J, Jahn K, Beerenwinkel N (2018) Single-cell mutation identification via phylogenetic inference. Nat Commun 9:1–8. https://doi.org/10.1038/s41467-018-07627-7
17. samtools Tutorial. http://quinlanlab.org/tutorials/samtools/samtools.html. Accessed 1 May 2021
18. GitHub - ncbi/sra-tools: SRA Tools. https://github.com/ncbi/sra-tools. Accessed 1 May 2021
19. GitHub—parklab/scan-snv: Single cell somatic genotyper. https://github.com/parklab/scan-snv. Accessed 1 May 2021

Chapter 18

Copy Number Variation Detection by Single-Cell DNA Sequencing with SCOPE

Rujin Wang and Yuchao Jiang

Abstract

Whole-genome single-cell DNA sequencing (scDNA-seq) enables the characterization of copy number profiles at the cellular level. This circumvents the averaging effects associated with bulk-tissue sequencing and has increased resolution yet decreased ambiguity in deconvolving cancer subclones and elucidating cancer evolutionary history. ScDNA-seq data is, however, sparse, noisy, and highly variable even within a homogeneous cell population, due to the biases and artifacts that are introduced during the library preparation and sequencing procedure. Here, we describe SCOPE, a normalization and copy number estimation method for scDNA-seq data. We give an overview of the methodology and illustrate SCOPE with step-by-step demonstrations.

Key words Single-cell DNA sequencing, Copy number variation, Data normalization, Segmentation, Latent factor

1 Introduction

Copy number variations (CNVs) refer to duplications and deletions that lead to gains or losses of large segments of the chromosomes. CNVs are an abundant source of variations and have been associated with diseases [1]. In cancer, somatic copy number aberrations are prevailing and have been associated with cancer progression and metastases [2, 3]. Recent advances in next-generation sequencing enable genome-wide CNV detection in a high-throughput manner. However, biases and artifacts are introduced during the library preparation and sequencing step, making data normalization crucial for accurate CNV profiling. Several algorithms have been developed to remove experimental noise from GC content bias [4], sequencing depth [5], mappability [6], amplification efficiency [7], and other latent systematic artifacts [8, 9]. Yet profiling of somatic CNVs in cancer remains challenging due to the heterogeneous nature of the tumor samples—the observed copy

Charlotte K. Y. Ng and Salvatore Piscuoglio (eds.), *Variant Calling: Methods and Protocols*,
Methods in Molecular Biology, vol. 2493, https://doi.org/10.1007/978-1-0716-2293-3_18,

number signals from bulk DNA sequencing (DNA-seq) are averaged and attenuated across multiple genotypically distinct cancer subpopulations [10, 11].

Whole-genome single-cell DNA-seq (scDNA-seq) enables the characterization of copy number profiles at the cellular level without the cell subpopulation confounding. This circumvents the averaging effects associated with bulk-tissue DNA-seq, increasing resolution while decreasing ambiguity in deconvolving cancer subclones and thus elucidating cancer evolutionary history [12, 13]. However, scDNA-seq data across all existing platforms and technologies is sparse, noisy, and highly variable, even within a homogeneous cell population. The extremely shallow and highly non-uniform depth of coverage, which is caused by nonlinear amplification and significant dropout events during the library preparation and sequencing step [14, 15], make detecting CNVs by scDNA-seq challenging. Furthermore, the cancer genomes undergo large chromosome or chromosome-arm level deletions or duplications, as well as changes in cellular ploidy, both of which lead to recurrent and frequent CNAs that disrupt a large proportion of the genome across multiple samples [16]. Existing methods for scDNA-seq either build an optimized normal/reference set for normalization [17] or adopt a cell-specific normalization procedure for removing systematic biases (e.g., those due to GC content), followed by a post hoc procedure for ploidy estimation and adjustment [18]. These methods, which are adapted from those that are developed for bulk DNA-seq data, do not address the challenges and complexities caused by aberrant copy number changes or the complicating factor of tumor ploidy; they also cannot adequately and correctly remove the artifacts, for they make the invalid assumption that all read counts are sampled from diploid regions under the null when estimating the noise terms. The recurrent CNV signals can either be accidentally removed during the normalization step or bias the correction step.

Here, we describe SCOPE [19] for *S*ingle-cell *COPy* number *E*stimation. The distinguishing features of SCOPE include: (1) utilization of cell-specific Gini coefficients for quality control and for identification of normal/diploid cells, which are then used as negative control samples in a Poisson latent factor model for read-depth normalization; (2) modeling of GC content bias using an expectation-maximization (EM) algorithm embedded in the Poisson regression models to account for the discretized copy number states along the genome; and (3) a cross-sample iterative segmentation procedure to identify breakpoints that are shared across cells with the same genetic background. In this chapter, we will further describe the SCOPE method and its software implementation.

2 Materials

In this section, we outline the required materials for CNV profiling by SCOPE, including data input, computational environment, and software packages. The bioinformatic pipeline and the R package for SCOPE are available at https://bioconductor.org/packages/SCOPE/ and https://github.com/rujinwang/SCOPE/.

2.1 Computational Environment

The upstream bioinformatic processing of raw sequencing data needs Linux or Unix platform in a high-performance (HPC) computing system, which will return assembled bam files for SCOPE. The CNV analysis can then be locally run on a Windows or Macintosh machine, where R is installed. For faster computation, SCOPE can be run in parallel on an HPC cluster.

2.2 Software Packages

For read alignment, BWA [20] is required for DNA sequencing (DNA-seq). Picard Tools (http://broadinstitute.github.io/picard) and SAMtools [21] are required for deduplication and quality controls. R package SCOPE, as well as its dependencies, is required for performing single-cell CNV analysis in R.

```
# Package installation
if (!requireNamespace("BiocManager", quietly = TRUE))
 install.packages("BiocManager")
BiocManager::install("WGSmapp")
BiocManager::install("SCOPE")
```

2.3 Data Input

The data input for SCOPE includes raw sequencing files from scDNA-seq studies (as fastq), as well as the corresponding genome assembly (as fasta). BWA [20] is used to align fastq files to the reference genome. For scDNA-seq datasets by the 10× Genomics, a python script (split_script.py) pre-stored in SCOPE's GitHub page can be used for demultiplexing. This is followed by sorting, adding read group, and deduplication on the aligned/demultiplexed bam files. The deduped bam files are used as the input for SCOPE's R package.

3 Methods

A typical CNV analysis using SCOPE consists of five steps: (1) pre-processing, (2) getting GC content, mappability, and coverage; (3) quality control; (4) normal cell identification and normalization; and (4) cross-sample segmentation and visualization.

3.1 Pre-processing

We start with loading the required packages, obtaining the indices for the bam files and the sample names, and reconstructing the bins along the genome. SCOPE processes the entire genome altogether and enables the reconstruction of user-defined genome-wide consecutive bins with a fixed interval length prior to downstream analysis. Make sure that all chromosomes are in the same format as in the bam files. By default, SCOPE is designed for hg19 with a fixed bin length of 500 kb (*see* **Note 1**).

```
library(SCOPE)
library(WGSmapp)
library(BSgenome.Hsapiens.UCSC.hg38)
bamfolder <- system.file("extdata", package = "WGSmapp")
bamFile <- list.files(bamfolder, pattern = '*.dedup.bam$')
bamdir <- file.path(bamfolder, bamFile) # Directories of bam
files
sampname_raw <- sapply(strsplit(bamFile, ".", fixed = TRUE),
"[", 1) # Cell names
bambedObj <- get_bam_bed(bamdir = bamdir, sampname = sampna-
me_raw, hgref = "hg38", resolution = 500)
ref_raw <- bambedObj$ref
```

3.2 Getting GC Content, Mappability, and Coverage

In this step, SCOPE computes GC content and mappability for each bin. To compute GC content, SCOPE calculates the frequencies of the four nucleotides within each bin using the Biostrings R package. To compute mappability, SCOPE employs the 100-mer mappability track from the ENCODE Project [22] and computes the weighted average of the mappability scores if multiple ENCODE regions overlap with the same bin (*see* **Note 1**).

```
mapp <- get_mapp(ref_raw, hgref = "hg38")
gc <- get_gc(ref_raw, hgref = "hg38")
values(ref_raw) <- cbind(values(ref_raw), DataFrame(gc, mapp))
```

SCOPE then computes the depth of coverage matrix from either single-end or paired-end sequencing. SCOPE, by default, adopts a fixed binning method to compute the depth of coverage while removing reads that are mapped to multiple genomic locations and to "blacklist" regions. This is followed by an additional step of quality control to remove bins with extreme mappability to avoid erroneous detections. Specifically, the "blacklist" bins include segmental duplication regions and gaps in reference assembly from telomere, centromere, and/or heterochromatin regions.

```
coverageObj <- get_coverage_scDNA(bambedObj, mapqthres =
40, seq = 'paired-end', hgref = "hg38")
Y_raw <- coverageObj$Y
```

3.3 Quality Control

For quality control, SCOPE first generates statistics including total number/proportion of reads, total number/proportion of mapped reads, total number/proportion of mapped non-duplicate reads, and total number/proportion of reads with mapping quality greater than 20. SCOPE then removes cells based on these computed cell-specific statistics, as well as bins that have extreme GC content (less than 20% and greater than 80%) and low mappability (less than 0.9).

```
QCmetric_raw <- get_samp_QC(bambedObj)
qcObj <- perform_qc(Y_raw = Y_raw, sampname_raw = sampname_r-
aw, ref_raw = ref_raw, QCmetric_raw = Qcmetric_raw)
## Removed 0 samples due to failed library preparation.
## Removed 0 samples due to failure to meet min coverage
requirement.
## Removed 0 samples due to low proportion of mapped reads.
## Excluded 216 bins due to extreme GC content.
## Excluded 355 bins due to low mappability.
## Removed 0 samples due to excessive zero read counts in
library size calculation.
## There are 3 samples and 5029 bins after QC step.
```

3.4 Normal Cell Identification and Normalization

One feature of SCOPE is to identify normal/diploid cells using the Gini coefficient. Gini coefficient is calculated for each cell as twice the area between the Lorenz curve and the diagonal. The value of the Gini index varies between 0 and 1, where 0 indicates the most uniformly distributed coverage and 1 indicates the most unevenly distributed coverage. Cells with extremely high Gini coefficients (greater than 0.5) are likely from failed library preparation and are recommended to be excluded. Set up a Gini threshold for identification of diploid/normal cells (e.g., Gini less than 0.12). We demonstrate the pre-stored toy dataset as follows.

```
Gini <- get_gini(Y_sim)
```

Normal cell index is determined by either the Gini coefficients or prior knowledge (e.g., from cell sorting). SCOPE by default utilizes ploidy estimates from a first-pass normalization run by CODEX2 [9] to ensure fast convergence and to avoid local optima. SCOPE then adopts a Poisson latent factor model with an embedded EM algorithm to capture both cell- and bin-specific systemic biases and artifacts, as well as GC content bias. Refer to Wang et al. [19] for the algorithmic details of SCOPE. Figure 1 gives the normalization results for three representative cells: diploid, hypodiploid, and hyperdiploid cells from a breast cancer patient [23]. Histogram of normalized z-scores (Fig. 2a), which are discretized based on the inferred copy numbers with integer values,

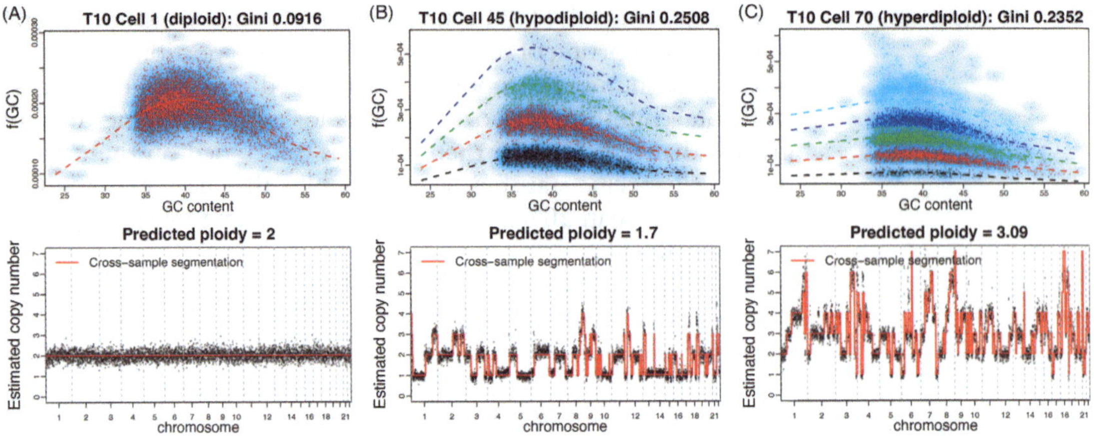

Fig. 1 SCOPE's normalization and segmentation results for three representative cells. GC content bias with EM fitting and cross-sample segmentation results of scDNA-seq data from breast cancer patient T10 are shown across three cells: (**a**) a diploid cell; (**b**) a hypodiploid cell with four copy number states; (**c**) a hyperdiploid cell with five copy number states

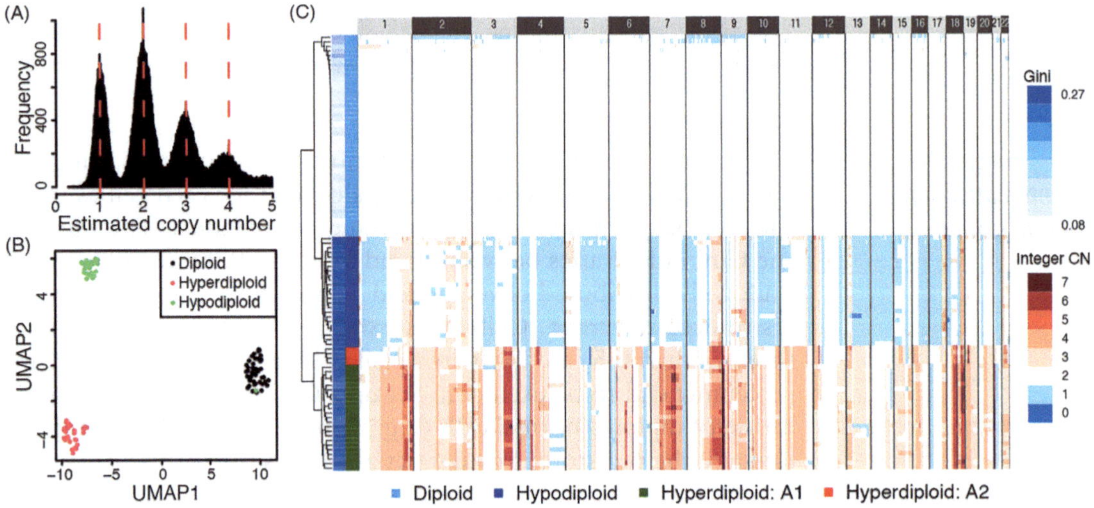

Fig. 2 Normalization results for scDNA-seq data of ploygenomic tumor T10. (**a**) Normalized z-scores from SCOPE, which automatically estimates discretized copy numbers and returns ploidy. (**b**) Dimension reduction of SCOPE's normalization results by UMAP separates the cells with different genetic backgrounds. (**c**) Heatmaps show post-segmentation copy number profiles of integer values. Hierarchical clustering separates normal cells from tumor cells and reveals cancer subclonal structure

demonstrate the effectiveness of SCOPE's normalization. Dimensionality reduction by UMAP successfully separates cells with different genetic backgrounds (Fig. 2b) (*see* **Note 2**).

```
# First-pass normalization for ploidy initialization
normObj.sim <- normalize_codex2_ns_noK(Y_qc = Y_sim, gc_qc =
ref_sim$gc, norm_index = which(Gini <= 0.12))
```

```
ploidy.sim <- initialize_ploidy(Y = Y_sim, Yhat = normObj.sim
$Yhat, ref = ref_sim)

# Normalization by SCOPE
normObj.scope.sim <- normalize_scope_foreach(Y_qc = Y_sim,
gc_qc = ref_sim$gc, K = 1, ploidyInt = ploidy.sim, norm_index
= which(Gini <= 0.12), T = 1:5, beta0 = normObj.sim$beta.hat,
nCores = 2)
# Model fitting with number of latent factors selected by the
Bayesian information criterion (BIC)
Yhat.sim <- normObj.scope.sim$Yhat[[which.max(normObj.scope.
sim$BIC)]]
fGC.hat.sim <- normObj.scope.sim$fGC.hat[[which.max(normObj.
scope.sim$BIC)]]

# Visualization of the EM fit and the selection of the number
of CNV groups
plot_EM_fit(Y_qc = Y_sim, gc_qc = ref_sim$gc, norm_index =
which(Gini<=0.12), T = 1:5, ploidyInt = ploidy.sim, beta0 =
normObj.sim$beta.hat, filename = "plot_EM_fit_demo.pdf")
```

3.5 Cross-Cell Segmentation and Visualization

SCOPE provides the cross-sample segmentation, which outputs shared breakpoints across cells from the same clone (*see* **Note 2**). This step processes the entire genome chromosome by chromosome. Shared breakpoints and integer copy number profiles will be returned. For visualization, SCOPE offers a heatmap of inferred integer copy number profiles with cells clustered by hierarchical clustering. Figure 2c shows the heatmap of the post-segmentation copy number profiles on a real breast cancer dataset (*see* **Note 3**).

```
# Cross-cell segmentation by chromosome
chrs <- unique(as.character(seqnames(ref_sim)))
segment_cs <- vector('list',length = length(chrs))
names(segment_cs) <- chrs
for (chri in chrs) {
 message('\n', chri, '\n')
 segment_cs[[chri]] <- segment_CBScs(Y = Y_sim, Yhat = Yhat.
sim,
 sampname = colnames(Y_sim), ref = ref_sim, chr = chri,
 mode = "integer", max.ns = 1)
}
iCN_sim <- do.call(rbind, lapply(segment_cs, function(z){z
[["iCN"]]}))
# Heatmap visualization
plot_iCN(iCNmat = iCN_sim, ref = ref_sim, Gini = Gini, filename
= "plot_iCN_demo")
```

3.6 Output

SCOPE returns three matrices as output. (1) Yhat.sim stores the normalized read depth with the number of latent factors selected by the Bayesian information criterion (BIC). (2) pool_sim aggregates the cross-sample CNV segmentation results, returning a set of genomic locations for shared breakpoints. This object is a matrix with two columns, "st" and "end," indicating the start and end genomic bin indices. (3) iCN_sim returns the inferred integer-valued copy numbers, which is used for visualization as shown in Fig. 2c.

```
# Normalized read depth matrix with EM algorithm
Yhat.sim <- normObj.scope.sim$Yhat[[which.max(normObj.scope.
sim$BIC)]]
# Cross-sample segmentation outputs
pool_sim <- do.call(rbind, lapply(segment_cs, function(z){z
[["poolcall"]]}))
# Inferred integer copy-number matrix by SCOPE
iCN_sim <- do.call(rbind, lapply(segment_cs, function(z){z
[["iCN"]]}))
```

4 Notes

1. SCOPE can also be applied to mouse scDNA-seq data. The correct mouse reference genome (mm10) needs to be specified for the computation of GC content and mappability. To compute mappability for mm10, there are two workarounds: (1) set all mappability to 1 to avoid extensive computation; (2) adopt QC procedures based on annotation results, e.g., filter out bins within the mouse "blacklist" regions, which generally have low mappability.

```
library(BSgenome.Mmusculus.UCSC.mm10)
mapp <- get_mapp(ref_raw, hgref = "mm10")
gc <- get_gc(ref_raw, hgref = "mm10")
```

2. Upon completion of normalization and segmentation at a first pass, SCOPE includes the option to cluster cells based on the matrix of normalized z-scores, estimated copy numbers, or estimated changepoints. Given the inferred subclones, SCOPE can opt to perform a second round of group-wise ploidy initialization and normalization.

```
# Group-wise ploidy initialization
# clones specifies the clonal assignment for each cell
clones <- c("normal", "tumor1", "normal", "tumor1", "tumor1")
ploidy.sim.group <- initialize_ploidy_group(Y = Y_sim, Yhat =
```

```
normObj.sim$Yhat, ref = ref_sim, groups = clones)
 # Group-wise normalization
 normObj.scope.sim.group <- normalize_scope_group(Y_qc =
Y_sim, gc_qc = ref_sim$gc, K = 1, ploidyInt = ploidy.sim.
group, norm_index = which(clones == "normal"), groups =
clones, T = 1:5, beta0 = normObj.sim$beta.hat)
```

3. SCOPE estimates total copy numbers. Recently developed methods, including CHISEL [24] and Alleloscope [25], characterize allele- and haplotype-specific copy numbers by aggregating sparse signals across loci and cells and can be used in a complementary allele-specific copy number analysis.

References

1. McCarroll SA, Altshuler DM (2007) Copy-number variation and association studies of human disease. Nat Genet 39(7 Suppl):S37–S42. https://doi.org/10.1038/ng2080
2. Jiang Y, Qiu Y, Minn AJ, Zhang NR (2016) Assessing intratumor heterogeneity and tracking longitudinal and spatial clonal evolutionary history by next-generation sequencing. Proc Natl Acad Sci U S A 113(37):E5528–E5537. https://doi.org/10.1073/pnas.1522203113
3. Urrutia E, Chen H, Zhou Z, Zhang NR, Jiang Y (2018) Integrative pipeline for profiling DNA copy number and inferring tumor phylogeny. Bioinformatics 34(12):2126–2128. https://doi.org/10.1093/bioinformatics/bty057
4. Benjamini Y, Speed TP (2012) Summarizing and correcting the GC content bias in high-throughput sequencing. Nucleic Acids Res 40(10):e72. https://doi.org/10.1093/nar/gks001
5. Sims D, Sudbery I, Ilott NE, Heger A, Ponting CP (2014) Sequencing depth and coverage: key considerations in genomic analyses. Nat Rev Genet 15(2):121–132. https://doi.org/10.1038/nrg3642
6. Iakovishina D, Janoueix-Lerosey I, Barillot E, Regnier M, Boeva V (2016) SV-Bay: structural variant detection in cancer genomes using a Bayesian approach with correction for GC-content and read mappability. Bioinformatics 32(7):984–992. https://doi.org/10.1093/bioinformatics/btv751
7. Jiang Y, Oldridge DA, Diskin SJ, Zhang NR (2015) CODEX: a normalization and copy number variation detection method for whole exome sequencing. Nucleic Acids Res 43(6):e39. https://doi.org/10.1093/nar/gku1363
8. Fromer M, Moran JL, Chambert K, Banks E, Bergen SE, Ruderfer DM, Handsaker RE, McCarroll SA, O'Donovan MC, Owen MJ, Kirov G, Sullivan PF, Hultman CM, Sklar P, Purcell SM (2012) Discovery and statistical genotyping of copy-number variation from whole-exome sequencing depth. Am J Hum Genet 91(4):597–607. https://doi.org/10.1016/j.ajhg.2012.08.005
9. Jiang Y, Wang R, Urrutia E, Anastopoulos IN, Nathanson KL, Zhang NR (2018) CODEX2: full-spectrum copy number variation detection by high-throughput DNA sequencing. Genome Biol 19(1):202. https://doi.org/10.1186/s13059-018-1578-y
10. Favero F, Joshi T, Marquard AM, Birkbak NJ, Krzystanek M, Li Q, Szallasi Z, Eklund AC (2015) Sequenza: allele-specific copy number and mutation profiles from tumor sequencing data. Ann Oncol 26(1):64–70. https://doi.org/10.1093/annonc/mdu479
11. Chen H, Jiang Y, Maxwell KN, Nathanson KL, Zhang N (2017) Allele-specific copy number estimation by whole exome sequencing. Ann Appl Stat 11(2):1169–1192. https://doi.org/10.1214/17-AOAS1043
12. Wang Y, Waters J, Leung ML, Unruh A, Roh W, Shi X, Chen K, Scheet P, Vattathil S, Liang H, Multani A, Zhang H, Zhao R, Michor F, Meric-Bernstam F, Navin NE (2014) Clonal evolution in breast cancer revealed by single nucleus genome sequencing. Nature 512(7513):155–160. https://doi.org/10.1038/nature13600
13. Kim C, Gao R, Sei E, Brandt R, Hartman J, Hatschek T, Crosetto N, Foukakis T, Navin NE (2018) Chemoresistance evolution in triple-negative breast cancer delineated by single-cell sequencing. Cell 173(4):879–893

e813. https://doi.org/10.1016/j.cell.2018.03.041
14. Liu J, Adhav R, Xu X (2017) Current progresses of single cell DNA sequencing in breast cancer research. Int J Biol Sci 13(8):949–960. https://doi.org/10.7150/ijbs.19627
15. Navin NE (2014) Cancer genomics: one cell at a time. Genome Biol 15(8):452. https://doi.org/10.1186/s13059-014-0452-9
16. Shlien A, Malkin D (2009) Copy number variations and cancer. Genome Med 1(6):62. https://doi.org/10.1186/gm62
17. Wang X, Chen H, Zhang NR (2018) DNA copy number profiling using single-cell sequencing. Brief Bioinform 19(5):731–736. https://doi.org/10.1093/bib/bbx004
18. Garvin T, Aboukhalil R, Kendall J, Baslan T, Atwal GS, Hicks J, Wigler M, Schatz MC (2015) Interactive analysis and assessment of single-cell copy-number variations. Nat Methods 12(11):1058–1060. https://doi.org/10.1038/nmeth.3578
19. Wang R, Lin DY, Jiang Y (2020) SCOPE: a normalization and copy-number estimation method for single-cell DNA sequencing. Cell Syst 10(5):445–452 e446. https://doi.org/10.1016/j.cels.2020.03.005
20. Li H, Durbin R (2009) Fast and accurate short read alignment with burrows-wheeler transform. Bioinformatics 25(14):1754–1760. https://doi.org/10.1093/bioinformatics/btp324
21. Li H, Handsaker B, Wysoker A, Fennell T, Ruan J, Homer N, Marth G, Abecasis G, Durbin R, Genome Project Data Processing S (2009) The sequence alignment/map format and SAMtools. Bioinformatics 25(16):2078–2079. https://doi.org/10.1093/bioinformatics/btp352
22. Derrien T, Estelle J, Marco Sola S, Knowles DG, Raineri E, Guigo R, Ribeca P (2012) Fast computation and applications of genome mappability. PLoS One 7(1):e30377. https://doi.org/10.1371/journal.pone.0030377
23. Navin N, Kendall J, Troge J, Andrews P, Rodgers L, McIndoo J, Cook K, Stepansky A, Levy D, Esposito D, Muthuswamy L, Krasnitz A, McCombie WR, Hicks J, Wigler M (2011) Tumour evolution inferred by single-cell sequencing. Nature 472(7341):90–94. https://doi.org/10.1038/nature09807
24. Zaccaria S, Raphael BJ (2021) Characterizing allele- and haplotype-specific copy numbers in single cells with CHISEL. Nat Biotechnol 39(2):207–214. https://doi.org/10.1038/s41587-020-0661-6
25. Wu CY, Lau BT, Kim HS, Sathe A, Grimes SM, Ji HP, Zhang NR (2021) Integrative single-cell analysis of allele-specific copy number alterations and chromatin accessibility in cancer. Nat Biotechnol 39(10):1259–1269. https://doi.org/10.1038/s41587-021-00911-w

Chapter 19

Variant Annotation and Functional Prediction: SnpEff

Pablo Cingolani

Abstract

Variant annotations, in general, refer to the process of information enrichment of genomic variants from a sequencing experiment. Typically these annotations include functional predictions, such as predicting the amino acid sequence changes from the DNA variant, predicting whether the variant will induce a splice anomaly, or predicting nonsense mediated decay. But other annotations also include combining with genomic databases, adding conservation scores, or comparing to allele frequencies from large population databases. Finally, all these annotations are combined to prioritize and filter variants into a reduced set of highly relevant variants for the study or clinical assay.

Key words Variant annotation, Functional annotations, Genomic variant prioritization, SnpEff, SnpSift

1 Introduction

Once DNA or RNA is sequenced, reads are aligned to the reference genome and genomic variants can be identified. The next step is to perform variant annotations in order to identify and prioritize variants, on the basis of their expected severity or impact over a specific phenotype.

Variant annotations are a broad category of analyses that add information to genomic variants. These include from the most fundamental aspects of "functional annotations," such as inferring how a coding variant could affect the protein sequence, to predictions, conservation scores, and merging information from other genomic databases (e.g. clinical information about established variants). But in all these cases, the fundamental goal is to add more information to make it easier to prioritize which variants can be considered likely to be "important" and which ones are likely to be less relevant in the context of our experiment or clinical assay.

SnpEff & SnpSift [5, 28] packages use highly optimized algorithms, to solve several variant annotation problems:

Charlotte K. Y. Ng and Salvatore Piscuoglio (eds.), *Variant Calling: Methods and Protocols*,
Methods in Molecular Biology, vol. 2493, https://doi.org/10.1007/978-1-0716-2293-3_19,

1. Standardizing the annotation process,
2. Calculate protein changes,
3. Estimate impact,
4. Estimate *loss of function* and *nonsense mediated decay*,
5. Add databases of genetic information,
6. Perform basic statistical analysis,
7. Facilitate variant filtering.

1.1 Reference Genome

Having a standard reference sequence is the key to establish comparisons and analysis. In order to compare DNA from different individuals (or samples), we need a *reference genome sequence* and *genomic annotations.*

Alignment and annotations must be based on the exact same reference genome sequence. Variants are called based on the reference genome, thus variant annotations must be performed using same reference genome. For instance, performing variant calling with respect to hg19 and then performing variant annotations using hg38 genome would result in completely erroneous results.

Genome annotation means identifying the functional elements within a reference genome sequence [2, 4, 26]. In a nutshell, this means locating the genomic coordinates of genes, transcripts, exons, introns, UTRs, and other functional elements of the genome [22, 24, 32]. For most studied organisms, reference genome sequences and genomic annotations are available, but the quality of both the reference sequence and the genomic annotations is highly variable depending on the organism. Another consideration is that sometimes a reference genome sequence has multiple genomic annotations. In this case, all variant annotations for a cohort must use the same genomic annotations in order to be comparable.

Genomic annotations identify multiple alternative transcripts per gene, also referred to as isoforms or splice variants. As we know that multiple transcripts exist in most genes in eukaryotic organisms. For instance, human genes have an average of 8.8 transcripts, according to Ensemble genomic annotations. But not all transcripts are equally important or equally well characterized. Some are far more abundant depending on the tissue and some have little evidence or are not well known, thus poorly annotated. Transcript Support Level (TSL), a score created by ENSEMBL, indicates a number from one (well characterized) to five (little evidence), helps to sort out the most relevant transcripts.

Alternative splicing also means that a variant in a genomic locus can have multiple effects on different transcripts for the same gene. For instance, a variant might be changing an amino acid in one transcript, but intronic in another transcript. Another complication raises when two genes overlap. In this case a variant may have multiple effects on each transcript for each gene. This sometimes makes it difficult to specify if a variant is "synonymous" or "non-synonymous" (e.g. when calculating Ks/Ka ratio), because the counts become somewhat ill-posed when multiple transcripts have different effects.

Canonical transcripts have emerged as a way to simplify the complexity created by isoforms. The main concept was to select one "representative" transcript per gene. Usually, the definition of Canonical Transcript is related to the longest coding sequence (CDS), the longest transcript, or the "most prevalent" transcript. The issue is that there is no uniform definition of canonical, since RefSeq, ENSEMBL, and Uniprot have inconsistent definitions. To make matters worse, "canonical transcript" definitions might even change in different versions or releases, for instance, hg19's definition of canonical is different than hg38's one. It is easy to see how arbitrary "canonical transcripts" are defined by just comparing the most common definitions. As a consequence, it is also easy to see that these definitions might not necessarily relate to the most biologically relevant transcript of a gene. Particularly when "most relevant" is intrinsically different for different studies or clinical assays.

1.2 Human Annotations

Human genomic variant annotation come with some complexities that might have profound implications in the results. Here we describe some common issues that we need to take into account since they can affect the results of genomic analyses.

The human genome is arguably the most studied genome, thus having the highest quality annotations. This also means that there are multiple sources of genomic annotations, NCBI and ENSEMBL being the most widespread. Furthermore, there are several *"transcript sets"* based on ENSEMBL and NCBI reference genome annotations, shaping a complex landscape. Here we enumerate some of the well known genome annotation transcript sets, as well as a summary of their characteristics, this is by no means exhaustive, it is just for illustrative purposes:

1. Ensembl: *Ensembl transcripts are products of the Ensembl automatic gene annotation system (a collection of gene annotation pipelines), termed the Ensembl annotation process. All Ensembl transcripts are based on experimental evidence and thus the automated pipeline relies on the mRNAs and protein sequences deposited into public databases from the scientific community. Manually curated transcripts are produced by the HAVANA group* [4].
2. RefSeq: *The Reference Sequence (RefSeq) collection provides a comprehensive, integrated, non-redundant, well-annotated set of sequences, including genomic DNA, transcripts, and proteins* [22]

3. LRG: *A Locus Reference Genomic (LRG) record is manually curated and contains stable genomic, transcript, and protein reference sequences for reporting clinically relevant sequence variants. LRGs are stable and therefore are not versioned, thus reducing ambiguity when reporting variants. When an LRG is established for any gene, the RefSeqGene and its annotation will be "frozen" to match that of the LRG* [6].
4. Gencode: *The GENCODE annotation is made by merging the manual gene annotation produced by the Ensembl-Havana team and the Ensembl-genebuild automated gene annotation. The GENCODE releases coincide with the Ensembl releases. In practical terms, the GENCODE annotation is essentially identical to the Ensembl annotation* [12].
5. GENCODE Basic: *It is a subset of the GENCODE gene set and is intended to provide a simplified, high-quality subset of the GENCODE transcript annotations that will be useful to the majority. This subset prioritizes full-length protein-coding transcripts over partial or non-protein-coding transcripts within the same gene* [11].
6. CCDS: *Consensus Coding Sequence Set is consistently annotated between Ensembl, MGI, HGNC, and NCBI* [25]
7. RefSeq select: *The RefSeq Select dataset consists of a representative or "Select" transcript for every protein-coding gene. The transcript is chosen by an automated pipeline based on multiple selection criteria, which include prior use in clinical databases (e.g., Locus Reference Genomic), transcript expression, conservation of the coding region, transcript and protein length and concordance with the Swiss-Prot canonical isoform. The RefSeq Select transcript is usually well-supported by archived data, well-expressed, conserved and represents the biology of the gene.*
8. KnownGene: *UCSC Known Genes dataset is constructed by a fully automated process, based on protein data from Swiss-Prot/TrEMBL (UniProt) and the associated mRNA data from Genbank*

As expected, this is somewhat confusing, particularly because the transcripts between the above-mentioned sets may often disagree. Unsurprisingly, as some studies point out, different genomic references affect the outcome of genomic annotations [20]. In some cases, low agreement was found between REFSEQ and ENSEMBL transcripts sets, even in high impact "loss of function" variants. A typical source of misunderstanding is when a scientist annotates variants with one transcript set and then uses a genome browser to visualize the variant, but the genome browser uses a different transcript set, so the variant annotation and the browser do not match. SnpEff's `show` command outputs a representation of transcripts or genes (*see* Notes 3.4), which can be helpful to understand the reasons for some of these "perceived mismatches."

An often overlooked aspect of the NCBI RefSeq transcripts is that RefSeq transcripts may not necessarily match the reference genome [24]. This is quite surprising to many researches and clinical professionals. The differences are usually limited to just one or few bases within the whole transcript, but the total number of transcripts affected is in the range from 3% to 7% (unpublished data). Since variant calling and variant annotations are performed with respect to a reference genome, both variant calls and annotations would be "erroneous" at specific locations where there are differences between a RefSeq transcript and the Reference Genome. SnpEff performs an analysis of the internal databases and flags those transcripts that do not match the reference genome, giving the user an options to filter them out during the analysis process. Furthermore, in some cases a RefSeq transcripts may be aligned to multiple genomic regions (*see* Notes 3.3), with each of these alignments having different mismatches with respect to reference genome. An option to avoid these problems altogether is to use ENSEMBL genomic annotations, since ENSEMBL transcripts do match the reference genome.

The need of a consistent genomic variant nomenclature is unquestionable. This would greatly facilitate searching variants in clinical databases as well as scientific journals. HGVS nomenclature from the Human Genome Variation Society is a "Sequence Variant Nomenclature" that is widely used, particularly in clinical environments. The nomenclature aims to have a consistent and unambiguous description of variants [8], for clinical diagnostics. Nevertheless the nomenclature has evolved throughout the years and there have been several different versions with inconsistent recommendations. The HGVS nomenclature can also be quite complex, to the point where it seems confusing even for experts. As an example, in a study [7] where a single variant was sent to 20 laboratories, there were 14 different HGVS syntaxes. Given these complexities, many clinical database have been found to have low concordance in HGVS notation, or even large number of errors in variants reported with HGVS notation [31].

On top of that HGVS is difficult to implement programmatically. As an example, HGVS "sometimes" recommends to shift the variants "right" with respect to the transcript, whereas VCF specification requires to always shift left with respect to the genome. These HGVS realignments add a layer of complexity to variant annotation programs. In order to follow HGVS recommendations, SnpEff must sometimes (depending on the variant type, transcript strand, and effect type) realign the variants and change their genomic coordinates. This can catch off guard many scientist who are unaware of this side effect of the HGVS notation and wonder why the variant annotation software is reporting some variants as if they were in "another location." In order to warn the users that such realignment occurred, an `INFO_REALIGN_3_PRIME` message is added to annotation in the VCF entry.

1.3 Clinical Annotations

Compliance with clinical requirements adds some subtleties to variant annotation processing. It is important to understand that clinical environments need to value consistency and compliance over innovation. That is why, despite human genome version 38 (GRCh38) being released in 2013 [29], many clinical services still rely on GRCh37/hg19. This is probably not only due to the costs of performing clinical validations when moving to a new genome reference but also due to a wealth of knowledge and experience developed on old reference genome. This experience includes clinical knowledge databases that companies developed during many years, as well as calibrated analysis pipelines that can filter out sequencing artifacts and noise.

The topic of which transcript sets (RefSeq or ENSEMBL) should be used in clinical environments was subject of different inter-related guidelines from different organizations. For example, old guidelines from ACMG (American College of Medical Pathology) and AMP (Association of Molecular Pathology) from 2015 recommended to use RefSeq transcript IDs [27]:

> The reference sequence should be complete and derived from either the National Center for Biotechnology Information RefSeq database ... Locus Reference Genomic database....

While CAP (College of American Pathologist) suggested to use either RefSeq or ENSEMBL for somatic variants but recommended to follow ACMG/AMP recommendations for germline variants.

These recommendations were corrected later in 2017 [18] and nowadays both Ensembl and RefSeq transcripts can be used:

> Human sequence variants must be reported using HGVS nomenclature and include the HUGO Gene Nomenclature Committee (HGNC) gene name, and a standard versioned reference identifier to the transcript/protein (e.g. RefSeq Accession Number, Ensembl Transcript/Protein ID, or CCDS ID) that allows unambiguous mapping of the variant.

But in practice, most labs were already following the old recommendation when this update was released, and it is now not worth to switch genomic references. As a consequence, clinical assays in the USA are almost exclusively based on RefSeq transcripts.

Fortunately both NCBI and ENSEMBL have been joining efforts in a new initiative to provide a joint transcript set compatible with both. *MANE* stands for "Matched Annotation from NCBI and EMBL-EBI." This is a relatively new effort, started in 2018, to converge into a common set of transcripts that has desirable characteristics:

(a) contains one well-supported transcript per protein-coding locus,
(b) perfectly align to the reference genome,
(c) has 100% match between RefSeq and ENSEMBL (including coding sequence and UTRs),
(d) has been manually curated by both groups,
(e) is versioned and largely stable,
(f) includes additional transcripts (i.e. more than one per gene) required to report variants of clinical interest, and
(g) will become the default transcripts shown in Genome Browsers.

As of this writing, version 0.93 was released in January 2021. Furthermore, the upcoming ENSEMBL release 104 is expected to switch the definition of "Canonical Transcript" to favor MANE transcripts.

2 Methods

In this protocol we show how to analyze genomic variants using the SnpEff [5] and SnpSift [28]. Both SnpEff and SnpSift have multiple "sub-commands" to help annotate and manipulate annotations. Some common sub-commands can be found in Tables 1 and 4.

We show three examples:

1. Functional annotations and Human databases,
2. Familial genomic analysis,
3. Build SnpEff database: Sars-Covid2.

In each step, an approximate running time is provided as a reference; this was measured by running command line on an OS. X laptop.

The dataset for these examples can be downloaded as:

```
wget -nc "https://snpeff.blob.core.windows.net/databases/dbs/protocols_202104.zip
unzip protocols_202104.zip
```

Requirements The computer hardware & software required for this protocol are:

1. *a computer running a Unix operating system (Linux, OS.X),*
2. *at least 8 GB of RAM,*
3. *at least 200 GB of free disk space,*
4. *Java version 1.12 or higher installed, and*
5. *a reasonably fast internet connection.*

Table 1
List of SnpEff commands

Command	Meaning
[eff—ann]	This is the default command. It is used for annotating variant filed (e.g. VCF files).
build	Build a SnpEff database from reference genome files (FASTA, GTF, etc.).
buildNextProt	Build NextProt database using XML files
cds	Compare CDS sequences calculated form a SnpEff database to FASTA file.
closest	Annotate the closest genomic region.
count	Count how many intervals (from a BAM, BED or VCF file) overlap with each genomic interval.
databases	Show currently available databases (from local config file).
download	Download a SnpEff database.
dump	Dump to STDOUT a SnpEff database (mostly used for debugging).
genes2bed	Create a bed file from a genes list.
len	Calculate total genomic length for each marker type.
protein	Compare protein sequences calculated form a SnpEff database to FASTA file.
spliceAnalysis	Perform an analysis of splice sites. Experimental feature.

Operating System Being a Java application, SnpEff runs on any modern version of OS.X, Unix, and Windows. All the steps in this tutorial will follow a Unix/OS.X command line style and file paths, but these can be easily adapted for Windows systems.

Download SnpEff can be downloaded from the web page *https://pcingola.github.io/SnpEff*, by clicking on the "Download" button/link. For these examples, we are using SnpEff & SnpSift versions 5.0, which is the latest versions at the time of writing. The package is downloaded as single a "Zip" file.

Installing SnpEff Installing SnpEff's package is just uncompressing in your computer's `HOME` directory. Typically OS.X and Windows system allow opening ZIP files by double clicking on them, and in Unix systems you can use the "unzip" shell command. The destination directory where SnpEff is installed is typically `$HOME/snpEff`, so we will follow that convention. If you decide to install somewhere else, you will have to adapt the paths shown in the commands accordingly, as well as update the `data_dir` parameter in your `snpEff.config` file, as described in SnpEff's documentation.

```
cd # Move to home directory
unzip path/to/snpEff_latest_core.zip # Unzip downloaded
SnpEff package
```

Note For SnpEff 5.0 you will need Java 12.0 or higher installed. This requirements will change for future SnpEff versions, so please check the requirements at *https://pcingola.github.io/SnpEff*. In order to check which Java version is installed in your system, you can run the command `java --version`

SNPEFF_HOME In all cases, the commands assume you are running from a terminal, within SnpEff's install directory, which we refer to as *SnpEff's home*. Assuming SnpEff was installed in the recommended path, you should run the command `cd $HOME/snpEff` to go to *SnpEff's home*. For convenience we assume you have a `SNPEFF_HOME` environment variable defined as follows:

```
export SNPEFF_HOME= "$HOME/snpEff" # Change to the appro-
priate path if necessary
```

SnpEff Databases In this tutorial, we will be using human genome, MANE transcripts from ENSEMBL (`GRCh38.mane.0.93.ensembl`). In order to download and install `GRCh38.mane.0.93.ensembl`, you can run the command:

```
java -jar snpEff download -v GRCh38.mane.0.93.ensembl
```

The "verbose" `-v` command line option shows download and install progress (see more command line options in Table 2).

SnpEff currently supports over 38,000 genomes, so there is a high chance that your organism is already supported. SnpEff databases are hosted by Microsoft Genomics https://www.microsoft.com/en-us/genomics/ on Azure Cloud infrastructure, which allowed to overcome several limitations related to the number and sizes of genomic databases. All SnpEff databases can be downloaded and installed using a `SnpEff download` command line similar to one shown above. A list of pre-built databases for all other species is available by running the following command:

```
java -jar snpEff.jar databases
```

Genomic Databases We will also be using variant databases: dbSnp, ClinVar, dbNSFP, and Gwas Catalog. As it often happens with genomic data, some of these databases are quite large (tens of GB) and might take a while to download.

dbSnp (Single-Nucleotide Polymorphism Database) Download from NCBI's DbSnp site, version: 151, size: 15 GB Warning: dbSnp is a large database and might take a while to download

Table 2
Snpeff: generic command line options

Option	Meaning
-c , -config	Specify config file
-d , -debug	Debug mode (very verbose).
-dataDir <path>	Override data_dir parameter from config file.
-download	Download a SnpEff database, if not available locally. Default: true
-nodownload	Do not download a SnpEff database, if not available locally.
-noShiftHgvs	Do not shift variants towards most 3-prime position (as required by HGVS).
-h , -help	Show this help and exit
-noLog	Do not report usage statistics to server
-t	Use multiple threads (implies '-noStats'). Default 'off'
-q , -quiet	Quiet mode (do not show any messages or errors)
-v , -verbose	Verbose mode

```
mkdir "$SNPEFF_HOME/db/GRCh38/dbSnp"
cd "$SNPEFF_HOME/db/GRCh38/dbSnp"
wget -nc "https://ftp.ncbi.nih.gov/snp/organisms/hu-
man_9606_b151_GRCh38p7/VCF/00-All.vcf.gz"
wget -nc "https://ftp.ncbi.nih.gov/snp/organisms/hu-
man_9606_b151_GRCh38p7/VCF/00-All.vcf.gz.tbi"
 # Create links for more intuitive names
ln -s 00-All.vcf.gz dbsnp.vcf.gz
ln -s 00-All.vcf.gz.tbi dbsnp.vcf.gz.tbi
```

ClinVar aggregates information about genomic variation and its relationship to human health [16]. Download from NCBI's ClinVar, version: 2021-05-01, size: 35 MB

```
mkdir "$SNPEFF_HOME/db/GRCh38/clinVar"
cd "$SNPEFF_HOME/db/GRCh38/clinVar"
wget "https://ftp.ncbi.nlm.nih.gov/pub/clinvar/vcf_GRCh38/
clinvar.vcf.gz"
wget "https://ftp.ncbi.nlm.nih.gov/pub/clinvar/vcf_GRCh38/
clinvar.vcf.gz.tbi"
```

dbNSFP is a database developed for functional prediction and annotation of all potential non-synonymous single-nucleotide variants (nsSNVs) in the human genome [19]. It includes many conservation scores and prediction algorithms, such as SIFT [21], Polyphen2 [3], CADD [15], PhyloP [23], phastCons [30], etc.

On top of that dbNSFP also provides allelic frequencies from "1000 Genomes Project" [1], ExAC [17], and GnomAD [13]. This database is transformed from the original format, so you must download it from SnpEff's site. Download from SnpEff's databases, version: 4.1b, size: 31 GB Warning: dbNSFP is a large database and might take a while to download

```
mkdir "$SNPEFF_HOME/db/GRCh38/dbNSFP"
cd "$SNPEFF_HOME/db/GRCh38/dbNSFP"
wget -nc "https://snpeff.blob.core.windows.net/databases/
dbs/GRCh38/dbNSFP_4.1a/dbNSFP4.1a.txt.gz"
wget -nc "https://snpeff.blob.core.windows.net/databases/
dbs/GRCh38/dbNSFP_4.1a/dbNSFP4.1a.txt.gz.tbi"
```

Gwas Catalog is a curated collection of human genome-wide association studies (GWAS). Download from EBI, version: 1.02, size: 35 MB

```
mkdir "$SNPEFF_HOME/db/GRCh38/gwasCatalog"
cd "$SNPEFF_HOME/db/GRCh38/gwasCatalog"
wget -nc -O gwascatalog.txt "https://www.ebi.ac.uk/gwas/
api/search/downloads/alternative"
```

2.1 Case 1: Functional Annotations and Human Databases

As an example, in this section we annotate a well know human whole genome NA12878 from Illumina's Platinum Genomes [10]. The original variant calls are in AWS S3 *s3://platinum-genomes/2017-1.0/hg38/small_variants/NA12878/NA12878.vcf.gz* The VCF contains approximately 4.2 million variants, for speed reasons, we will use a subset of the variants consisting of 50,000 exonic variants.

We will annotate with functional annotations and well known databases: dbSnp, ClinVar, dbNSFP, and Gwas Catalog.

Step 1 Annotate using GRCh38 MANE transcripts (time: 1 min):

```
java -Xmx8g -jar snpEff.jar ann -v \
    GRCh38.mane.0.93.ensembl \
    protocols/NA12878.vcf.gz \
    > NA12878.ann.vcf
```

For more command line options *see* Tables 3 and 2

Step 2 Annotate using DbSnp annotations (time: 3 min). Note that when performing this annotations, the ID field in the VCF file will be populated with dbSnp information (i.e. an "rs number") for the variants matching a dbSnp entry (Table 4).

Table 3
Database command line options

Option	Meaning
-canon	Only use canonical transcripts.
-interval	Use a custom intervals. You may use this option many times)
-motif	Annotate using motifs (requires Motif database).
-nextProt	Annotate using NextProt (requires NextProt database).
-noGenome	Do not load any genomic database (e.g. annotate using custom files).
-noMotif	Disable motif annotations.
-noNextProt	Disable NextProt annotations.
-onlyReg	Only use regulation tracks.
-onlyProtein	Only use protein-coding transcripts. Default: false
-onlyTr <file.txt>	Only use the transcripts in this file. Format: One transcript ID per line.
-reg <name>	Regulation track to use (this option can be used add several times).
-ss , -spliceSiteSize <int>	Set size for splice sites (donor and acceptor) in bases. Default: 2
-strict	Only use 'validated' transcripts (i.e. sequence has been checked). Default: false

```
java -Xmx8g -jar SnpSift.jar ann -v \
    db/GRCh38/dbSnp/dbsnp.vcf.gz \
    NA12878.ann.vcf \
    > NA12878.dbsnp.ann.vcf
```

Step 3 Add ClinVar annotations (time: less than a minute)

```
java -Xmx8g -jar SnpSift.jar ann -v \
    db/GRCh38/clinVar/clinvar.vcf.gz \
    NA12878.dbsnp.ann.vcf \
    > NA12878.clinvar.dbsnp.ann.vcf
```

Step 4 Add dbNSFP annotations (time: 25 min)

```
java -Xmx8g -jar SnpSift.jar dbnsfp -v \
    -db db/GRCh38/dbNSFP/dbNSFP4.1a.txt.gz \
    NA12878.clinvar.dbsnp.ann.vcf \
    > NA12878.dbnsfp.clinvar.dbsnp.ann.vcf
```

Step 5 Add Gwas Catalog annotations (time: less than a minute)

Table 4
SnpSift commands

Command	Meaning
alleleMat	Create an allele matrix output.
annotate	Annotate 'ID' from a database (e.g. dbSnp). Assumes entries are sorted.
caseControl	Compare how many variants are in 'case' and in 'control' groups; calculate p-values.
ccs	Case control summary.
concordance	Concordance metrics between two VCF files.
covMat	Create an covariance matrix output (allele matrix as input).
dbnsfp	Annotate with multiple entries from dbNSFP.
extractFields	Extract fields from VCF file into tab separated format.
filter	Filter using arbitrary expressions
geneSets	Annotate using MSigDb gene sets (MSigDb includes: GO, KEGG, Reactome, BioCarta, etc.)
gt	Add Genotype to INFO fields and remove genotype fields when possible.
gtfilter	Filter genotype using arbitrary expressions.
gwasCat	Annotate using GWAS catalog
hwe	Calculate Hardy-Weimberg parameters and perform a goodness of fit test.
intersect	Intersect intervals (genomic regions).
intervals	Keep variants that intersect with intervals.
intIdx	Keep variants that intersect with intervals (index-based)
join	Join files by genomic region.
op	Annotate using an operator.
phastCons	Annotate using conservation scores (phastCons).
private	Annotate if a variant is private to a family or group.
rmRefGen	Remove reference genotypes.
rmInfo	Remove INFO fields.
sort	Sort VCF file/s (if multiple input VCFs, merge and sort).
split	Split VCF by chromosome.
tstv	Calculate transition to transversion ratio.
varType	Annotate variant type (SNP,MNP,INS,DEL or MIXED).
vcfCheck	Check that VCF file is well formed.
vcf2tped	Convert VCF to TPED.

```
java -Xmx8g -jar SnpSift.jar gwasCat -v \
    -db db/GRCh38/gwasCatalog/gwascatalog.txt \
    NA12878.dbnsfp.clinvar.dbsnp.ann.vcf \
    > NA12878.gwascat.dbnsfp.clinvar.dbsnp.ann.vcf
```

All in One SnpEff and SnpSift accept input from standard input (STDIN), so all the previous annotations can be combined in one step, using pipes (time 25 min):

```
 time java -Xmx8g -jar snpEff.jar ann -v GRCh38.mane.0.93.
ensembl protocols/NA12878.vcf.gz \
      | java -Xmx8g -jar SnpSift.jar ann -v db/GRCh38/dbSnp/
dbsnp.vcf.gz \
      | java -Xmx8g -jar SnpSift.jar ann -v db/GRCh38/clinVar/
clinvar.vcf.gz \
      | java -Xmx8g -jar SnpSift.jar dbnsfp -db db/GRCh38/
dbNSFP/dbNSFP4.1a.txt.gz - \
      | java -Xmx8g -jar SnpSift.jar gwasCat -db db/GRCh38/
gwasCatalog/gwascatalog.txt - \
      > NA12878.ann.vcf
```

2.2 Case 2: Family Tree

The goal in this example is to use SnpEff to find a mutation causing a Mendelian recessive trait. The data is a contrived example from a publicly available "CEPH 1463" family dataset, sequenced by Complete Genomics [9], consisting of 4 grandparents, 2 parents, and 11 siblings.

From this pedigree of healthy individuals, a variant causing cystic fibrosis was artificially introduced in three siblings. For the purpose of this example, we assume that we do not know the causative variant, but that we know that we are dealing with a Mendelian recessive disorder (*see* Fig. 1), where the three siblings are affected (cases), but the 14 other siblings, parents, and grandparents are not (controls).

Genomic variants are usually provided in a VCF file containing variant information of all the samples; storing the variant data in a single VCF file is the standard practice, not only because variant calling algorithms have better accuracy when run on all samples simultaneously, but also because it is much easier to annotate, manipulate, and compare individuals when the data is stored and transferred together. A caveat of this approach is that VCF files can become very large when performing experiments with thousands of samples (from several Gigabytes to Terabytes in size).

First, we use SnpEff to add annotation fields to each variant record in the input VCF file. Then SnpSift is used to filter the most significant variants having annotations meeting certain criteria.

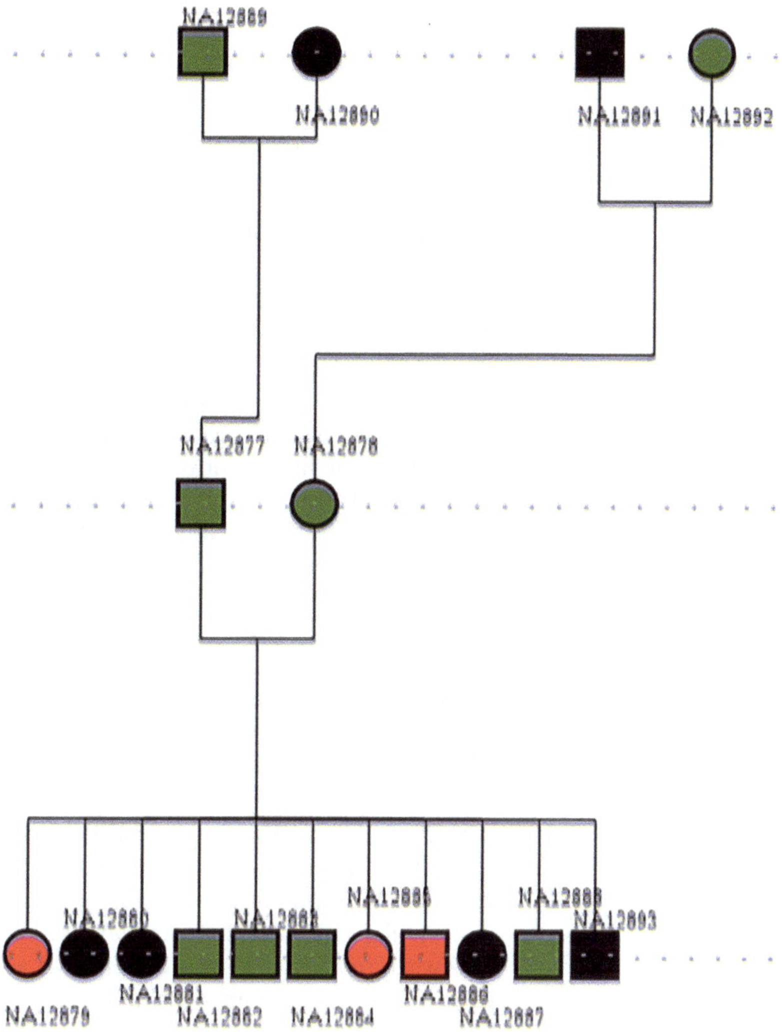

Fig. 1 Family tree for CEPH 1463. In this dataset, artificially generated "affected" individuals (red) were introduced for illustrative purposes

Functional Variant Annotations Our first step is to annotate each of the 500,000 variants contained in the VCF file. Note that we are using GRCh37, because the variant calling was performed using GRCh37/hg19. The first time we run the command, SnpEff will attempt to download and install the database, if it is not already installed.

```
java -Xmx4g -jar snpEff.jar -v \
    -stats ex1.html \
    GRCh37.75 \
    protocols/ex1.vcf \
    > protocols/ex1.ann.vcf
```

SnpEff produces three output files :

1. the HTML file containing summary statistics about the variants and their annotations,
2. an annotated VCF file, and
3. a text file summarizing the number of variant types per gene.

Quality Control SnpEff creates summary files for quality control. The summary file "ex1.html" is a standard HTML file that can be opened in any web browser to view quality control (QC) metrics. It can also be created in comma-separated values format (CSV) to be used by downstream processing programs as part of an automated pipeline. In our example, the summary file contains basic quality control statistics calculated from the variant file. One of the first obvious QC metrics is the number of errors and warnings. Having too many `ERROR_CHROMOSOME_NOT_FOUND` or `WARNING_REF_DOES_NOT_MATCH_GENOME` usually indicates a mismatch between variant calling and variant annotation genome versions (*see* **Notes 3.1** for common errors and warnings).

A typical statistics is the T_s/T_v ratio, which in case of human data should be close to 2.0 (Fig. 2c), similarly the missense/silent ratio should be around 1.0 (Fig. 2d), both of which are within expected ranges for this type of human variant analyses. Note that expected values for Whole Genome Sequencing (WGS) are slightly different than Whole Exome Sequencing (WES) or Panel. Large deviations from the expected values for the organism being sequenced might indicate problems with either the sequencing or variant calling pipelines. The summary file also contains QC information for the gene annotation used as input. In this example, 829 warnings (Fig. 1a) were identified as a result of possible genomic annotation errors or small inconsistencies identified in the reference genome so we have to carefully analyze those genes/transcripts. Other summary statistics are available, such as variant types (Fig. 1e), variants effects (Fig. 1d and g), and primary impacts (Fig. 1b and g).

In this step of our protocol, SnpEff created a VCF file with half million annotated variants. Obviously scanning variants manually are not feasible, so we next use the SnpSift to add statistics on cases and controls, and then filter variants to a small subset of variants with interesting high functional impact. Our approach is to create filters using an increasingly more restrictive set of criteria.

Cases and Controls In our example the pedigree is consistent with a Mendelian recessive disease, so we will first use SnpSift to find high impact variants that are homozygous in cases and either absent or heterozygous in controls. This genetic rationale to select the promising variants will be translated as the corresponding filter expression in the next stage. To do this, we will identify cases and control samples by providing SnpEff with pedigree information using a "TFAM" file, a standard file format used to describe

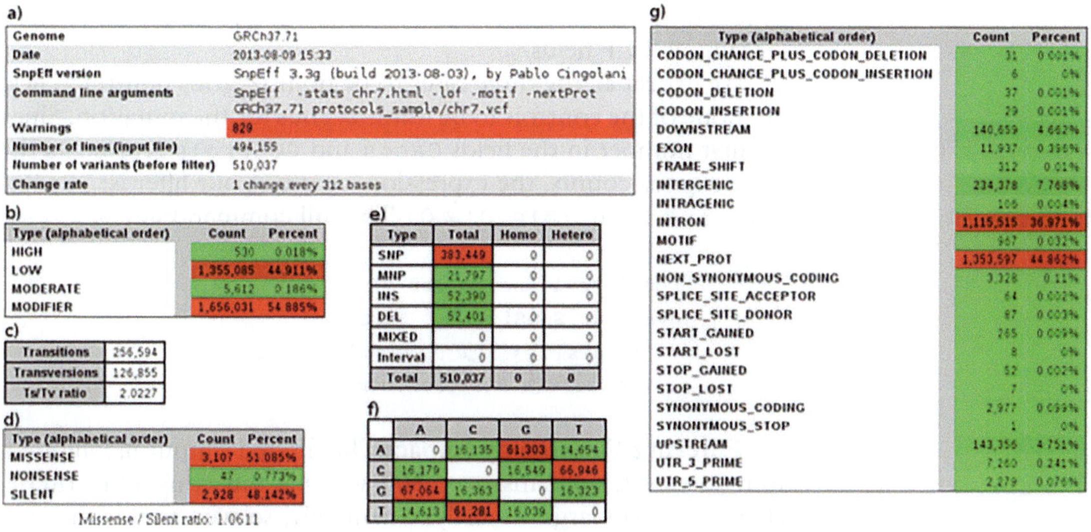

Fig. 2 SnpEff summary (HTML) report

pedigrees. In our example, the TFAM file (`pedigree.tfam`) identifies the three cases (NA12879, NA12885, NA12886) and lists the other family members as controls.

The `SnpSfit caseControl` command counts the number homozygous non-reference, heterozygous, and allele count (i.e. number of non-reference alleles in each DNA sample) for both cases and controls groups (running time: 60 min):

```
java -Xmx1g -jar SnpSift.jar caseControl -v \
    -tfam protocols/pedigree.tfam \
    protocols/ex1.ann.vcf \
    > protocols/ex1.ann.cc.vcf
```

This analysis creates an output VCF file (`ex1.ann.cc.vcf`) with new information in the `INFO` field for each variant. The `INFO` fields called `Cases` and `Controls`, alternative alleles, contain counts (i.e. number of non-reference alleles). Each field has three numbers representing the counts of homozygous non-reference, heterozygous, and total non-reference allele counts. For instance, `Cases=1,2,4` means that there is: 1 homozygous non-reference, 2 heterozygous, and a total of 4 non-reference alleles ($2*1+2=4$). SnpSift also calculates basic statistics for each variant based on the allele frequencies in the two groups: Cochran-Armitage, ChiSquaredDistribution, and Fisher exact tests for allelic, dominant, and recessive models. These can be useful as a starting point for more in-depth statistical analysis or simply to help filtering variants.

Filtering Variants by Cases & Control Counts We can use the `SnpSift filter` command to reduce the number of candidate loci base on alleles in cases and controls. SnpSift filter allows users to create powerful filters that select variants using arbitrary expressions containing VCF fields.

We expect all the three cases to be homozygous non-reference, and none of the controls to be homozygous for the mutation. Since the first number in the fields `Cases` and `Controls` represents the homozygous counts, the expression we use in our filter is: `(Cases[0] = 3) & (Controls[0] = 0)` The full command line is:

```
cat protocols/ex1.ann.cc.vcf \
    | java -jar SnpSift.jar filter \
        "(Cases[0]␣=␣3)␣&␣(Controls[0]␣=␣0)" \
    > protocols/ex1.filtered.hom.vcf
```

Filtering Variants by Impact The filtered output file, filtered.hom_cases.vcf, contains over 400 variants satisfying our criteria. This is still too large to analyze manually, so we can add another filter to narrow down the criteria. We can filter variants that are expected to have a high impact.

SnpEff adds a simple "impact" assessment for each variant:

1. `HIGH`: The variant is assumed to have high (disruptive) impact in the protein, probably causing protein truncation, loss of function or triggering nonsense mediated decay. For example, `stop_gained`, `frameshift_variant`
2. `MODERATE`: A non-disruptive variant that might change protein effectiveness. For example, `missense_variant`, `inframe_deletion`
3. `LOW`: Assumed to be mostly harmless or unlikely to change protein behavior. For example, `synonymous_variant`
4. `MODIFIER`: Usually non-coding variants or variants affecting non-coding genes, where predictions are difficult or there is no evidence of impact. For example, `exon_variant`, `downstream_gene_variant`

To identify variants classified as `HIGH` impact, we use the SnpSift expression `ANN[*].IMPACT = HIGH`. The new filtering commands become:

```
cat protocols/ex1.ann.cc.vcf \
    | java -jar SnpSift.jar filter \
        ''(Cases[0] = 3) & (Controls[0] = 0) & (ANN[*].IMPACT
= 'HIGH')'' \
    > protocols/ex1.filtered.vcf
```

After filtering, only two variants satisfy our criteria, one of them is a `stop_gained` loss of function variant, also a known Cystic fibrosis variant [14].

```
$ cat protocols/ex1.filtered.vcf | ./scripts/vcfInfoOnePer-
Line.pl
 7 117227832 . G T . .
 AC 14
 AN 22
 ANN T|stop_gained|HIGH|CFTR|ENSG00000001626|transcript|
ENST00000003084|protein_coding|12/27|c.1624G>T|p.Gly542*|
1756/6128|1624/4443|542/1480||
 ANN T|stop_gained|HIGH|CFTR|ENSG00000001626|transcript|
ENST00000454343|protein_coding|11/26|c.1441G>T|p.Gly481*|
1573/5949|1441/4260|481/1419||
 ANN T|stop_gained|HIGH|CFTR|ENSG00000001626|transcript|
ENST00000426809|protein_coding|11/26|c.1534G>T|p.Gly512*|
1534/4316|1534/4316|512/1437||WARNING_TRANSCRIPT_INCOMPLETE
 ANN T|sequence_feature|LOW|CFTR|ENSG00000001626|topologi-
cal_domain:Cytoplasmic|ENST00000003084|protein_coding||
c.1624G>T|||||
 ANN T|sequence_feature|LOW|CFTR|ENSG00000001626|domain:
ABC_transporter_1|ENST00000003084|protein_coding||
c.1624G>T|||||
 ANN T|sequence_feature|LOW|CFTR|ENSG00000001626|beta_-
strand|ENST00000003084|protein_coding|12/27|c.1624G>T|||||
 ANN T|sequence_feature|LOW|CFTR|ENSG00000001626|beta_-
strand|ENST00000454343|protein_coding|11/26|c.1441G>T|||||
 ANN T|upstream_gene_variant|MODIFIER|AC000111.5|
ENSG00000234001|transcript|ENST00000448200|processed_pseudo-
gene||n.-1C>A||||1362|
 ANN T|downstream_gene_variant|MODIFIER|CFTR|
ENSG00000001626|transcript|ENST00000472848|processed_tran-
script||n.*148G>T||||29|
 LOF (CFTR|ENSG00000001626|11|0.27)
 NMD (CFTR|ENSG00000001626|11|0.27)
 Cases 3
 Cases 0
 Cases 6
 Controls 0
 Controls 8
 Controls 8
 CC_TREND 9.111e-04
 CC_GENO NaN
 CC_ALL 4.025e-02
 CC_DOM 6.061e-03
 CC_REC 1.000e+00
```

A chart showing how the variant propagates across the pedigree structure can be created as follows (note that this is how Fig. 1 was generated):

```
java -jar SnpSift.jar pedShow \
    protocols/pedigree.tfam \
    protocols/ex1.filtered.vcf \
    protocols/chart
```

Filters Using ClinVar Annotations So far, since the purpose of the example was to show how annotations and filtering are performed to uncover new variants, we assumed that the causative variant was not known. Since the variant is known ClinVar has this information in convenient VCF format that can be used for annotations. We can annotate using ClinVar by using the following command:

```
java -Xmx1g -jar SnpSift.jar annotate -v \
    protocols/db/clinvar_00-latest.vcf \
    protocols/ex1.ann.cc.vcf \
    > protocols/ex1.ann.cc.clinvar.vcf
```

Our variant of interest is then annotated as "Cystic Fibrosis." To find the variant, we filter for variants having ClinVar annotation `CLNDBN` that are in CFTR gene and have a `stop_gained` annotation:

```
$ cat protocols/ex1.ann.cc.clinvar.vcf \
    | java -jar SnpSift.jar filter \
        "(exists CLNDBN) & (ANN[*].EFFECT has 'stop_gained') & 
(ANN[*].GENE = 'CFTR')" \
    > protocols/ex1.ann.cc.clinvar.filtered.vcf

$ cat protocols/ex1.ann.cc.clinvar.filtered.vcf | ./scripts/
vcfInfoOnePerLine.pl
7 117227832 rs113993959 G T . .
AC 14
AN 22
ANN  T|stop_gained|HIGH|CFTR|ENSG00000001626|transcript|
ENST00000003084|protein_coding|12/27|c.1624G>T|p.Gly542*|
1756/6128|1624/4443|542/1480||
...
Cases 3
Cases 0
Cases 6
Controls 0
Controls 8
Controls 8
...
CLNACC RCV000007535.6|RCV000058931.3|RCV000119041.1
CLNALLE 1
CLNDBN Cystic_fibrosis|not_provided|Hereditary_pancreatitis
```

```
  CLNDSDB GeneReviews:MedGen:OMIM:Orphanet:SNOMED_CT|MedGen|
GeneReviews:MedGen:OMIM:Orphanet:SNOMED_CT
  CLNDSDBID NBK1250:C0010674:219700:ORPHA586:190905008|
CN221809|NBK84399:C0238339:167800:ORPHA676:68072000
  CLNHGVS NC_000007.13:g.117227832G>T
  CLNORIGIN 1
  CLNREVSTAT prof|single|single
  CLNSIG 5|5|5
  CLNSRC CFTR2|HGMD|OMIM_Allelic_Variant|OMIM_Allelic_Variant
  CLNSRCID G542X|CM900049|602421.0009|602421.0095
  ...
```

Once these variants have been filtered to a handful of interesting variants, we can use the `-fastaProt` command line option to output create the resulting protein sequences from these variants (*see* **Notes 3.2**).

2.3 Build SnpEff Database: Sars-Covid2

In this part we show an example on how to build a SnpEff database for the Sars-Covid2 genome. Note that this database is already available for download, so there is no need to build it from scratch, this exercise is to show how to build one database.

In order to build a database, we need the genome reference sequence and genome annotations. Typically genome annotations are distributed in genome annotation files using standard formats, such as GTF, GFF, GenBank, or Embl. None of these formats is perfect, and there are limitations making them more or less suitable depending on the organism. As a general rule, simple organisms tend to use GenBank/EMBL files, but for complex organisms GTF and GFF are preferred.

Also, GTF should be favored over GFF. The two most prominent reasons are: (1) GFF is too flexible allowing for too many ways to make mistakes, and (2) GFF files are notoriously difficult to validate. There are some validators but they are usually unable to catch many of the genomic annotation issues commonly found. So, it is often better to avoid GFF files.

Other typical limitations of genome annotation file formats relate to unusual events, such as *rare amino acids* and *ribosomal slippage* events.

Rare amino acids, such as Selenocystine, may occur only in 25 genes in the whole human genome. Likewise, Pyrrolysine occurs in archaea and bacteria. These rare amino acids are encoded by a stop codon, so they are not part of the codon table. Genome annotation formats do not provide a standard way to encode rare amino acids, so they have to be inferred using other mechanisms. For example, SnpEff can infer them if a protein FASTA file is used during the database build process, and the FASTA file properly encodes Selenocystines.

Ribosomal slippage means that the ribosomal machinery jumps, typically one base back, when translating the sequence. This event is not well represented in GTF or GFF files. Nevertheless, GenBank provides a standard mechanism for representing it using the "`/ribosomal_slippage`" tag in the CDS. Typical examples of ribosomal slippage are Sars-Covid2 (ORF1ab gene) and Ebola virus (GP gene).

Sars-Covid2 has a small genome and one of its genes has a known ribosomal slippage event. In light of the above-mentioned guidelines, we should use GenBank genome annotation files that correctly represent the genome and must not use neither GTF nor GFF files that would introduce errors in our database.

Building the Database We need to perform three steps: Note: We assume the commands are executed within `SNPEFF_HOME` dir.

(1) Configure SnpEff to add the genome entry

```
export GENOME_ID= "NC_045512.2"
echo  "$GENOME_ID.genome␣:␣Sars-COVID19" >> snpEff.config
```

(2) Download the GenBank genome annotation file

```
mkdir data/$GENOME_ID
cd data/$GENOME_ID
curl  "https://eutils.ncbi.nlm.nih.gov/entrez/eutils/efetch.
fcgi?db=nucleotide&id=$GENOME_ID&rettype=gbwithparts&retmo-
de=text" \
      > genes.gbk
```

(3) Build the database

```
java -Xmx2g -jar snpEff.jar build -v $GENOME_ID
```

All these steps are conveniently wrapped in the script `buildDbNcbi.sh` from SnpEff's package, so this command is equivalent to the three steps shown above:

```
./scripts/buildDbNcbi.sh NC_045512.2
```

3 Notes

3.1 SnpEff Error and Warning Messages

ERROR_CHROMOSOME_NOT_FOUND This error is due to a difference between the chromosome names in input VCF file and the chromosome names in SnpEff's database. The chromosome name does not exist in the reference database. Typically this means that there is a mismatch between the chromosome names in your input VCF file and the chromosome names used in the reference genome to build SnpEff's database.

This error could be caused because you are trying to annotate using a reference genome that is different than the one you used for sequence alignment. Obviously doing this makes the annotation information almost certainly incorrect. That is why SnpEff shows an error message instead of a warning.

SnpEff can report chromosome names used in the database by using -v (verbose) option:

```
$ java -Xmx4g -jar snpEff.jar -v hg19 my.vcf > my.ann.vcf
...
 # Chromosomes : Format 'chromo_name size codon_table'
 # '1' 249250621 Standard
 # '2' 243199373 Standard
 # '3' 198022430 Standard
...
```

WARNING_REF_DOES_NOT_MATCH_GENOME If the genome reference sequence does not match the variant's reference (i.e. `REF` field in the VCF file), SnpEff will emit a warning. For example, if the VCF file indicates that the reference at a certain location is `A`, while SnpEff's database indicates that the reference should be `C`. Under normal circumstances, there should be none of these warnings (or at most just a handful).

If too many of these warnings are seen, this indicates a severe problem. A typical case is when trying to annotate using a different genome than the one used for alignment (e.g. reads are aligned to hg19 but variants are annotated to using hg38).

WARNING_TRANSCRIPT_INCOMPLETE The number of coding bases is NOT multiple of 3, so there is missing information for at least one codon. This indicates an error in the reference genome gene and/or transcript definition. This can happen in transcripts that are poorly annotated.

WARNING_TRANSCRIPT_MULTIPLE_STOP_CODONS Multiple STOP codons found in a CDS. There should be only one STOP codon at the end of the transcript, but in this case, the transcript has multiple STOP codons, which is unlikely to be real. This usually indicates an error on the reference genome annotation. Could for, for example, indicating frame errors in the reference genome file for one or more exons in this transcript.

3.2 Creating a Protein Sequence FASTA File

SnpEff has a command line option (`-fastaProt`) to output the "reference" and "alternative" protein sequences for each variant into a FASTA file. This means that for each variant, the output FASTA file will have an entry with protein sequence resulting from applying that variant to the reference sequence. For example, here we create a "one variant VCF input":

```
echo "1␣889455␣.␣G␣A␣.␣.␣." \
    | tr ' ' '\t' \
     | java -Xmx6g -jar snpEff.jar ann -fastaProt z.prot.fa
hg19 \
    > z.ann.vcf
```

The resulting protein FASTA file looks like this (lines edited for readability):

```
>NM_015658.3 Ref
MAAAGSR...LLFGKVAKDSSRMLQPSSSPLWGKLRVDIKAYLGS...

>NM_015658.3 Variant 1:889455-889455 Ref:G Alt:A HGVS.p:p.
Gln236*
MAAAGSR...LLFGKVAKDSSRML*PSSSPLWGKLRVDIKAYLGS...
```

3.3 Multiple Version of RefSeq Transcripts

When using RefSeq transcripts, for instance in the human genome versions hg38 or hg19, can lead to some confusion due to multiply mapped transcripts. For example, transcript `NM_001135865.1` from hg38 is mapped four times. From the original RefSeq data, you can see that there are actually four mappings of NM_001135865.1:

```
# Note: Output edited for readability
\$ zgrep NM_001135865.1 ~/snpEff/data/hg38/genes.refseq.gz
751 NM_001135865.1 chr16 - 21834582 21857657 21834717
21857378 11 ...
756 NM_001135865.1 chr16 + 22513522 22536520 22513801
22536385 7 ...
597 NM_001135865.1 chr16_KV880768v1_fix + 1679394 1702742
1679673 1702607 11 ...
589 NM_001135865.1 chr16_KV880768v1_fix - 568516 591514
568651 591235 7 ...
```

Not only NM_001135865.1 maps twice to regions in chr16 but also one is mapped in the forward strand and the other on the reverse strand (notice the + and – signs). In these cases, SnpEff disambiguates the transcripts IDs. When there are multiple mappings for a transcript SnpEff will make sure each mapping is uniquely identified by appending a number to the original transcript ID. So the transcript IDs are renamed (notice that the first one is not changed):

```
NM_001135865.1
NM_001135865.1.2
NM_001135865.1.3
NM_001135865.1.4
```

Even though they are mapped to different chromosomes and strands in chr16, the protein sequence will be very similar. This is why RefSeq has multiple mappings of the same transcript.

3.4 Show Gene or Transcript Information

SnpEff can show information of genes and transcripts, as well as a text "graphical representation" of the transcript.

```
java -Xmx8g -jar snpEff.jar show hg38 NM_001135865.1 >
NM_001135865.1.txt
```

This shows detailed information about the transcript, sequences. The "graphical" representation of the gene/transcript can have lines with thousands of characters, depending on the length of the transcript. Also, it should be noted that SnpEff is showing internal data structures, so the coordinates are zero-based, instead of one-based.

References

1. 1000 Genomes Project Consortium et al (2015) A global reference for human genetic variation. Nature 526(7571):68
2. Abril JF, Castellano Hereza S (2019) Genome annotation. Elsevier, Amsterdam
3. Adzhubei I, Jordan DM, Sunyaev SR (2013) Predicting functional effect of human missense mutations using PolyPhen-2. Curr Protoc Hum Genet 76(1):7–20
4. Aken BL et al (2016) The Ensembl gene annotation system. Database. https://doi.org/10.1093/database/baw093
5. Cingolani P et al (2012) A program for annotating and predicting the effects of single nucleotide polymorphisms, SnpEff: SNPs in the genome of Drosophila melanogaster strain w1118; iso-2; iso-3. Fly 6(2):80–92
6. Dalgleish R et al (2010) Locus Reference Genomic sequences: an improved basis for describing human DNA variants. Genome Med 2(40):1–7
7. Deans ZC, Fairley JA, den Dunnen JT, Clark C (2016) HGVS nomenclature in practice: an example from the United Kingdom National External Quality Assessment Scheme. Hum Mutat 37(6):576–578
8. den Dunnen JT et al (2016) HGVS recommendations for the description of sequence variants: 2016 update. Hum Mutat 37(6): 564–569
9. Drmanac R et al (2010) Human genome sequencing using unchained base reads on self-assembling DNA nanoarrays. Science 327(5961):78–81
10. Eberle MA et al (2017) A reference data set of 5.4 million phased human variants validated by genetic inheritance from sequencing a three-generation 17-member pedigree. Genome Res 27(1):157–164
11. Frankish A et al (2015) Comparison of GENCODE and RefSeq gene annotation and the impact of reference geneset on variant effect prediction. BMC Genomics 16(8):1–11
12. Harrow J et al (2012) GENCODE: the reference human genome annotation for The ENCODE Project. Genome Res 22(9): 1760–1774
13. Karczewski KJ et al (2019) Variation across 141,456 human exomes and genomes reveals the spectrum of loss-of-function intolerance across human protein-coding genes. BioRxiv, p 531210

14. Kerem, B-s et al (1990) Identification of mutations in regions corresponding to the two putative nucleotide (ATP)-binding folds of the cystic fibrosis gene. Proc Natl Acad Sci 87(21):8447–8451
15. Kircher M et al (2014) A general framework for estimating the relative pathogenicity of human genetic variants. Nat Genet 46(3):310–315
16. Landrum MJ et al (2014) ClinVar: public archive of relationships among sequence variation and human phenotype. Nucleic Acids Res 42(D1):D980–D985
17. Lek M et al (2016) Analysis of protein-coding genetic variation in 60,706 humans. Nature 536(7616):285–291
18. Li MM et al (2017) Standards and guidelines for the interpretation and reporting of sequence variants in cancer: a joint consensus recommendation of the Association for Molecular Pathology, American Society of Clinical Oncology, and College of American Pathologists. J Mol Diagn 19(1):4–23
19. Liu X, Jian X, Boerwinkle E (2011) dbNSFP: a lightweight database of human nonsynonymous SNPs and their functional predictions. Hum Mutat 32(8):894–899
20. McCarthy DJ et al (2014) Choice of transcripts and software has a large effect on variant annotation. Genome Med 6(3):1–16
21. Ng PC, Henikoff S (2003) SIFT: predicting amino acid changes that affect protein function. Nucleic Acids Res 31(13):3812–3814
22. O'Leary NA et al (2016) Reference sequence (RefSeq) database at NCBI: current status, taxonomic expansion, and functional annotation. Nucleic Acids Res 44(D1):D733–D745
23. Pollard KS, Hubisz MJ, Rosenbloom KR, Siepel A (2010) Detection of nonneutral substitution rates on mammalian phylogenies. Genome Res 20(1):110–121
24. Pruitt KD, Tatusova T, Maglott DR (2007) NCBI reference sequences (RefSeq): a curated non-redundant sequence database of genomes, transcripts and proteins. Nucleic Acids Res 35 (Suppl_1):D61–D65
25. Pruitt KD et al (2009) The consensus coding sequence (CCDS) project: identifying a common protein-coding gene set for the human and mouse genomes. Genome Res 19(7): 1316–1323
26. Ranganathan S, Nakai K, Schonbach C (2018) Encyclopedia of bioinformatics and computational biology: ABC of bioinformatics. Elsevier, Amsterdam
27. Richards S et al (2015) Standards and guidelines for the interpretation of sequence variants: a joint consensus recommendation of the American College of Medical Genetics and Genomics and the Association for Molecular Pathology. Genet Med 17(5):405–423
28. Ruden DM et al (2012) Using Drosophila melanogaster as a model for genotoxic chemical mutational studies with a new program, SnpSift. Front Genet 3:35
29. Schneider VA et al (2017) Evaluation of GRCh38 and de novo haploid genome assemblies demonstrates the enduring quality of the reference assembly. Genome Res 27(5): 849–864
30. Siepel A et al (2005) Evolutionarily conserved elements in vertebrate, insect, worm, and yeast genomes. Genome Res 15(8):1034–1050
31. Yen JL et al (2017). A variant by any name: quantifying annotation discordance across tools and clinical databases. Genome Med 9(1):1–14
32. Zerbino DR et al (2018) Ensembl 2018. Nucleic Acids Res 46(D1):D754–D761

Chapter 20

Annotating Cancer-Related Variants at Protein–Protein Interface with Structure-PPi

Miguel Vazquez and Tirso Pons

Abstract

A comprehensive analysis of germline and somatic variants requires complex computational approaches that combine next-generation sequencing (NGS)-based omics data with curated annotations from public repositories. Here, we describe Structure-PPi, which facilitates the analysis of cancer-related variants onto protein 3D structures, interaction interfaces, and other important functional sites (i.e., catalytic, ligand-binding, posttranslational modification). Our approach relies on features extracted from Interactome3D, UniProtKB, InterPro, APPRIS, dbNSFP, and COSMIC databases and provides complementary information to pathogenicity prediction methods. Thus, Structure-PPi helps in the discrimination of false-positive predictions and adds both mechanistic and biological insights into the role of variants in a given cancer. An online version of the tools is available at https://rbbt.bsc.es/StructurePPI/.

Key words Cancer mutations, Variants prioritization, Rbbt workflows, Protein–protein interactions, 3D structure analysis, Protein isoforms

1 Introduction

Understanding the clinical pathogenic effects of variants is crucial for the development of precision medicine approaches. However, our interpretation about the functional and biological consequences of genetic variants identified by human genome sequencing projects is limited. Recent efforts use integrated analysis by merging genomic, protein interactions, and 3D structural data in the prioritization of variants from large-scale sequencing studies [1, 2].

Several studies have suggested that disease-associated germline and somatic variants selectively target protein–protein interaction (PPI) interfaces [3–6]. In particular, cancer variants affecting different interfaces on the same protein could result in different phenotypes. So, interaction perturbation ("edgetic") variants lead to the loss or gain of specific interactions, and, consequently, their

Charlotte K. Y. Ng and Salvatore Piscuoglio (eds.), *Variant Calling: Methods and Protocols*, Methods in Molecular Biology, vol. 2493, https://doi.org/10.1007/978-1-0716-2293-3_20,

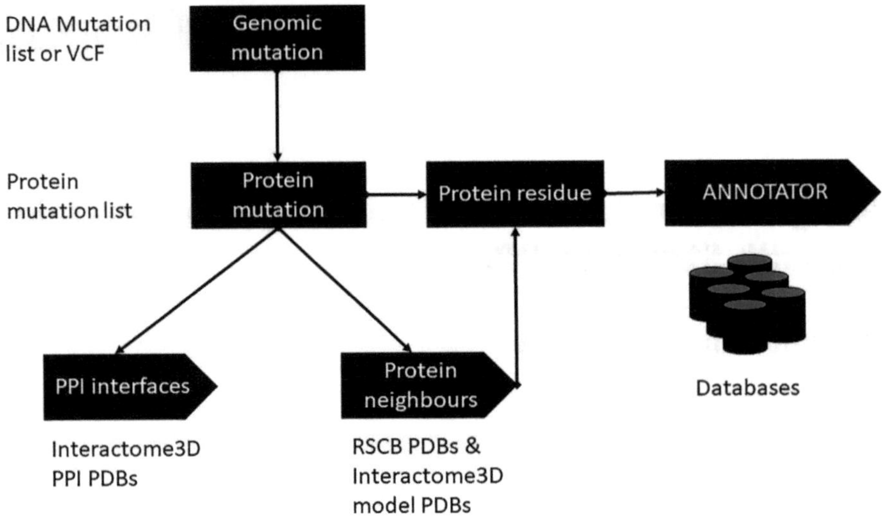

Fig. 1 Overview of the Structure-PPi tool. Inputs are mutations on DNA or over proteins. The mutated protein residues or their neighbors can be annotated with information from several databases. Also, residues in PPI interfaces are detected

analysis can improve our understanding of the heterogeneity in genotype–phenotype relationships.

Different databases with high-quality 3D structurally resolved protein interactome networks are available [7, 8]. As well, many bioinformatics tools are available to explore and predict the potential impact of human variants perturbing PPI networks [9–14]. While those methods might seem straightforward, they usually are time-consuming and costly if we submit batches of tens of thousands of mutations. Structure-PPi takes care of file retrieval, processing, and caching on demand to provide a platform to serve these functionalities as swiftly as possible without redoing work.

Figure 1 shows an overview of the Structure-PPi tool. It can take in a list of protein mutations, or a list of genomic mutations (or a VCF file) which will be transformed into their consequence as protein mutations. For each protein mutation, Structure-PPi can annotate the residue it affects with features from several databases. Using PDBs from RSCB and from Interactome3D, it can find other residues in close proximity and annotate those with the different databases. It can also use PDBs of PPIs to find if the mutation affects an interface residue.

2 Materials

2.1 Annotation Databases

Structure-PPi annotates protein amino acids with information coming from several databases.

1. Appris: catalytic sites, ligand-binding sites, signal peptides, transmembrane domains.

2. InterPro: protein domains.
3. UniProtKB: protein domains, secondary structures, relevant variants (i.e., sites which have been experimentally altered by mutagenesis).
4. Variants: human variants curated from literature and annotated in UniProtKB/SwissProt.
5. COSMIC: mutations occurring in tumor samples.
6. Drug resistance mutations: COSMIC selection of variants that confer tumor resistance to drug therapies.
7. PDB and Interactome3D: three-dimensional structure of proteins and protein–protein complexes.

Each of these resources associates single amino acids, or continuous stretches of them, with one or several fields of information, depending on the database. For instance, the Appris database associates three fields: the type of feature, the location of the feature on the protein sequence, and an identifier that can be used to find information about that feature. Protein sequences are also gathered from the UniProt Knowledgebase (UniProtKB) entries in order to resolve discrepancies between UniProtKB annotations and the Ensembl sequences used by Rbbt.

2.2 Protein Structure Files

In order to determine neighborhoods of residues, Structure-PPi makes use of protein structure files in PDB format. It finds PBD files associated to each protein from UniProtKB which are retrieved from RSCB. Interactome3D assigns proteins to models of PDB structures adapted from other organisms. In addition to those, protein–protein interaction complexes are also available as PDB files at Interactome3D.

2.3 Organisms Builds and Gene Sets

The low-level functionalities of Structure-PPi are compatible with other organisms in addition to human. All the basic data is downloaded from Ensembl for any organism that Ensembl supports with little configuration; out of the box it supports mouse, rat, and yeast, but support for others is easy and can be done by the user. Of course, functional annotations and protein 3D structures will be scarce or not available for other organisms, so Structure-PPi is in practice a human-centric tool.

The data on human evolves in time, both at the level of the chromosome sequences (the genome build; e.g., hg38 or hg19), as well as the gene-set. In Structure-PPi, we use the following convention: organisms are identified by a three-letter code based on the scientific name (a convention borrowed from the KEGG database; e.g., *Hsa* for ***homo sapiens***, *Mmu for* ***mus musculus***, *Sce for saccharomyces* ***cerevisiae***) and is followed by the Ensembl *archive date* that determines the build and gene-set to be used. For instance, the default code is "Hsa/feb2014" which consists of the latest available gene-set for the hg19 human build. For a hg38 version of the data use "Hsa/may2017." These codes ensure that all the information gathered from Ensembl is stable and correctly versioned. Using the right code is the most important when the input are genomic mutations, i.e., mutations in DNA, since the genome build is critically important, but can also have important effects in protein sequences or identifier equivalence mappings in particular cases. When imputing protein point mutations using gene symbol identifiers, for instance, having the wrong archive data in the code might not have much importance. Pairwise sequence alignment is used to resolve discrepancies between protein sequences, which also helps alleviate issues with differences in gene sets.

2.4 Inputs and Output Formats, and a Word on Multiplicity

Structure-PPi operates primarily with two file formats: a list of entities and a TSV file. In addition to these two, it also supports VCF files to specify genomic mutations. Outputs are generally TSV files, with each row representing an entity, specified as the first column, associated with a number of fields, one per column. Multiple values inside a column are separated by the character "|".

Many parts of Structure-PPi present one-to-many relationships. For instance, a protein mutation might overlap with several regions in the same database, or a protein location might have several neighboring residues that each have their own annotations. Depending on the analysis, these multiplicities pile up and challenge the expressiveness of the TSV format. In Structure-PPi, we tried to retain the use of TSV files, but not lose the details, which forced us to introduce two additional separators ";" and "," to represent the second and third level of multiplicities. When one analysis builds on another and introduces multiplicities, for instance annotating neighbors instead of the actual residues, the separators used in the innermost analyses are downgraded one level, turning "|" into ";" and ";" into ",". However, to make use of the tool easier for casual use, there is an option to simplify the output so that all values inside a field are joined and repetitions removed.

3 Methods

3.1 Resolving Differences in Identifiers and Sequences

Structure-PPi accepts specifying protein mutations using different formats for identifying the proteins. The preferred one is "Ensembl Protein ID" (ENSP), since this ties then unequivocally and unambiguously with the right protein information from Ensembl (especially if the exact archive date is used). When gene symbols are used, we resolve it to a principal isoform (as defined by Appris), with a preference to one that is also associated with a UniProtKB entry. Annotation sources like Appris and InterPro associate entries with ENSP identifiers, while others are associated with UniProtKB entries. UniProtKB and ENSP ids are translated using a translation table based on the equivalence tables from Ensembl BioMart but taking special consideration to match the UniProtKB entry with the ENSP isoform that best matches its sequence. Often differences still appear between UniProtKB entries and the best ENSP that we can find, so all the annotations of resources using UniProtKB entries are mapped back into the ENSP isoforms using an alignment map build using Smith-Waterman pair sequence alignment [15]. Alignment maps are also used to resolve discrepancies between the sequences of PDB chains and ENSP isoforms.

3.2 Annotation Engine

A core engine is in charge of annotating protein positions with features, which is customized for each database to provide the resulting TSV file with the right column headers and content. Each protein mutation is processed to extract the protein identifier and the mutations position. The identifier is translated into ENSP and then annotated using the annotation engine, which might involve translating in turn into UniProtKB and using alignment maps to resolved discrepancies.

3.3 Turning DNA Mutations into Protein Mutations

Structure-PPi uses the Rbbt workflow "Sequence" to turn DNA mutations into protein mutations, which can work with VCF files or with mutations specified in the succinct format "<chr>:<position>:<alt allele>." Since genes often present multiple possible isoforms, the DNA mutations might have consequences in several different isoforms. To simplify the output, only principal isoforms are considered (as defined by Appris), but even then, some genes might have multiple principal isoforms. This is one source of multiplicity, like we mentioned before, so the "DNA" version of the functionalities shifts all separator characters one level when compared with the protein mutation counterparts.

3.4 Mapping to Protein Sequence Locations to Structures

Each UniProtKB entry may be associated with a number of PDBs from RSCB. In addition, Interactome3D provides PDB structures of pairs of proteins forming complexes, which are also assigned UniProtKB identifiers. To identify the position of a residue in our isoform inside the structure described in a PDB, one must identify the chain and the position inside the chain that this residue occupies, which may be more than one. Instead of trusting the annotations inside the PDB file, we perform sequence alignment of the sequence of each chain with the sequence of our isoform to identify that best matching chain that overlaps our residue of interest, if any.

3.5 Calculating Neighborhoods and Interactions Surfaces

Residue neighborhoods are calculated inside the PDB using a matrix with all the distances between all PDB atoms belonging to different residues. Distances between contiguous backbone atoms are not considered. The pairwise distance matrix is used to extract all atoms at a particular distance. A distance of 5 Å is used by default when looking for neighboring residues; this distance is increased to

8 Å when looking for interaction surfaces. In the case of neighboring residues, these are required to be on the same PDB chain, while the opposite is required when detecting PPI interaction surfaces.

When looking for neighbors of a particular residue in a particular protein, all PDB files are gathered from UniProtKB and Interactome3D and traversed in order until one is found that covers the query residue. This first PDB file is used to determine the neighborhood of the residue. This behavior can be overridden so that all PDB files are examined, in which case the neighborhood is the union of all. This can make the process much lengthier.

3.6 Caching

An installation of Structure-PPi (i.e., the Proteomics Rbbt workflow) will initially feature just the code and some of the databases like Appris or InterPro. Other resources like UniProtKB entries and PDB files will be downloaded on demand and cached. Sequence maps between isoforms and UniProtKB entries or PDB chains, or atom distance matrices will also be computed on demand and cached. This means that when looking at a new isoform for the first time, the process might slow down due to the gathering and preparation of the necessary resources. Individual analysis jobs are also cached (*see* **Note 1**).

3.7 URLs

1. Appris: http://appriws.bioinfo.cnio.es
2. COSMIC: https://cancer.sanger.ac.uk/cosmic
3. dbNSFP: https://sites.google.com/site/jpopgen/dbNSFP
4. Interactome3D: https://interactome3d.irbbarcelona.org/
5. InterPro: https://www.ebi.ac.uk/interpro/
6. RSCB PDB: https://www.rcsb.org
7. Swiss-Prot Variant Pages: https://www.uniprot.org/docs/humsavar
8. UniProtKB: https://www.uniprot.org

4 Structure-PPI Functionalities

4.1 Low-Level Tasks

There are a number of low-level functionalities (tasks on Rbbt lingo) on which the annotation tasks (Subheading 4.2) build and which are also made accessible on their own. These are not the usual entry points, but can be used to develop other functionalities (*see* **Note 2**), such as the JsMOL visual interface available in the Structure-PPI web.

1. "pdb_alignment_map": uses pairwise alignment to return a coordinate map between an amino acid sequence provided as input and the different chains of a given PDB.

2. "pdb_chain_position_in_sequence" and "sequence_position_in_pdb": can translate from coordinates in a given PDB chain to coordinates in a given amino acid sequence and conversely.
3. "neighbour_map": for each residue in a given PDB (format <chain>:<position>) returns all the neighboring residues (at a given distance).
4. "neighbours_in_pdb": uses a given PDB to find the neighbors of different positions inside a given amino acid sequence.

4.2 Annotation Tasks

There are four annotation tasks, two for protein mutations and two for genomic mutations. The two functionalities in each case are to annotate features overlapping just the mutated residues ("annotate_mi" and "annotate_dna") or to annotate features overlapping also their neighbors ("annotate_mi_neighbours" and "annotate_dna_neighbours"). In addition to the annotation tasks, there are four that make 3D calculations to calculate neighboring residues and residues in interface positions, again two for protein mutations and two for DNA mutations. The tasks for finding neighboring residues are used by the corresponding annotation tasks, unlike the tasks for identifying interface residues, which have functional value on their own. In the case when the input is genomic mutations (or VCF files), these are translated into protein mutations using a different rbbt workflow. Functionalities accepting genomic mutations are indicated with the term "dna" in their name, while those accepting protein mutations are indicated with the term "mi" (for *mutated isoform*). These corresponds to the left-hand side of Fig. 2. *See* **Note 3** for some tips on formatting the protein mutations.

1. "mi_neighbours" and "dna_neighbours": detect residues neighboring the mutated residue.
2. "annotate_mi" and "annotate_dna": annotate the mutated residues with functional information from the different annotations databases described in Subheading 2.
3. "annotate_mi_neighbours" and "annotate_dna_neighbours": annotate neighbors of the mutated residues with functional information from the different annotations databases described in Subheading 2.
4. "mi_interfaces" and "dna_interfaces": report PPI interfaces that overlap with the mutations.

4.3 High-Level Tasks

There are a few tasks that help run more comprehensive analyses, represented on the right-hand side of Fig. 2:

1. "mi_wizard," "dna_wizard," and "wizard": these tasks run all the annotation tasks for a list of protein mutations or DNA

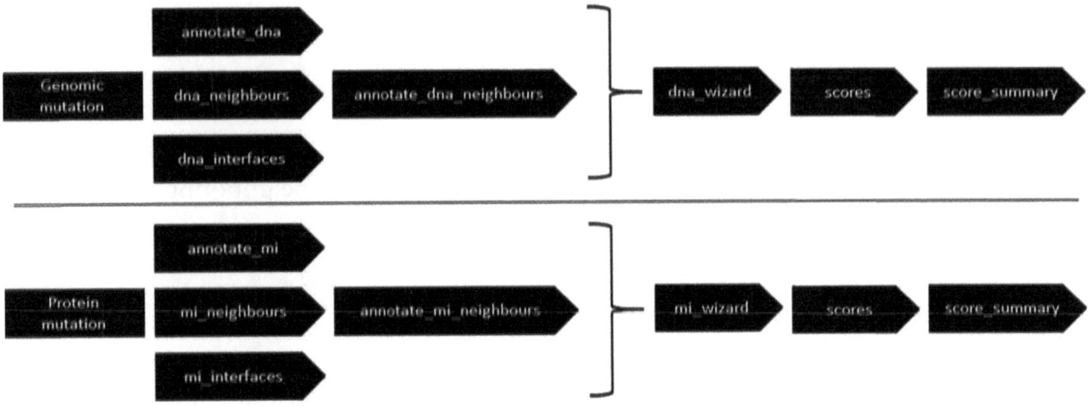

Fig. 2 Workflow dependencies for jobs using DNA (top) and Protein (bottom) mutations. The left-hand side are basic annotation tasks; note how "dna_neighbours" is used for "annotate_dna_neighbours" and likewise for the "mi" version. The right-hand side are high-level tasks, "score_summary" being the one-stop-shop task that runs all analyses

mutations. The third form autodetects the input and calls one of the other automatically.

2. "scores" and "score_summary": these tasks use the wizard to prepare all the results and then scores the different variants according to some hand-written rules regarding how disruptive each variant could potentially be based on some of the possible annotations. In addition to these scores, Structure-PPI includes damage predictions from dbNSFP.

The score calculation consists of accumulating points for certain features like having affected interface residues, mutations over ligand binding domains, highly recurrently mutated positions in COSMIC, or UniProtKB variants of penetrance. The scores are calculated first for features overlapping the queried position and then the summed by the same scores calculated for features overlapping neighboring positions, but halved. They are only intended to highlight feature rich mutations which might merit additional inspection, not as a replacement for proper damage predictions, which are also provided separately. The "scores_summary" task is a one-stop shop to all functionalities.

5 Availability

5.1 Software Package

Structure-PPi is implemented in the Proteomics Rbbt workflow, available at https://github.com/Rbbt-Workflows/Proteomics. To use this, you will need a working installation of Rbbt. Installing this workflow will give access to the command line interface to the Proteomics workflow and the Structure-PPi web interfaces, which also serves as a REST web service end-point.

Please note that the Proteomics workflow only includes the code, not the necessary underlying datasets. Rbbt will take care of gathering and preparing all the datasets and indices as they are required, as well as additional workflows that support some of its functionalities. Notably, the COSMIC dataset requires setting up an access token to be able to download the files. All these preparations might add up to a few hours. After that there are still files that will be cached through the use of the application. It is recommended initially to set a verbose log level to monitor these processes. Please consult this page to learn more about rbbt: http://mikisvaz.github.io/rbbt/.

5.2 Containers

To simplify the deployment on a new infrastructure, a container has been created with the Rbbt base system, the Proteomics workflow and other workflows and data resources needed to begin work. You should be able to find an up-to-date version of this container at: http://mikisvaz.github.io/rbbt/tutorial/containers/.

5.3 Command Line Tool

Structure-PPI can be used from the command line using the "rbbt" command, which controls everything in the Rbbt framework. In particular, "rbbt workflow task Proteomics" is used for enacting tasks from the Proteomics workflow, which holds the Structure-PPI backend functionalities. Typing just "rbbt workflow task Proteomics" lists the available functionalities; for instance: "annotate_mi." Typing "rbbt workflow task Proteomics annotate_mi -h" will display the documentation for that task, which includes the different inputs that it accepts as well as the location of examples that you can use to test the functionalities. To run a task, just include the relevant inputs in the command line like this "rbbt workflow task Proteomics annotate_mi -m SYT8:A55T."

5.4 Structure-PPi Web Interface

A working instance of the Structure-PPi software is available at https://rbbt.bsc.es/StructurePPI/. From there, you can access all the different tasks in the Proteomics workflow. It also links the results with specific report pages for protein mutations and protein isoforms that help put them in context. These report interactive visualizations of proteins sequences and PDB structures. Structure-PPi has also been embedded into other applications (*see* **Note 4**).

5.5 REST Web Services

The functionalities of the Proteomics workflow are also accessible as web services at the end-point https://rbbt.bsc.es/StructurePPI/Proteomics, or at another end-point setup by the user. Please consult the rbbt documentation at http://mikisvaz.github.io/rbbt/ to learn more about REST web services in Rbbt.

5.6 Performance

The performance of the Proteomics workflow is dominated by downloading files and caching resource. For tasks involving 3D calculations, the throughput can be initially of about a couple of mutations per second and of about a few dozen for normal annotations jobs. Once the system has been used for a while and resources have been prepared for the proteins being queried the throughput rises to about fifty to a hundred mutations in tasks involving 3D calculations and about a couple of thousand mutations per second for normal annotation jobs. Normal memory usage is about 300–500 MB, but can raise to a couple GB when processing new PDB files. Throughput can be controlled in the command line by specifying the number of concurrent workers to use, the above values are for using three workers, which is the default on an average workstation.

6 Use Cases

This protocol describes a series of tasks that enable the annotation of genomic and protein mutations with information from several databases. These tasks include: (1) annotating a vcf file, (2) prioritizing protein mutations, (3) driver genes scores, and (4) protein centric investigation.

6.1 Annotating a VCF File

Annotating variants from a VCF file is one of the most common use cases. The tasks "annotate_dna," "annotate_dna_neighbours," and "dna_interfaces" all work with genomic mutations as input. From a command line interface, you can see the documentation for these tasks, as shown in Fig. 3. A typical command would be "rbbt workflow task Proteomics annotate_dna --vcf --mutations <vcf.file> --database Appris --simplify --organism Hsa/may2017." The same inputs can be specified from the web interface at https://rbbt.bsc.es/StructurePPI. This command will produce a TSV file like the one in Table 1.

6.2 Prioritizing Protein Mutations

Here we are interested in the prioritization of three novel mutations identified in the fibroblast growth factor receptor 2 (FGFR2) in Keratinocytic epidermal nevi [16]. The p.S252W (c.755C>G), p.P286S (c.857C>T), and p.Y376C (c.1127A>G) mutations were studied with the Structure-PPi tool [17]. FGFR2 amino acid positions for these novel mutations were first annotated using the "annotate_mi" task (Fig. 4a). The Structure-PPi "annotate_mi_neighbours" task was also used to annotate neighbor positions (less than 5 Å or adjacent in the sequence if no PDB covers that area) (Fig. 4b). In particular, "annotate_mi" and "annotate_mi_neighbours" tasks indicate that: (1) these novel mutations are located in the FGFR2 extracellular domain, (2) amino acid position 252 accommodate variants in Apert syndrome (VAR_004114) and Pfeiffer syndrome (VAR_004116), and (3) adjacent residue to the amino acid position 376 accommodate a variant in Beare-Stevenson cutis gyrata and Pfeiffer syndromes (VAR_017275).

```
Proteomics
==========

## annotate_dna:

-o,--organism=<string> (default: 'Hsa/feb2014') Organism code
-d,--database=<select> (default: 'UniProt') Database of annotations
-p,--principal[=false] (default: true) Consider only principal isoforms
-s,--simplify[=false] (default: false) Simplify output lists

Inputs from dependencies:

  [Sequence] mutated_isoforms_fast:
-m,--mutations=<list|file|->    Mutations Chr:Position:Allele (e.g. 19:54646887:T).
Extra fields are ignored but kept
-w,--watson[=false] (default: true) Mutations all reported on the watson (forward)
strand as opposed to the gene strand
-v,--vcf[=false] (default: false) Is the input an VCF file instead of genomic
mutations?
-c,--coding[=false] (default: false) Consider only transcripts with protein_coding
biotype

Returns: tsv

Input select options

database: UniProt, variants, Appris, InterPro, COSMIC, COSMIC_resistance_mutations

## DEPENDENCY GRAPH (abridged)

Proteomics#annotate_dna
 Sequence#mutated_isoforms_fast
```

Fig. 3 Extract from command-line documentation for the task "annotate_dna" from the Proteomics rbbt workflow

Table 1
Extract of result from task "annotate_dna" over a VCF file using database "Appris"

Genomic mutation	Mutated isoform	Appris features	Appris feature locations	Appris feature descriptions
1:113212635:G	(CAPZA1) ENSP00000263168:Q248E	Domain	17-283	F-actin_cap_A
10:7755183:C	(ITIH2) ENSP00000351190:E134D	Domain	72-183	VIT
13:20717179:C	(GJA3) ENSP00000241125:I83M	Domain	3-227	Connexin
15:75981700:A	(CSPG4) ENSP00000312506:T569M	Domain	538-645	Cadherin_3
17:28618492:G	(BLMH) ENSP00000261714:Q16H	No_pep_signal	1-28	
16:23400369:A	(COG7) ENSP00000305442:P729S	Domain	2-767	COG7
19:11917376:A	(ZNF491) ENSP00000313443:R203H	Domain	187-209	zf-C2H2
19:40408268:T	(FCGBP) ENSP00000221347:T1524N	Domain	1461-1528	C8
16:23400369:A	(COG7) ENSP00000305442:P729S	Domain	2-767	COG7
2:238280897:C	(COL6A3) ENSP00000295550:I1255V	Domain	1233-1402	VWA
20:8769120:A	(PLCB1) ENSP00000338185:A1046T	Domain	1003-1174	PLC-beta_C
3:20019844:A	(RAB5A) ENSP00000273047:E161K	Domain	22-181	Ras

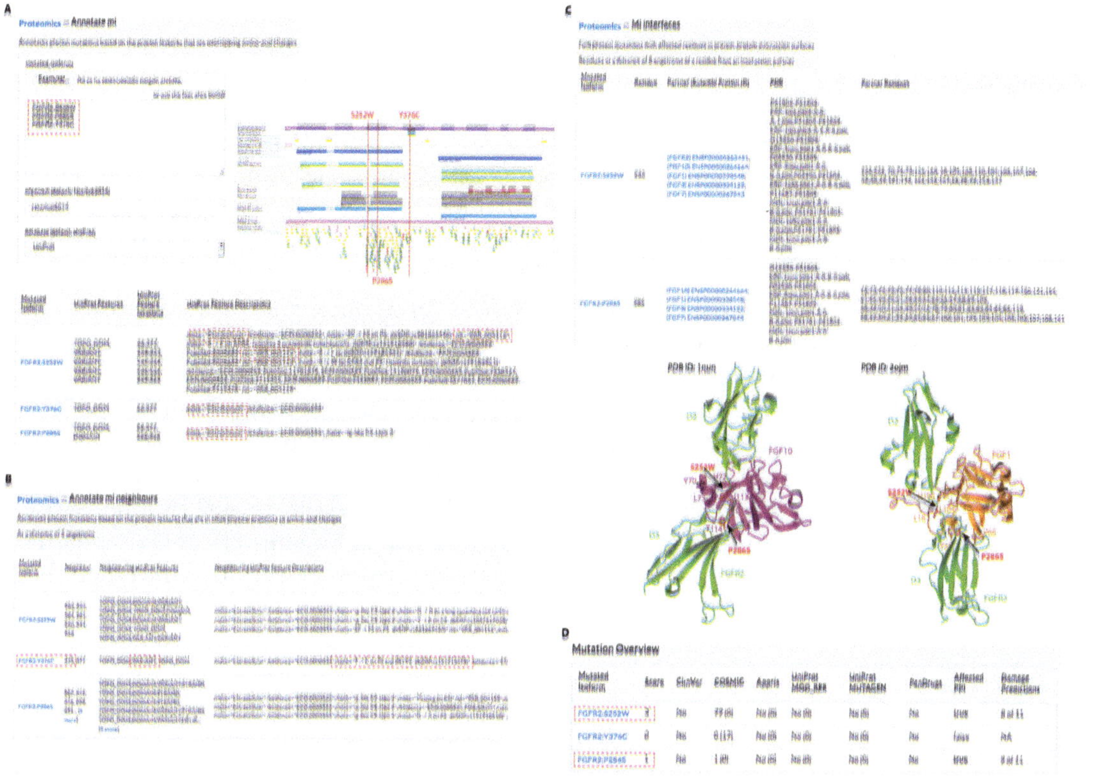

Fig. 4 Prioritization of novel mutations in FGFR2. Screenshots of the input/output options in the "annotate_mi" and "annotate_mi_neighbours" tasks are shown in panels (**a**) and (**b**), respectively. An output of the "mi_interfaces" task is shown in (**c**). A mutation overview with the "score_summary" task is shown in (**d**). The identification codes "VAR_xxxxxx" correspond to entries in the file humsavar.txt (release 2021_02 of Apr 27, 2021) from the Swiss-Prot Variant Pages. The amino acid numbering Y376 refers to the FGFR2IIIb isoform, RefSeq ID: NM_022970.3, and Ensembl Transcript ID: ENST00000457416, which is the main transcript expressed in Keratinocytic epidermal nevi

FGF ligands interact with the second and third Ig domains (hereafter referred to as D2 and D3) and with the interconnecting D2-D3 linker region of FGFR2. Then we applied the "mi_interfaces" task to identify which mutations are located at interfaces (Fig. 4c). The p.S252W (c.755C>G) mutation is localized in the D2-D3 linker and can alter the ligand-binding specificity. The p. P286S (c.857C>T) mutation is also positioned in D3 in the ligand-binding site (for more details, *see* Supplementary Fig. 4 in Toll et al. [16]). Both the mutations show the highest score according to the "score_summary" task (Fig. 4d).

We also used Structure-PPi for the prioritization of variants across several cancers in large cohorts of patients. Indeed, Structure-PPi functionalities have been implemented in PCAWG-Scout (http://pcawgscout.bsc.es) as part of The Cancer Genome Atlas (TCGA) and International Cancer Genome Consortium (ICGC) Drivers and Functional Interpretation Working Group [18].

6.3 Protein Centric Investigation

Structure-PPi has previously been used in the interpretation of pathogenic variants in human protein kinases [19]. The "annotate_mi," "annotate_mi_neighbours," "mi_interfaces," and "score_summary" tasks provide the wKinMut-2 tool [19] with complementary information about the functional role of the affected residues. To illustrate this, we used the B-Raf proto-oncogene p.V600E mutation recurrently found in the literature associated with certain types of cancer including colorectal cancer, sarcoma, melanoma, thyroid carcinoma, and ovarian serous carcinoma. Reports about the pathogenicity of B-raf p.V600E are coherent with the information obtained from "annotate_mi" (Fig. 5a). According to "annotate_mi_neighbours," p.V600E could affect the ATP binding due to the proximity to F468, localized in the ATP-binding region (PDB ID: 5jsm) (Fig. 5b, c). Moreover, p.V600E received a high score of 4.0 by the "score_summary" task, which suggested a pathogenic effect (Fig. 5d). This output from "score_summary" of the p.V600E mutation is in

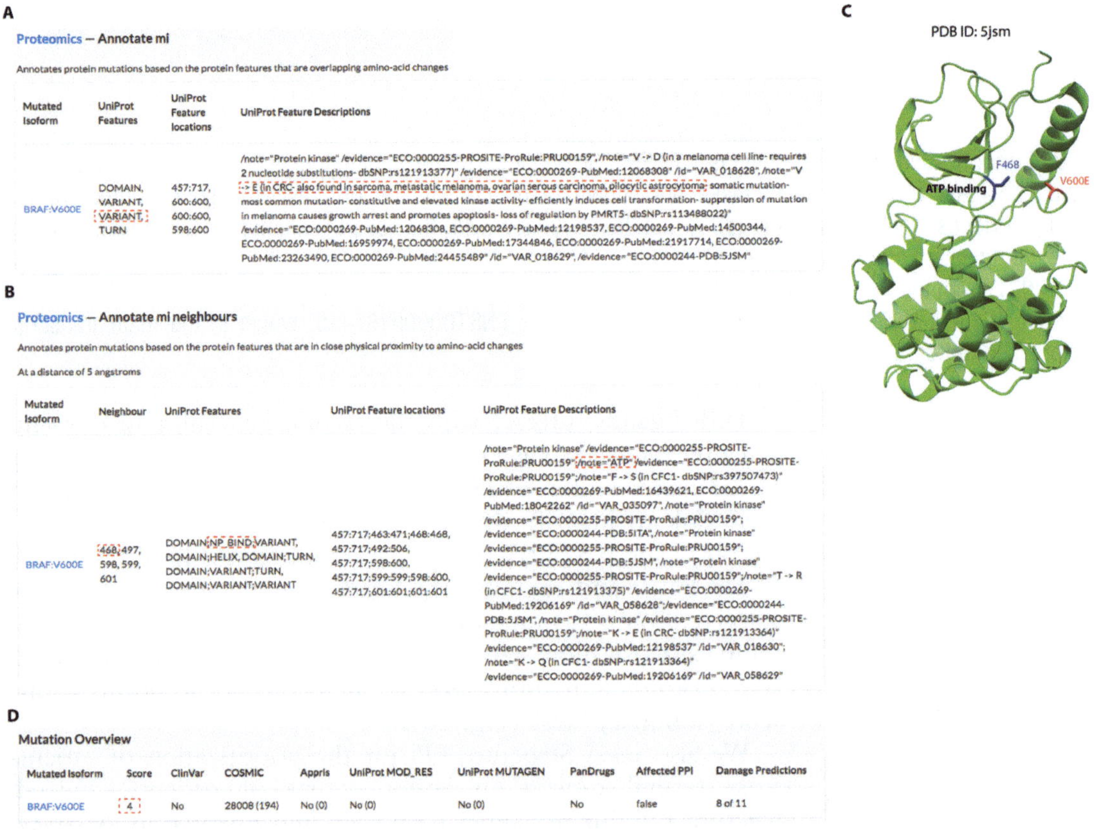

Fig. 5 Structure-PPi annotations for the BRAF p.V600E mutation. Screenshots of the outputs from each task: "annotate_mi" (**a**), "annotate_mi_neighbours" (**b**), and "score_summary" (**d**). Mapping of BRAF mutation onto the 3D crystal structure (**c**)

complete agreement with several records in the literature that suggest a decreased overall survival, poor treatment responses, and different patterns of metastatic spread [20, 21].

7 Notes

1. When issuing a job, a custom name can be specified; otherwise, the name 'Default' will be used. Changing the name of the job avoids reusing existing job results.
2. For a selected PDB file, users can apply the "neighbour_map" functionality to generate a contact matrix with all pairs of residues that fall within a given "distance" of each other.
3. In Structure-PPi, the functionalities "annotate_mi," "mi_neighbours," "annotate_mi_neighbours," "mi_interfaces," and "mi_wizard" allow annotation of stop-gain variants and either in-frame or frameshift indels. For example, input variants TP53:E68*, BARD1:L359_P365del, PTEN:Y76del, or SMARCA4:T296Rfs*7 must be edited and annotated as TP53:E68A, BARD1:L359A, PTEN:Y76A, and SMARCA4: T296A, respectively.
4. Structure-PPi is currently integrated in the PCAWG-Scout online browser (http://pcawgscout.bsc.es) for data access, visualization, and analysis of the Pan-Cancer data [18].

References

1. Yi S, Lin S, Li Y, Zhao W, Mills GB, Sahni N (2017) Functional variomics and network perturbation: connecting genotype to phenotype in cancer. Nat Rev Genet 18(7):395–410
2. Capriotti E, Ozturk K, Carter H (2019) Integrating molecular networks with genetic variant interpretation for precision medicine. WIREs Syst Biol Med 11(3):e1443
3. Zhong Q, Simonis N, Li Q-R, Charloteaux B, Heuze F, Klitgord N et al (2009) Edgetic perturbation models of human inherited disorders. Mol Syst Biol 5(1):321
4. Guo Y, Wei X, Das J, Grimson A, Lipkin SM, Clark AG et al (2013) Dissecting disease inheritance modes in a three-dimensional protein network challenges the "guilt-by-association" principle. Am J Hum Genet 93(1):78–89
5. Chen S, Fragoza R, Klei L, Liu Y, Wang J, Roeder K et al (2018) An interactome perturbation framework prioritizes damaging missense mutations for developmental disorders. Nat Genet 50(7):1032–1040
6. Cheng F, Zhao J, Wang Y, Lu W, Liu Z, Zhou Y et al (2021) Comprehensive characterization of protein–protein interactions perturbed by disease mutations. Nat Genet 53(3):342–353
7. Meyer MJ, Das J, Wang X, Yu H (2013) INstruct: a database of high-quality 3D structurally resolved protein interactome networks. Bioinformatics 29(12):1577–1579
8. Mosca R, Céol A, Aloy P (2013) Interactome3D: adding structural details to protein networks. Nat Methods 10(1):47–53
9. Betts MJ, Lu Q, Jiang Y, Drusko A, Wichmann O, Utz M et al (2015) Mechismo: predicting the mechanistic impact of mutations and modifications on molecular interactions. Nucleic Acids Res 43(2):e10–e10
10. Mosca R, Tenorio-Laranga J, Olivella R, Alcalde V, Céol A, Soler-López M et al (2015) dSysMap: exploring the edgetic role of disease mutations. Nat Methods 12(3): 167–168
11. Céol A, Müller H (2015) The MI bundle: enabling network and structural biology in

genome visualization tools. Bioinformatics 31(22):3679–3681

12. Lu H-C, Herrera Braga J, Fraternali F (2016) PinSnps: structural and functional analysis of SNPs in the context of protein interaction networks. Bioinformatics 32(16):2534–2536
13. Ozturk K, Carter H (2019) Identifying driver interfaces enriched for somatic missense mutations in tumors. In: Starr TK (ed) Cancer driver genes: methods and protocols [internet]. Springer, New York, NY, pp 51–72. (methods in molecular biology). Available from: https://doi.org/10.1007/978-1-4939-8967-6_4
14. Sedova M, Iyer M, Li Z, Jaroszewski L, Post KW, Hrabe T et al (2019) Cancer3D 2.0: interactive analysis of 3D patterns of cancer mutations in cancer subsets. Nucleic Acids Res 47 (D1):D895–D899
15. Smith TF, Waterman MS (1981) Identification of common molecular subsequences. J Mol Biol 147(1):195–197
16. Toll A, Fernández LC, Pons T, Groesser L, Sagrera A, Pau ECS et al (2016) Somatic embryonic FGFR2 mutations in Keratinocytic epidermal nevi. J Invest Dermatol 136(8): 1718–1721
17. Vázquez M, Valencia A, Pons T (2015) Structure-PPi: a module for the annotation of cancer-related single-nucleotide variants at protein–protein interfaces. Bioinformatics 31(14):2397–2399
18. Campbell PJ, Getz G, Korbel JO, Stuart JM, Jennings JL, Stein LD et al (2020) Pan-cancer analysis of whole genomes. Nature 578(7793): 82–93
19. Vazquez M, Pons T, Brunak S, Valencia A, Izarzugaza JMG (2016) wKinMut-2: identification and interpretation of pathogenic variants in human protein kinases. Hum Mutat 37(1):36–42
20. Grothey A, Fakih M, Tabernero J (2021) Management of BRAF-mutant metastatic colorectal cancer: a review of treatment options and evidence-based guidelines. Ann Oncol 32(8): 959–967. https://www.annalsofoncology.org/article/S0923-7534(21)01112-1/abstract
21. Morris VK, Bekaii-Saab T (2020) Improvements in clinical outcomes for BRAFV600E-mutant metastatic colorectal cancer. Clin Cancer Res 26(17):4435–4441

Chapter 21

Preanalytical Variables and Sample Quality Control for Clinical Variant Analysis

Ilaria Alborelli and Philip M. Jermann

Abstract

Broad molecular profiling by next-generation sequencing of solid tumors has become a critical tool for clinical decision-making in the era of precision oncology. In addition to many already approved targeted therapies, more than half of ongoing oncology-related clinical trials are biomarker-driven. Therefore, accurate and reliable assays are needed to assess the genetic make-up of tumor cells and guide clinicians in the therapy decision process. In order to obtain high-quality NGS data for variant detection, certain preanalytical steps and quality metrics should be followed. These include assessment of sample types, choice of extraction method, library preparation technology, sequencing platform, and finally sequencing quality control. Each of these steps has certain challenges and pitfalls that need to be addressed and overcome, respectively. In this chapter, we address the preanalytical quality control and how each of the involved steps may influence the final result. Following these guidelines and QC metrics may help in obtaining optimal results that will allow the precise and robust assessment of genetic variants in a clinical setting.

Key words Preanalytical variable, FFPE, Liquid biopsy, ctDNA, Next-generation sequencing, Molecular pathology, Variant calling

1 Introduction

In the era of precision oncology, biomarker-driven targeted therapies have become a viable and often preferred treatment option. Due to the increasing number of targeted therapies available across different tumor entities, accurate molecular diagnostics is a key factor in enabling state-of-the-art patient care. In particular, DNA- and RNA-based analyses are now performed routinely to assess the tumor's genetic makeup and enable access to these novel drugs. Next-generation sequencing (NGS) plays a pivotal role in this process and has become the method of choice for routine molecular profiling. The application of NGS in a clinical setting may range from targeted sequencing of a few genes to whole exome or even whole-genome sequencing. For routine cancer diagnostics, however, it is currently recommended to limit sequencing to

Charlotte K. Y. Ng and Salvatore Piscuoglio (eds.), *Variant Calling: Methods and Protocols*,
Methods in Molecular Biology, vol. 2493, https://doi.org/10.1007/978-1-0716-2293-3_21,

clinically relevant targets, including those for which approved targeted therapies exist or those that may allow access to novel drugs in the setting of a clinical trial [1]. Thus, current solutions in clinical laboratories range from 10- to 500-gene panels.

Importantly, NGS in routine diagnostic laboratories lacks standardization in regard to target enrichment strategies, as well as sequencing technologies. Each of the currently used approaches has its own capabilities and challenges that need to be taken into consideration when trying to determine the presence of genetic variations with high sensitivity and specificity. While some of these challenges are technology specific, others are universal and should be addressed in a platform-independent manner.

Here, we describe the preanalytical steps that should be considered and optimized in a laboratory performing targeted NGS-based genomic analyses. We will focus on gene-panel sequencing of formalin-fixed paraffin-embedded (FFPE) tissue specimens as well as blood plasma samples used for cell-free DNA (cfDNA) analysis as these are most representative of a clinical molecular pathology setting. Following these steps and optimizing each of them can maximize the success rate of NGS-based molecular profiling.

The following quality control (QC) steps within the NGS end-to-end workflow are critical for obtaining high-quality results and will be discussed in detail in the following sections of this chapter:

1. Sample.
   (a) Origin (topology, size).
   (b) Type (FFPE, blood, fresh frozen).
   (c) Processing (Tissue fixation, grossing, dehydration, embedding, cutting).
   (d) Storage (Temperature, time).
   (e) Tumor cell content.
   (f) Tumor area size.
2. Nucleic acid extraction.
   (a) Purity.
   (b) Nucleic acid type (DNA, cfDNA, RNA).
   (c) Fragmentation.
   (d) Chemical modifications (deamination, abasic sites).
   (e) Yield.
3. Target enrichment (NGS library preparation).
   (a) Input requirements.
   (b) Fragment size requirements.
   (c) Yield.

   (d) Quality.
   (e) Complexity.
4. Sequencing.
   (a) Read quality.
   (b) Read depth.
   (c) Read uniformity.

2 Materials

2.1 Sample Types: General Considerations and Challenges

The first thing to consider when planning a next-generation sequencing experiment is the sample type (Fig. 1a). The source of nucleic acid (NA) will strongly influence critical parameters such as nucleic acid yield and quality. Cell lines and whole-blood samples usually yield high amounts of high-quality nucleic acids. In contrast, other sample types, such as clinical FFPE tissue, may offer only a minimal number of cells and generally contain compromised and fragmented DNA and RNA. Cell fixation interferes not only with nucleic acid quality but also with yield. DNA yield is on average 6 pg/cell in unfixed material, but decreases to 3 pg/cell in fixed cells [2]. In addition, for somatic mutations analysis in tumor biopsy samples, sample purity and heterogeneity must be considered. Most often, clinical FFPE tissue blocks obtained from tumor biopsies contain both normal and tumor tissue. Thus, it is important that a trained pathologist determines the area in which tumor cells are enriched and estimates the tumor cell content within that area. This process is commonly done on hematoxylin-eosin-stained tissue sections. The section containing the region of interest marked by the pathologist is then used as a reference for additional unstained sections, from which that area is dissected and used for nucleic acid isolation. Similarly, when working with blood plasma samples used to extract cfDNA, the ratio between tumor-derived and normal circulating cfDNA can vary significantly between patients or even between samples taken from the same patient at different time points [3]. Additionally, both NA derived from FFPE and cfDNA are fragmented with an average size of 180 bp for cfDNA and very variable in the case of FFPE (Fig. 1b, c).

In conclusion, it is crucial to choose a library preparation method compatible with the type of sample used since current technologies differ significantly in minimum requirements for NA input and quality (e.g., fragment length) [4, 5].

2.2 Formalin-Fixed Paraffin-Embedded Tissue Samples

2.2.1 Tissue Fixation

The most common tissue storage method in routine clinical pathology is fixation of the tissue in formalin, followed by embedding in paraffin [6]. Traditionally, this type of preservation has been introduced for the purposes of preserving tissue structure and

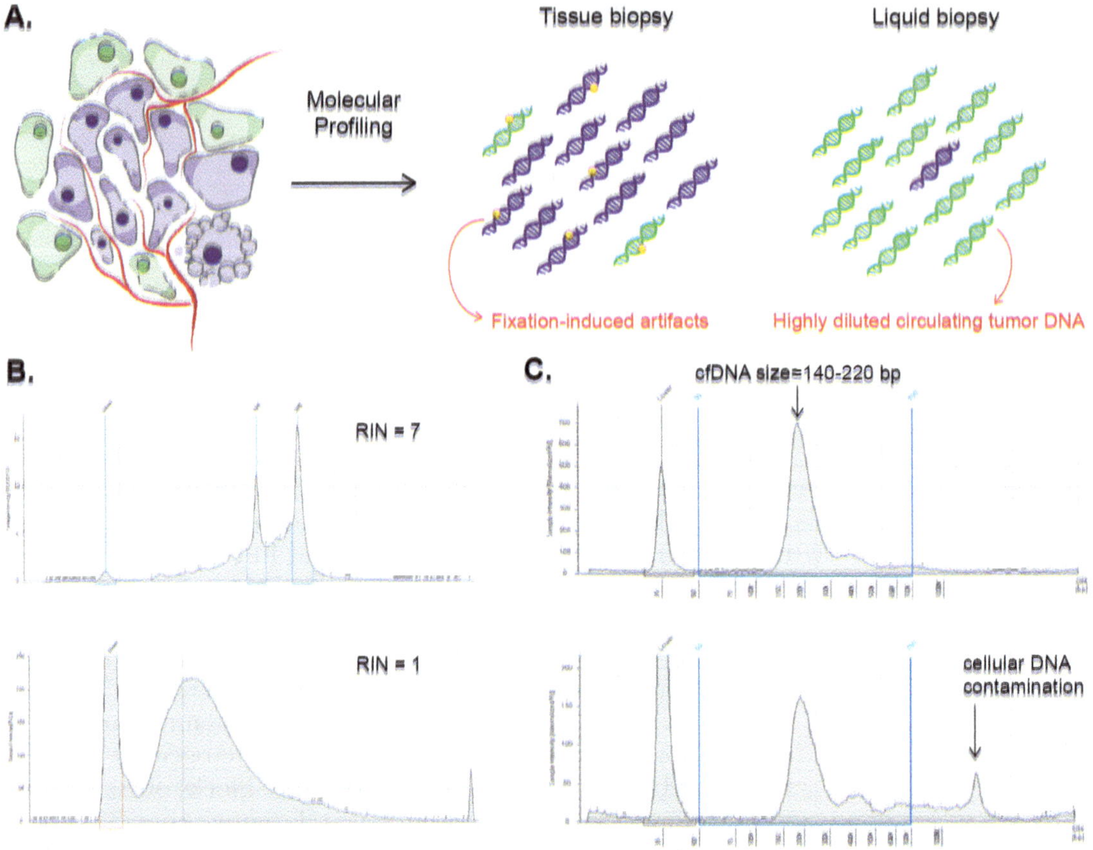

Fig. 1 (**a**) Sample type overview for molecular profiling of cancer cells. The most common sample types used for clinical genomic profiling are tissue and liquid biopsy. Highlighted in the figure are challenges associated with these sample types, such as fixation-artifacts and high dilution of molecules of interest. (**b**) NA derived from FFPE sample is highly fragmented, due to fixation induced damage and NA-protein cross-linking events. Top section: Capillary electrophoresis profile of RNA derived from cell line, showing high quality (RNA Integrity Number (RIN) score = 7). The two peaks corresponding to 18S and 28S ribosomal RNA are clear and visible. Bottom section: Capillary electrophoresis profile of FFPE-derived RNA shows high-level degradation. (**c**) Capillary electrophoresis profiles of cfDNA samples. cfDNA arises from cell death and tissue remodeling of normal and cancerous cells. Due to the nucleases activated during necrosis and apoptosis, cfDNA is highly fragmented. Top section: cfDNA sample passing quality control, with a clear peak around 180 bp. Bottom section: cfDNA sample presenting cellular DNA contamination, visible as a higher peak >1000 bp

morphology and allowing subsequent assessment of hematoxylin-eosin and immunohistochemical-based staining. With the introduction of genetic analysis as part of routine molecular profiling, FFPE tissue specimens have been shown to be compatible with downstream NGS analyses, including both DNA- and RNA-based assays [7, 8]. Since this first demonstration, NGS protocols have been successfully adapted for short-read sequencing platforms such as Illumina and Ion Torrent [9, 10]. However, while formalin fixation helps preserve tissue structure and epitope integrity, working with FFPE samples for NGS-based genetic analyses comes with

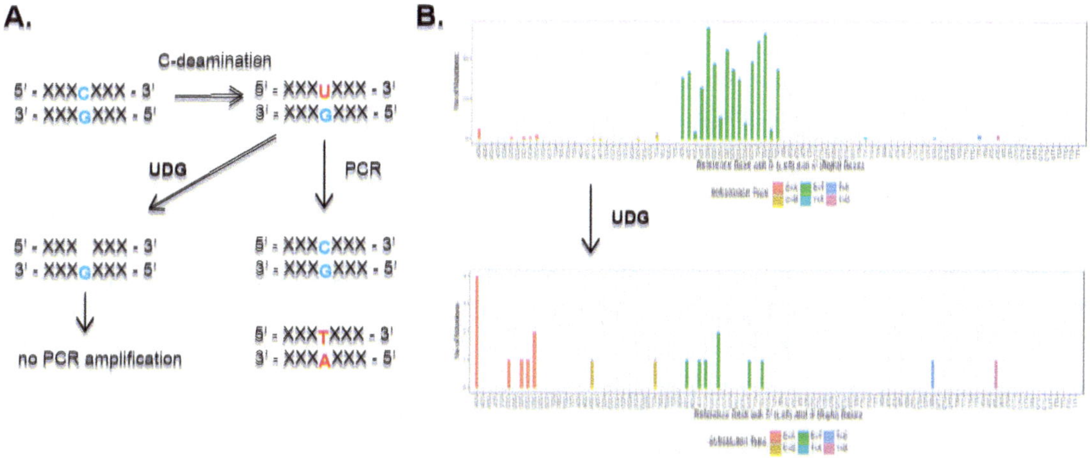

Fig. 2 Formalin-fixation causes deamination artifacts. (**a**) Deamination of cytosine causes conversion to uracil and subsequent PCR-induced C>T/G>A transition. By treating compromised DNA with UDG, uracil is eliminated from the original strand, removing the chance of detecting false positive mutations. (**b**) The mutational signature of a highly compromised FFPE sample is shown before (top section) and after UDG treatment (bottom section). UDG causes a specific decrease in the number of C>T mutations, consistent with deamination artifacts

several caveats related mainly to over- or under-fixation of the tissue in formaldehyde. Under-fixation can lead to DNA damage, resulting in over-fragmented DNA which may be incompatible with downstream target enrichment and sequencing. On the other hand, over-fixation can result in formaldehyde-induced crosslinks, DNA fragmentation, deamination of cytosine bases leading to C→T transitions, and generation of abasic sites [11, 12], all of which will strongly affect sample processing and variant calling. *See* **Note 1** for practical recommendations to maximize quality and retrievable quantity of FFPE-derived nucleic acids [13].

Uracil-DNA glycosylase (UDG) treatment can be used to reduce formalin-fixation artifacts. Fixation-induced cytosine deamination is responsible for turning cytosine into uracil, which in turn results in C>T (and G>A) transitions after PCR amplification (Fig. 2a). UDG treatment can significantly reduce deamination artifacts by removing unwanted uracil from DNA [11, 14, 15] (Fig. 2b). In addition, unique molecular identifiers (UMIs, *see* Subheading 3.1.2) can also be employed to analytically remove deamination artifacts since they allow for PCR-error correction by tagging individual template molecules [16].

2.2.2 Tumor Cell Content of Clinical FFPE Samples

An important aspect when working with FFPE samples from tumor biopsies in a clinical setting is the accurate estimation of tumor cell content present in the sample. NGS in these samples is commonly performed to assess the clinically relevant somatic mutations present in the tumor cells. In order to robustly detect such mutations, it is critical to ensure presence of sufficient tumor cells within the tissue sample. Tumor cell content is usually estimated by trained pathologists, who determine the area enriched for tumor cells and estimate the tumor cell content within these areas (*see* **Note 2**).

2.2.3 Nucleic Acid Extraction and Purification

Nucleic acid extraction from tissue samples can be performed using different methods, and the appropriate technology should be chosen for the types of samples from which nucleic acids are to be extracted. In the case of clinical FFPE tissue sections used to assess the genetic variants present in tumor cells, samples first need to be deparaffinized, most commonly in Xylene, which dissolves the paraffin wax and leaves the tissue intact. Next, the tissue is washed in ethanol and air-dried. It can then be microdissected to enrich tumor cells, which are then transferred into a lysis buffer containing proteinase K. The sample is then incubated, commonly at 56 °C between 1 h (RNA) and 24 h (DNA) to reverse the crosslinks between nucleic acids and proteins that were induced during the formalin-fixation process.

Downstream nucleic acid purification can be done with different methods, the most common being either spin column-based and nucleic acid binding bead-based technologies. Both methods have proven viable for performing downstream NGS [17]. Spin column-based methods utilize membranes that contain silica or glass fiber to bind nucleic acids. They offer a simple and straightforward procedure that can be run manually, but also in an automated way, which is especially important when scaling up sample numbers. However, automation methods for column-based methods can be expensive and complex to set up. Also, starting with excessive input sample may clog the membrane and result in inefficient purification. In contrast, bead-based extraction utilizes beads with a paramagnetic core most commonly coated with a matrix of silica for binding nucleic acids. In this method, tissue lysate is incubated with the magnetic beads, allowing the particles to bind DNA or RNA molecules. This type of nucleic acid purification can be easily automated and therefore offers a high degree of scalability and requires less hand-on resources. However, viscous samples can reduce migration of magnetic beads, and manual performance of this technique can be tedious with a large number of samples. In any case, the following points should be considered when choosing the extraction method for a particular sample type:

1. Scalability: Depending on the throughput, a method should be chosen that can be scaled up to the number of samples that are run by a laboratory. While in a research setting, turnaround times are not as crucial, they can be critical in a clinical setting. Therefore, being able to multiplex samples can be of critical importance.

2. Automation: Nucleic acid extraction automation may help not only with optimizing turnaround times but also with generating robust and reproducible nucleic acids.
3. Minimum and maximum elution volume: The final elution volume strongly influences the final concentration of the extracted nucleic acids. The lower the final volume, the higher the concentration will be. This is especially important when working with very sparse input material where total yield will be limited. Some downstream library preparation assays require certain minimal input DNA/RNA. Therefore, it is important to make sure the final concentration obtained from the purification kit is compatible with the library preparation method that will be used.
4. Relative yield (% recovery of total input): When working with very limited input material, it is important to make sure that one can extract most, if not all of the DNA and/or RNA from the sample, to be able to run downstream library preparation. It may be necessary to run comparisons between different extraction methods to assess what works best for the specific samples that will be run.
5. Quality vs. Quantity: While yield is certainly an important factor, quality of the obtained nucleic acids may be just as crucial, especially when working with FFPE samples. In contrast to nucleic acids obtained from whole blood, sample to sample variation in nucleic acid quality can vary substantially with FFPE samples, depending on fixation conditions as described above in Subheading 2.2.1. It is therefore important to not only quantify the obtained nucleic acids but also qualify them to determine the amount of functional DNA or RNA.

2.2.4 Nucleic Acid Quantitation

NGS-targeted library preparation protocols require between 10 and 200 ng of input DNA or RNA, depending on the type of enrichment technology [9, 10, 18, 19]. Amplicon-based methods generally require less input and can yield high-quality NGS libraries with as little as 1 ng input DNA. This is particularly useful when handling samples with limited material. On the other hand, hybridization-capture techniques require between 50 and 100 ng of input DNA, but can offer better uniformity.

The most commonly used techniques for accurate quantitation of extracted nucleic acids are either fluorescence or UV-absorbance based. They both have unique features that make them the preferred methods in specific situations (*see* **Note 3**):

1. Fluorometric quantitation:
   (a) Assay dyes bind selectively to DNA, RNA, or protein, making it more specific than UV absorbance.

   (b) More sensitive than UV absorbance, detecting as little as 10 pg/μL of DNA.
   (c) Useful when quantifying purified nucleic acids, even at low concentrations.

2. UV-absorbance:
   (a) Measures anything above 260 nm, including intact nucleic acids, but also free nucleotides, proteins, and even excessive salts.
   (b) Good for detecting potential contaminants.
   (c) Not suitable for low concentrations.

2.2.5 Nucleic Acid Qualification

While NA quantitation is a key factor for the success of an NGS experiment, assessment of NA quality is just as important. In particular, when working with FFPE samples, one must expect high variability in sample quality due to potential DNA and RNA damage induced during tissue fixation and sample storage (*see* Subheadings 2.2.1 and 2.3.2). In particular, over-fragmentation of nucleic acids may not be detectable with the quantitation methods mentioned above, especially when using UV absorbance. Even fluorometric methods will provide no information about the length of nucleic acid fragments present in the sample. This may lead to overestimation of "functional" nucleic acids and potentially result in failed library preparation or low complexity libraries, limiting the sensitivity of genetic variant detection.

To circumvent this issue, there are different ways to assess the quality of purified nucleic acids. The most common methods include gel electrophoresis and quantitative real-time PCR-based techniques. Advanced automated gel electrophoresis platforms allow for quick assessment of nucleic acid fragment length as well as concentration with little hands-on time. However, despite getting a read-out on fragment length, it is not a direct measure of functional DNA or RNA. This is particularly important when working with amplicon-based target enrichment methods for library preparation (*see* Subheading 3.2.1), as the input DNA or RNA needs to be PCR amplifiable, and depending on the amplicon length, this requires a minimum fragment length of the purified input nucleic acids. When working with NAs extracted from cell lines or whole blood, this is not much of an issue. However, working with FFPE tissue-derived NAs, this is an important factor to consider, as these NAs are often heavily degraded and fragmented, limiting their ability to be amplified using PCR. To account for this, qPCR-based methods can be employed to determine the amount of amplifiable NAs in a sample. Often, these assays amplify a single target such as the *RNaseP* gene [20] with a specific amplicon length that will somewhat reflect the length of the library prep amplicons. Additionally, other qPCR assays, often used in forensic settings, where NA fragmentation is even more advanced, will amplify and quantify several targets of different amplicon lengths to determine the amount of degradation within the sample. Both of these methods will allow the user to directly measure the amplifiable NA present in the sample and adjust input for target enrichment accordingly.

2.3 Liquid Biopsy Samples

In the field of oncology, detecting and monitoring oncogenic variants is of primary importance for patient management. Clinicians can employ liquid biopsy when tissue biopsy is not feasible or to follow-up response to therapy and track resistance mechanisms. Liquid biopsy generally refers to sampling body fluids to gain insight into molecular alterations occurring in the tumor cells. This technology prides itself on being fast and noninvasive compared to conventional tissue biopsy. Several body fluids can be used for the analysis, as cerebrospinal fluid, saliva, pleural fluid, urine, and ascites. However, the most commonly used material in diagnostics is plasma derived from peripheral blood. The main biological entities examined in liquid biopsy are cfDNA/circulating-tumor DNA (ctDNA), circulating tumor cells, and exosomes. Because variant calling is mainly performed on cfDNA/ctDNA, we will focus on this entity during this chapter.

Standardization of preanalytical conditions for liquid biopsy is of utmost importance and has been recommended by several organizations, including the European Liquid Biopsy Society (ELBS), the National Cancer Institute (NCI), and the Blood Profiling Atlas in Cancer (BloodPAC) [21]. As a consequence, guidelines have been proposed for sample storage and transport conditions, plasma isolation, cfDNA extraction, and quantification [22], with initiatives such as the Biospecimen Evidence-Based Practices [23], and CANCER-ID [24] aiming to harmonize and define best practices for liquid biopsy assay implementation. We will go through these standard procedures in the following paragraphs.

2.3.1 Types of Liquid Biopsy

The type of body fluid used has a substantial impact on the sensitivity of the analysis. In particular, the choice of body fluid to collect and analyze should be driven by tumor location and vascularization. Many studies have reported effective detection of ctDNA from cerebrospinal fluid (CSF), saliva, pleural effusion, ascites, urine, and other bodily fluids [25–29]. CSF has shown high sensitivity and clinical utility for genotyping brain tumors and metastatic brain lesions [30, 31]. Because of the blood–brain barrier, CSF should be preferred over blood for these indications. Peripheral blood can be used for most other cancers. However, tumor type and stage strongly influence sensitivity rates. For example, bladder and colorectal cancers have the highest reported fraction of cases with detectable ctDNA in plasma, while brain and thyroid cancers are on the lower end of the spectrum [3]. While 80% of metastatic tumors release ctDNA in plasma, detecting oncogenic mutations in early stage tumors remains challenging [3]. Methylation profiling of cfDNA might provide more informative results for early cancer detection and population screening [32, 33], but this method will not be covered here since it is outside of the scope of this chapter.

2.3.2 Sample Collection and Storage

The vessel used for sample collection can drastically influence cfDNA quality. Several commercial tubes are available for peripheral blood collection, and they differ mainly in the type of preservative substance they use. If plasma and not serum is the body fluid of choice, the tubes used during blood drawing should contain an anticoagulant (heparin) and need to be inverted several times to ensure proper mixing and avoid clotting. In addition, regular tubes for blood collection contain EDTA to increase DNA stability. Because cellular DNA contamination represents the highest risk for decreased sensitivity of cfDNA analyses, plasma must be isolated from EDTA tubes within 4 h following phlebotomy [34–36]. When prolonged storage is needed, blood should be collected in tubes containing fixatives that prevent cell lysis and dilution of cfDNA with cellular DNA derived from nucleated blood cells. Several commercial solutions are available (e.g., Streck Cell Free BCT®, PAXgene Blood ccfDNA Tubes or Roche Cell Free DNA Collection Tubes®), and they enable up to 2 weeks of storage [35].

2.3.3 Plasma Isolation

For blood-based ctDNA assays, the non-cellular fraction needs to be isolated by centrifugation (plasma) or blood clotting (serum). Plasma should be preferred to serum since serum ctDNA is highly diluted by cfDNA of non-cancerous origins arising from leukocyte lysis [37]. For plasma isolation, whole blood should be centrifuged at low speed to avoid cellular lysis. Conventional protocols include two centrifugation steps to separate plasma and then eliminate cell debris.

Optimal centrifugation conditions for plasma isolation are as follows [22, 36, 38–40]:

1. Speed settings: 800–2000 × g is recommended for the first spin, whereas 1500–16,000 × g is recommended for the second spin.
2. Centrifugation time: 10–20 min is required for each spin.
3. Centrifugation temperature: can be set at 4 °C or room temperature, with no reported impact on ctDNA quality.

2.3.4 Nucleic Acid Extraction

Generic DNA extraction methods should not be used for cfDNA isolation since they are designed to remove fragmented DNA and have also been shown to lead to loss of cfDNA integrity and increase fragmentation [41].

Dedicated cfDNA extraction kits differ in:

1. Extraction method: magnetic beads, silica gel membrane columns, or polymer-mediated enrichment.
2. Scalability of input volumes: different protocols can allow cfDNA isolation starting from 1 mL up to 6 mL of blood plasma.
3. Scalability of output volumes: from 400 μL to 15 μL.
4. Workflow: manual or automated solutions.
5. Relative yield and fragment size: variable, smaller DNA fragments tend to have a lower yield [42].

Different extraction methods have shown similar performance [43, 44] (*see* **Note 4**).

2.3.5 Nucleic Acid Quantitation and Qualification

Standard methods for NA quantitation (*see* Subheading 2.2.4) can theoretically be used for cfDNA. However, since yield is considerably lower for cfDNA than for cellular DNA, high-sensitivity methods should be used for quantification. Depending on physical conditions, cfDNA concentration in healthy donors is on average 5 ng per mL plasma [45–47], which would lead to a total output of 25 ng if 5 mL of plasma is used for extraction and yield is 100%. Considering 70–80% recovery rate and using an elution volume of 20 μL, the cfDNA concentration to be measured will be in the range of 0.5–1.5 ng/μL. In order to obtain precise measures at such concentrations, high-sensitivity fluorometric assays must be used (PicoGreen or Qubit Fluorometer). In addition, if carrier RNA is added during extraction and present in the eluate, it will interfere with the UV absorbance readings of spectrophotometers.

Finally, the use of capillary electrophoresis is recommended to ensure the quality and purity of the extracted cfDNA. Agilent Bioanalyzer or TapeStation can be used for this purpose. Since cfDNA is mostly derived from cell death, it is highly fragmented with a distribution peak of sizes ranging from 140 to 200 bp (Fig. 1c). Fragment lenght analysis is particularly useful for detecting cellular DNA contamination in hemolytic plasma samples (Fig. 1c).

3 Methods

3.1 NGS Library Preparation QC

3.1.1 Input Requirements

DNA and RNA input requirements differ greatly between different library preparation technologies. While whole-genome sequencing requires between 100 ng and 1 μg of DNA, targeted sequencing methods may yield functional libraries with as little as 1 ng input DNA or RNA. This is particularly important when working with clinical samples that often offer limited material. Based on the points discussed in Subheadings 2.2.5 and 2.3.5, the NA input needs to be adjusted to ensure that sufficient functional molecules are being provided for successful library preparation. In particular, when performing amplicon-based target enrichment on FFPE samples, it may be crucial to determine the fragment length of the isolated nucleic acids and make sure it is compatible with the enrichment assay. In the following section, we will focus on NGS library preparation for targeted sequencing and the steps that have to be taken to ensure optimal library and sequencing quality.

3.1.2 Library Preparation Technologies

Successful targeted NGS library preparation can be achieved with different methods using two main technologies, namely hybrid capture and amplicon-based target enrichment [19, 48]. Both the technologies have their advantages and disadvantages and should be considered depending on the specific needs of a laboratory. Hybrid capture-based methods work through using long, biotinylated oligonucleotide probes to hybridize to the regions of interest. The captured target regions can then be pulled out through streptavidin beads and eluted. In contrast, amplicon-based target enrichment utilizes multiplex PCRs to simultaneously amplify hundreds to thousands of genomic regions in one PCR. This has the advantage of working with very little DNA and RNA input, often as low as 1 ng. Hybrid-capture methods instead often require >50 ng input material. However, hybridization methods tend to have more uniform enrichment, resulting in an overall more even read distribution across the enriched regions. Also, due to the nature of this type of target enrichment, final sequencing reads will have unique start and end positions, allowing for higher sensitivity when performing variant calling. Lastly, amplicon-based methods require less hands-on work in the laboratory and offer a faster turnaround time than hybrid-capture methods.

A variation of these technologies includes the use of unique molecular identifiers (UMIs), enabling error correction and higher overall sensitivity for detecting genetic variants [49]. This is achieved through the use of random nucleotide sequences barcoding on each DNA fragment prior to PCR amplification during the target enrichment process. After sequencing, this approach allows for distinguishing reads amplified from the same original DNA molecule based on identical UMI, enabling the detection of genetic variants on a single-molecule level. This technology is mainly used for clinical applications requiring ultra-sensitive variant detection, such as liquid biopsy on the basis of ctDNA [50].

3.1.3 Library QC

Qualification of the final NGS library is an important QC step before proceeding with sequencing. In particular, when working with library preparation technologies that involve fragmentation of input DNA, it is important to determine the distribution of the final library fragment length. Depending on the sequencing technology used further downstream, read-length may be limited, so it is important to be in the range of these limitations with the fragment length distribution of the libraries. In the case of amplicon-based libraries, the required fragment length is determined by the size of the PCR amplicon. There, it is important to choose a panel that is compatible with the sample type from which nucleic acids are extracted. In order to assess the fragment length distributions of the finished libraries, automated electrophoresis is commonly used. This method requires minimal hands-on time and is sensitive enough to detect even small amounts of final library molecules. This step is important in particular for libraries where hybrid-capture technologies are used for target enrichment as their length can vary depending on the initial DNA shearing conditions.

3.1.4 Library Concentration

Before proceeding with sequencing, the final libraries need to be accurately quantified. This is a key step to ensure optimal performance in cluster generation (Illumina) or template preparation (Ion Torrent) reactions. Both overloading and underloading these reactions can result in decreased read quality and reduction of overall data output. For library quantification, different methods can be used. The most common are:

1. Fluorometric DNA quantification (Qubit/PicoGreen): measures total DNA present in the sample. Does not distinguish between library molecules and unspecific molecules.
2. qPCR: the most accurate method for library quantification. It relies on primers and/or probes binding adapter sequences present on functional library fragments. This method allows for accurate and reliable quantification of NGS libraries.
3. Capillary electrophoresis (TapeStation/Bioanalyzer): electrophoresis methods can be used not only for library qualification but also for their quantification. As for fluorometric DNA measurement, this method does not distinguish between functional library molecules (e.g., fragments containing ligated adapter molecules) and other DNA present in the sample. However, it does allow for an estimation of library to primer/adapter dimer ratio based on the size of the fragments.

3.2 NGS Post-sequencing Data QC

Quality control of raw sequencing data should be performed prior to downstream data analysis and variant calling. We will focus here mainly on QC performed after short-read sequencing as it is employed in most clinical laboratories. We will cover the main steps performed on two main sequencing platforms used in a clinical setting, namely Ion Torrent semiconductor sequencing (Thermo Fisher Scientific) platform and Illumina sequencing by synthesis technology. While the main steps are very similar between the two platforms, they vary in detail and some of these steps can be automated post-sequencing.

3.2.1 Sequencing Run QC

The first step is to get a general overview of the sequencing run by looking at the following platform-specific key metrics given by the sequencing software provided by the manufacturers of the sequencing platforms:

Illumina Sequencing

1. Cluster density (K/mm^2): This metric indicates the density of clusters across the flow cell. It is a critical value that influences overall run quality, Q30 scores (described below), and total yield. Under-clustering can lead to poor performance, including lower quality scores and lower data output. In contrast, over-clustering can lead to the introduction of sequencing artifacts, lower overall quality, and therefore lower total data output. The ideal value depends on the instrument/kit combination and is given by the manufacturer. For example, for MiSeq and NextSeq instruments, the density should ideally be between 170 and 220 K/mm^2. The most common factors that influence cluster density are library quality and library quantification. It is important that libraries are denatured in freshly prepared NaOH with a pH >12.5. Also, to accurately quantify libraries, qPCR is the most effective method when a suitable standard is used. Other compatible methods are fluorescent methods such as Qubit/Picogreen or Bioanalyzer/TapeStation, although in case of the latter, these methods are recommended primarily for library quality control rather than quantitation.
2. Yield: this is the number of bases generated in the sequencing run. A specification regarding expected yield for a given instrument/kit combination is given by the manufacturer. It is important to maximize yield in particular when the full capacity of the flow cell has been used. Lower than expected yields may result in insufficient coverage of sequenced libraries and may affect variant calling sensitivity and specificity in downstream analysis.
3. %Q30: This value indicates the percentage of sequenced bases with a quality score of 30 or higher. This quality score (q) is an estimate of the probability (p) of a particular base being called wrongly and is calculated as $q = -10 \times \log10(p)$. For example, a q score of 10 indicates that there is a 10% probability that the called base is a sequencing error. Correspondingly, Q30 refers to 0.1% probability that a given base is called wrongly. Most Illumina runs should generate a minimum of 70–80% Q30 bases. Importantly, the per base quality scores usually drop toward the end of a read. This is important to consider, in particular when the region of interest is located at the end of the reads.

Ion Torrent Sequencing

1. ISP Loading: This number indicates the number of wells on a chip with ion-sphere particles (ISP) that are loaded with library DNA fragments. Ideally, values >80% should be aimed for. ISP loading greatly depends on the accurate quantification and dilution of libraries for template preparation. A lower ISP loading value indicates that an insufficient number of library molecules were present in the template preparation reaction, resulting in a lower number of DNA loaded ISPs on the sequencing chip. In contrast, too much input library DNA can result in high chip loading, but will also cause higher polyclonality and a higher number of low-quality reads, resulting in an overall reduced total data output. The optimal input concentrations for template preparation depend on the chip type and kit combination and are given by the manufacturer.
2. Polyclonality: Due to the way that ISPs are loaded with library DNA during the template preparation process, it is inevitable that some ISPs will be loaded with copies of more than one library molecule. Reads originating from these ISPs will result in so-called polyclonal reads as they are essentially a mix of more than one sequenced library template. These reads are detected by the sequencing software and filtered out. Normal values for polyclonality range between 20% and 35%, depending on the chip and library type. The value is primarily influenced by the library concentration that was used as input for the template preparation reaction. Therefore, if values >35% are observed constantly across runs, library input concentration should be reduced. While it is possible to get polyclonality values <20% by dramatically lowering the input amount, this will also result in much lower ISP loading, causing an overall reduced total data output.
3. Enrichment: This value indicates how many of the detected ISPs on the chip were loaded with library DNA. This value should be 99–100%. Values lower than this indicate an issue with the template preparation reaction prior to sequencing.
4. Low quality/adapter dimers: This value indicates the number of reads filtered due to low quality and/or present adapter dimer sequences. It usually varies between 1 and 20%, depending on various factors such as library type, library complexity, and library quality.
5. Total reads/bases: This value indicates the total number of quality filter-passing (as described above) reads and bases that can be used for downstream data analysis. The Ion Torrent platform automatically performs read quality assessment and read trimming based on default parameters, yielding ready-to-use reads for downstream data analysis.

6. Read-length profile: The read-length profile plot can be used to quickly assess whether a sequencing run yielded reads with the expected length distribution based on the type of libraries. For hybrid-capture libraries, one would expect a normal distribution with a peak at the size that was selected during library preparation. In contrast, for amplicon libraries, read-length distributions are determined by the size of the sequenced amplicons. Therefore, each library panel will result in a unique read-length distribution pattern.

3.2.2 Per-Sample Sequencing QC

Once overall run QC has been performed, the next step is to assess the data quality for each sequenced sample. We will focus here on target-enrichment panel sequencing and the key metrics that should be assessed before moving to downstream variant calling analysis:

1. Read mapping: Once overall read quality has been assessed and reads have been filtered with the desired quality criteria, they are usually mapped to a reference genome (exception for de novo whole-genome sequencing, which will not be covered here). There are different mapping algorithms available, each with the aim to align the obtained sequencing reads to a given reference sequence. These algorithms usually allow flexible parameter settings, i.e., tolerate a certain number and type of mismatches to account for the presence of genetic variants present in the sample.
2. Uniformity: This metric is used to describe the distribution of reads across the enriched genomic regions. The higher the uniformity value, the more evenly the reads are distributed across the different targets. Ideally, each sequenced target region should be sequenced to the same depth. However, due to biases introduced during library preparation and sequencing, this is rarely the case. Reasons for this can include sequence composition, amplicon length (amplicon libraries), capture efficiency (hybrid capture libraries), and quality of template DNA. Generally, it has been shown that hybrid-capture-based library preparation technologies result in better read uniformity than PCR-based enrichment technologies [19].
3. Coverage and mean read depth: Read depth describes the number of times that a specific region or base is sequenced. This metric is critical as it directly determines the sensitivity of variant detection at each sequenced genomic position. Commonly, a mean read depth of 1000–2000× is aimed for when trying to identify somatic mutations in clinical tumor specimens. Depending on the type of target enrichment that was used for library preparation, read uniformity (described above) may vary [19]. Therefore, in order to reach a minimum read

depth threshold in all sequenced regions, it may be necessary to sequence to a higher mean read depth to ensure that the regions with the lowest read depth is sequenced deeply enough to allow for sensitive variant calling.

4. Molecular Coverage: When working with liquid biopsy samples and, in general, aiming for high sensitivity for mutation detection, it is recommended to use UMIs. UMIs enable the distinction of reads originating from the same template molecule allowing for increased sensitivity and specificity of variant detection [50]. The overall limit of detection (LoD) of these assays is mainly dependent on the amount of input material, the conversion rate of this material into functional library molecules, and sequencing depth. To determine the LoD of a sample after sequencing, one can assess the number of UMIs covering a specific region along with overall read depth, in contrast to simply measuring the total reads covering that region (described above). Usually, a UMI is considered as functional if it is supported by at least three reads, but this number can vary based on the library preparation kit and manufacturer's recommendation. Overall, UMI-based libraries used for ctDNA analysis require substantially higher sequencing read depth to allow for sufficient functional UMI-based read families and can achieve variant detection LoDs down to 0.1% allelic frequency.

4 Notes

1. To maximize quality and retrievable quantity of FFPE-derived nucleic acids, the following are suggested:
   (a) Fixative type: 10% neutral buffered formalin is recommended for tissue fixation.
   (b) Fixative time: Tissue fixation should be done between 18 h and 48 h, depending on tissue size. Over-fixation can lead to DNA damage such as cytosine deamination or abasic sites. Under-fixation can lead to nucleic acid degradation and result in lower overall yield of functional DNA and RNA.
   (c) Fixative temperature: Fixation should be performed at room temperature.
   (d) FFPE tissue storage temperature: FFPE blocks can be stored at room temperature but storage at 4 °C will preserve better DNA and RNA quality even after long-term storage.
   (e) FFPE tissue storage time: Functional nucleic acids can be extracted from FFPE tissue that have been stored for

several decades. Proper quality assessment as described further below should be performed when extracting nucleic acids from FFPE blocks that have been stored over an extended period.

2. In FFPE samples, the minimal tumor cell content required for accurate identification of heterozygous somatic variants is usually between 10% and 20%, corresponding to a theoretical limit of detection of 5%–10% allelic frequency. This limit is employed in FFPE samples due to the fact that below this threshold, deamination artifacts (Subheading 2.2.1) may occur, making it difficult to distinguish true variants from sequencing artifacts.
3. For the accurate quantitation of extracted nucleic acids in a clinical routine laboratory setting, fluorometric quantitation is the preferred method as it offers more reproducible and more sensitive results, unaffected by contaminants. Having a standard operating protocol in place to extract and purify NAs from different input materials, one can generally exclude presence of such contaminants.
4. When choosing a cfDNA extraction kit, particular attention should be reserved to the scalability of input and output volume in order to obtain the maximum yield. In particular, smaller elution volumes should be preferred since a higher concentration of cfDNA results in increased sensitivity for mutation analysis. Interestingly, fragment size-specific yields have been shown to differ between isolation kits [42].

5 Conclusions

This chapter provides an overview of current challenges and general practices for molecular profiling of clinical samples. We highlighted relevant tools for ensuring the highest sequencing standard and best results for variant calling. Performing genomic analysis of challenging sample types has now become part of routine diagnostics and clinical research. By following guidelines and recommended procedures, molecular biologists and pathologists can ensure the highest analytical sensitivity and specificity while reducing sample attrition rate. Finally, preanalytical variables significantly impact subsequent data analysis and should be carefully considered to support clinical and analytical decisions and ensure reliable variant detection.

References

1. Mosele F, Remon J, Mateo J et al (2020) Recommendations for the use of next-generation sequencing (NGS) for patients with metastatic cancers: a report from the ESMO precision medicine working group. Ann Oncol 31:1491–1505
2. Liu HE, Triboulet M, Zia A et al (2017) Workflow optimization of whole genome amplification and targeted panel sequencing for CTC mutation detection. NPJ Genom Med 2:34
3. Bettegowda C, Sausen M, Leary RJ et al (2014) Detection of circulating tumor DNA in early- and late-stage human malignancies. Sci Transl Med 6:224ra24
4. Goodwin S, McPherson JD, McCombie WR (2016) Coming of age: ten years of next-generation sequencing technologies. Nat Rev Genet 17:333–351
5. Hess JF, Kohl TA, Kotrová M, Rönsch K, Paprotka T, Mohr V, Hutzenlaub T, Brüggemann M, Zengerle R, Niemann S, Paust N (2020) Library preparation for next generation sequencing: A review of automation strategies. Biotechnol Adv 41:107537
6. Kofanova O, Bellora C, Garcia Frasquilho S et al (2020) Standardization of the preanalytical phase of DNA extraction from fixed tissue for next-generation sequencing analyses. New Biotechnol 54:52–61
7. Kerick M, Isau M, Timmermann B et al (2011) Targeted high throughput sequencing in clinical cancer settings: formaldehyde fixed-paraffin embedded (FFPE) tumor tissues, input amount and tumor heterogeneity. BMC Med Genet 4:68
8. Ludyga N, Grünwald B, Azimzadeh O et al (2012) Nucleic acids from long-term preserved FFPE tissues are suitable for downstream analyses. Virchows Arch 460:131–140
9. Zhang L, Chen L, Sah S et al (2014) Profiling cancer gene mutations in clinical formalin-fixed, paraffin-embedded colorectal tumor specimens using targeted next-generation sequencing. Oncologist 19:336–343
10. Chung J, Son D-S, Jeon H-J et al (2016) The minimal amount of starting DNA for Agilent's hybrid capture-based targeted massively parallel sequencing. Sci Rep 6:26732
11. Kim S, Park C, Ji Y et al (2017) Deamination effects in formalin-fixed, paraffin-embedded tissue samples in the era of precision medicine. J Mol Diagn 19:137–146
12. Do H, Dobrovic A (2015) Sequence artifacts in DNA from formalin-fixed tissues: causes and strategies for minimization. Clin Chem 61: 64–71
13. Arreaza G, Qiu P, Pang L et al (2016) Pre-analytical considerations for successful next-generation sequencing (NGS): challenges and opportunities for formalin-fixed and paraffin-embedded tumor tissue (FFPE) samples. Int J Mol Sci 17:1579. https://doi.org/10.3390/ijms17091579
14. Alborelli I, Leonards K, Rothschild SI et al (2020) Tumor mutational burden assessed by targeted NGS predicts clinical benefit from immune checkpoint inhibitors in non-small cell lung cancer. J Pathol 250:19–29
15. Do H, Dobrovic A (2012) Dramatic reduction of sequence artefacts from DNA isolated from formalin-fixed cancer biopsies by treatment with uracil- DNA glycosylase. Oncotarget 3: 546–558
16. Peng Q, Xu C, Kim D et al (2019) Targeted single primer enrichment sequencing with single end duplex-UMI. Sci Rep 9:4810
17. Seiler C, Sharpe A, Barrett JC et al (2016) Nucleic acid extraction from formalin-fixed paraffin-embedded cancer cell line samples: a trade off between quantity and quality? BMC Clin Pathol 16:17
18. de Leng WWJ, Gadellaa-van Hooijdonk CG, Barendregt-Smouter FAS et al (2016) Targeted next generation sequencing as a reliable diagnostic assay for the detection of somatic mutations in Tumours using minimal DNA amounts from formalin fixed paraffin embedded material. PLoS One 11:e0149405
19. Samorodnitsky E, Jewell BM, Hagopian R et al (2015) Evaluation of hybridization capture versus amplicon-based methods for whole-exome sequencing. Hum Mutat 36:903–914
20. Dang J, Mendez P, Lee S et al (2016) Development of a robust DNA quality and quantity assessment qPCR assay for targeted next-generation sequencing library preparation. Int J Oncol 49:1755–1765
21. Febbo PG, Martin A, Scher HI et al (2020) Minimum technical data elements for liquid biopsy data submitted to public databases. Clin Pharmacol Ther 107:730–734
22. Meddeb R, Pisareva E, Thierry AR (2019) Guidelines for the Preanalytical conditions for analyzing circulating cell-free DNA. Clin Chem 65:623–633
23. Greytak SR, Engel KB, Parpart-Li S et al (2020) Harmonizing cell-free DNA collection and processing practices through evidence-

based guidance. Clin Cancer Res 26:3104–3109

24. Lampignano R, Neumann MHD, Weber S et al (2020) Multicenter evaluation of circulating cell-free DNA extraction and downstream analyses for the development of standardized (pre)analytical work flows. Clin Chem 66: 149–160
25. Husain H, Melnikova VO, Kosco K et al (2017) Monitoring daily dynamics of early tumor response to targeted therapy by detecting circulating tumor DNA in urine. Clin Cancer Res 23:4716–4723
26. Wang Y, Springer S, Mulvey CL et al (2015) Detection of somatic mutations and HPV in the saliva and plasma of patients with head and neck squamous cell carcinomas. Sci Transl Med 7:293ra104
27. Li YS, Jiang BY, Yang JJ et al (2018) Unique genetic profiles from cerebrospinal fluid cell-free DNA in leptomeningeal metastases of EGFR-mutant non-small-cell lung cancer: a new medium of liquid biopsy. Ann Oncol 29: 945–952
28. Han M-R, Lee SH, Park JY et al (2020) Clinical implications of circulating tumor DNA from ascites and serial plasma in ovarian cancer. Cancer Res Treat 52:779–788
29. Villatoro S, Mayo-de-Las-Casas C, Jordana-Ariza N et al (2019) Prospective detection of mutations in cerebrospinal fluid, pleural effusion, and ascites of advanced cancer patients to guide treatment decisions. Mol Oncol 13: 2633–2645
30. Zheng M-M, Li Y-S, Jiang B-Y et al (2019) Clinical utility of cerebrospinal fluid cell-free DNA as liquid biopsy for leptomeningeal metastases in ALK-rearranged NSCLC. J Thorac Oncol 14:924–932
31. Escudero L, Llort A, Arias A et al (2020) Circulating tumour DNA from the cerebrospinal fluid allows the characterisation and monitoring of medulloblastoma. Nat Commun 11: 5376
32. Liu J, Zhao H, Huang Y et al (2021) Genome-wide cell-free DNA methylation analyses improve accuracy of non-invasive diagnostic imaging for early-stage breast cancer. Mol Cancer 20:36
33. Shen SY, Singhania R, Fehringer G et al (2018) Sensitive tumour detection and classification using plasma cell-free DNA methylomes. Nature 563:579–583
34. van Dessel LF, Beije N, Helmijr JCA et al (2017) Application of circulating tumor DNA in prospective clinical oncology trials - standardization of preanalytical conditions. Mol Oncol 11:295–304
35. Kang Q, Lynn Henry N, Paoletti C et al (2016) Comparative analysis of circulating tumor DNA stability In K3EDTA, Streck, and Cell-Save blood collection tubes. Clin Biochem 49: 1354–1360
36. Volckmar A-L, Sültmann H, Riediger A et al (2018) A field guide for cancer diagnostics using cell-free DNA: from principles to practice and clinical applications. Genes Chromosom Cancer 57:123–139
37. Pittella-Silva F, Chin YM, Chan HT et al (2020) Plasma or serum: which is preferable for mutation detection in liquid biopsy? Clin Chem 66:946–957
38. Sherwood JL, Corcoran C, Brown H et al (2016) Optimised pre-analytical methods improve KRAS mutation detection in circulating tumour DNA (ctDNA) from patients with non-small cell lung cancer (NSCLC). PLoS One 11:e0150197
39. Page K, Guttery DS, Zahra N et al (2013) Influence of plasma processing on recovery and analysis of circulating nucleic acids. PLoS One 8:e77963
40. Page K, Powles T, Slade MJ et al (2006) The importance of careful blood processing in isolation of cell-free DNA. Ann N Y Acad Sci 1075:313–317
41. Malentacchi F, Pizzamiglio S, Verderio P et al (2015) Influence of storage conditions and extraction methods on the quantity and quality of circulating cell-free DNA (ccfDNA): the SPIDIA-DNAplas external quality assessment experience. Clin Chem Lab Med 53:1935–1942
42. Cook L, Starr K, Boonyaratanakornkit J et al (2018) Does size matter? Comparison of extraction yields for different-sized DNA fragments by seven different routine and four new circulating cell-free extraction methods. J Clin Microbiol 56:e01061-18. https://doi.org/10.1128/JCM.01061-18
43. Pérez-Barrios C, Nieto-Alcolado I, Torrente M et al (2016) Comparison of methods for circulating cell-free DNA isolation using blood from cancer patients: impact on biomarker testing. Transl Lung Cancer Res 5:665–672
44. Wolf A, Beller K, Groemminger S et al (2016) Purification of circulating cell-free DNA from plasma and urine using the automated large-volume extraction on the QIAsymphony® SP instrument. Adv Exp Med Biol 924:179–185
45. Mouliere F, Robert B, Arnau Peyrotte E et al (2011) High fragmentation characterizes

tumour-derived circulating DNA. PLoS One 6: e23418

46. Alborelli I, Generali D, Jermann P et al (2019) Cell-free DNA analysis in healthy individuals by next-generation sequencing: a proof of concept and technical validation study. Cell Death Dis 10:534

47. Mehra N, Dolling D, Sumanasuriya S et al (2018) Plasma cell-free DNA concentration and outcomes from Taxane therapy in metastatic castration-resistant prostate cancer from two phase III trials (FIRSTANA and PROSELICA). Eur Urol 74:283–291

48. Jennings LJ, Arcila ME, Corless C et al (2017) Guidelines for validation of next-generation sequencing–based oncology panels. J Mol Diagn 19:341–365

49. Kivioja T, Vähärautio A, Karlsson K et al (2011) Counting absolute numbers of molecules using unique molecular identifiers. Nat Methods 9:72–74

50. Vollbrecht C, Lehmann A, Lenze D, Hummel M (2018) Validation and comparison of two NGS assays for the detection of EGFR T790M resistance mutation in liquid biopsies of NSCLC patients. Oncotarget 9:18529–18539

Index

Charlotte K. Y. Ng and Salvatore Piscuoglio (eds.), *Variant Calling: Methods and Protocols*,
Methods in Molecular Biology, vol. 2493, https://doi.org/10.1007/978-1-0716-2293-3,

MIX
Papier aus verantwortungsvollen Quellen
Paper from responsible sources
FSC® C105338
FSC
www.fsc.org

If you have any concerns about our products,
you can contact us on
ProductSafety@springernature.com

In case Publisher is established outside the EU,
the EU authorized representative is:
Springer Nature Customer Service Center GmbH
Europaplatz 3, 69115 Heidelberg, Germany

Printed by Libri Plureos GmbH
in Hamburg, Germany